Calving Problems and Early Viability of the Calf

Current Topics in Veterinary Medicine and Animal Science

Volume 4

Formerly known as Current Topics in Veterinary Medicine, this series is expanding to include topics of interest to those working in the fields of animal science and veterinary medicine.

other titles in this series

Volume 1
Control of Reproduction in the Cow
edited by J.R. Sreenan

Volume 2
Patterns of Growth and Development in Cattle
edited by H. de Boer and J. Martin

Volume 3
Respiratory Diseases in Cattle
edited by W.B. Martin

Calving Problems and Early Viability of the Calf

A Seminar in the EEC Programme of Coordination of Research on Beef Production held at Freising, Federal Republic of Germany, May 4-6, 1977

Sponsored by the Commission of the European Communities, Directorate-General for Agriculture, Coordination of Agricultural Research

Edited by

B. Hoffmann

Institut für Physiologie,
Südd. Versuchs- und Forschungsanstalt für
    Milchwirtschaft Weihenstephan
Technische Universität München,
Freising

I. L. Mason

FAO,
Rome

J. Schmidt

Institut für Physiologie,
Südd. Versuchs- und Forschungsanstalt für
    Milchwirtschaft Weihenstephan
Technische Universität München,
Freising

Martinus Nijhoff Publishers - The Hague/Boston/London   1979
for
The Commission of the European Communities

Publication arranged by

Commission of the European Communities,
Directorate-General Scientific and Technical Information
and Information Management, Luxembourg.

EUR 6178 EN

For further information:
Martinus Nijhoff Publishers b.v., P.O.B. 566, 2501 CN
The Hague, The Netherlands.

LEGAL NOTICE

ISBN-13: 978-94-009-9317-4   e-ISBN-13: 978-94-009-9315-0
DOI: 10.1007/978-94-009-9315-0

Table of contents

	Page
Preface	IX
Objective and Background	X

SURVEY OF THE PRESENT SITUATION IN THE EEC

Survey of the incidence of calving problems, calf mortality, and their economic importance: dairy and dual-purpose cattle
D. Smidt and F. W. Huth 3

Present situation of calving problems in the EEC: incidence of calving difficulties and early calf mortality in beef herds
F. Ménissier and J. L. Foulley 30

Discussion 86

GENETIC FACTORS AND BREEDING FOR CALVING PERFORMANCE. PART 1

Selection for double-muscling and calving problems
R. Hanset and M. Jandrain 91

Reproductive performance in crossbreeding: results from a current experiment in the Federal Republic of Germany
H.-J. Langholz, R. F. Diehl and W. Pabst 105

Influence of sire breed on calving performance, perinatal mortality and gestation length
T. Liboriussen 120

Breeding considerations for minimising difficult calving
R. T. Berg 133

Discussion 141

GENETIC FACTORS AND BREEDING FOR CALVING PERFORMANCE. PART 2

A breeding strategy for reducing perinatal calf mortality in heifer calvings
R. Bar-Anan 149

Selection for calving ability in French beef breeds
J. L. Foulley and F. Ménissier 159

Relationship between performance test data and calving performance of test bulls
K. Osterkorn, H. Kräusslich, G. Averdunk and A. Göttschalk 177

Breeding for calving performance
J. Philipsson 189

Sire evaluation for dystocia in Dutch cattle breeds
R. D. Politiek . . . . 206 |

Investigations on the relationships of body measurements and weight of heifer and calf to calving difficulties in German Simmental (Fleckvieh) cattle: preliminary results of EEC project no. 320 of the beef production programme
W. Schlote and H. Hässig 220

Comparison of the main European cattle breeds used in industrial crossing on French Friesian dairy cows: preliminary results on calving difficulties
F. Ménissier, J. Sapa, J. Gogue, B. Bonaiti and J. Frebling 230

Discussion 245

PHYSIOLOGICAL ASPECTS OF PARTURITION

Hormonal mechanism involved in control of parturition in the cow
B. Hoffmann, J. Schmidt and E. Schallenberger 263

The influence of the sire on the oestrogen production of the bovine foetus-placental unit
A. Osinga and W. Hazeleger 282

Neural control of the reproductive tract in the cow as it relates to parturition
M. Rüsse 293

Preliminary observations on myometrial electrical activity before, during and after parturition in the cow
M. A. N. Taverne, G. C. van der Weyden and P. Fontijne 297

Discussion 312

INDUCED PARTURITION

Betamethasone induced calving: a comparison between induced and non-induced dairy cows
K. J. O'Farrell 325

Some results with induced parturition in cows and heifers
H. O. Gravert and E. Kordts 338

Induction of parturition in the bovine
R. W. J. Plenderleith 341

Use of prostaglandins for induction of parturition in the cow
M. J. Bosc 353

Discussion 366

NUTRITION AND MANAGEMENT OF THE DAM IN RELATION TO CALVING PROBLEMS. PART 1

Effect of rearing intensity and age at calving on calving performance
J. Brolund Larsen and K. Sejrsen 375

Pre-calving management and feeding of the beef cow in relation to calving problems and viability of the calf
B. G. Lowman 392

Problems associated with the calving and neonatal period in beef cattle
W. Oxender and W. Adams 408

Discussion 423

NUTRITION AND MANAGEMENT OF THE DAM IN RELATION TO CALVING PROBLEMS. PART 2

Effect of plane of nutrition during late pregnancy on the incidence of calving problems in beef cows and heifers
M. J. Drennan 429

The influence of pre-partal feeding on energy metabolism in early lactation
E. Farries 444

The influence of pre-calving feeding and management of the cow on ease of calving and calf viability
K. Sejrsen and A. Neimann-Sørensen 456

Clinical aspects of the nutritional status of the dam and parturition
E. Grunert 468

Discussion 478

STATUS, NUTRITION AND MANAGEMENT OF THE NEWBORN CALF. PART 1

Conclusions from the EEC seminar on perinatal ill health in calves
H. Thornberry 487

Immune mechanisms in the newborn calf
D. K. Hammer 494

The effect of different methods of feeding colostrum on calf blood serum immunoglobulin levels
R. J. Fallon 507

Discussion 518

STATUS, NUTRITION AND MANAGEMENT OF THE NEWBORN CALF. PART 2

Management of the newborn calf: an attempt at an economic analysis
O. Aalund 525

Treatment of the newborn calf
P. Larvor 539

Acidosis and clinical state in depressed calves
K. Walser and H. Maurer-Schweizer 551

Discussion 564

GENERAL DISCUSSION WITH PANEL OF SESSION CHAIRMEN 569

List of participants 589

PREFACE

This publication contains the proceedings of a seminar held in Germany (Fed. Rep.) on May 4 - 6, 1977, under the auspices of the Commission of the European Communities, as part of the EEC programme of co-ordination of research on beef production.

The programme was drawn up by a combined scientific working group on Genetics and Selection and on Nutrition and Management on behalf of the Beef Production Committee. The working group consisted of Professor Dr. Neimann-Sørensen (Denmark, Chairman), Dr. Brolund Larsen (Denmark), Mr. Boccard (France), Dr. H. de Boer (Netherlands), Priv.-Doz. Dr. B. Hoffmann (Germany, Fed. Rep.), Professor Dr. H.J. Langholz (Germany, Fed. Rep.), Dr. J.W.B. King (UK), Mr. R. Jarrige (France), Mr. B. Vissac (France), Professor Dr. A. Romita (Italy), Professor Dr. E.P. Cunningham (Ireland), Mr. P. L'Hermite (CEC) and Dr. J.C. Tayler (Scientific Adviser to CEC).

The subject chosen for this seminar was drawn from the list of priorities in research objectives drawn up in 1973 by members of a committee (now the Standing Committee on Agricultural Research, (CPRA). One of the functions of this series of seminars was to summarise and update the information available on the selected subjects and to discuss future needs for research, so as to assist the Commission in evaluating the probable impact of research on agricultural production within the Community.

The Commission wishes to thank those representatives of the Member States who took responsibility in the organisation and conduct of this seminar; notably, Priv.-Doz. Dr. B. Hoffmann (Chairman and local organiser) and Dr. J. Schmidt (local organisation) and their other colleagues at the Institute of Physiology, Südd. Versuchs- und Forschungsanstalt für Milch-wirtschaft, Weihenstephan.

Thanks are also accorded to the Chairmen of sessions, Mr. I.L. Mason, Professor R.D. Politiek, Professor Dr. R. Hanset, Dr. M. Bosc, Professor Dr. H. Karg, Dr. R. Bar-Anan, Professor Dr. D. Smidt, Dr. H. Thornberry and Professor W. Oxender.

OBJECTIVE OF SEMINAR

To discuss in an interdisciplinary group of scientists consisting
of geneticists, physiologists, clinicians and animal scientists, problems
related to parturition and early viability of the calf in dairy, beef and
dual-purpose cattle; to review the current situation; to consider practical,
economic and scientific aspects, and to examine objectives and priorities
for future research.

BACKGROUND

In order to overcome the tremendous losses in agriculture due to
perinatal death of the calf, to dystocia, stillbirth and subsequent
fertility problems in the cow, the need has developed to analyse critically
all the factors involved.

Only from conclusions derived from such an effort is it possible to
find adequate ways for improvement. In consequence any improvements made
in that fundamental basis of cattle breeding would certainly greatly speed
up the rate of progress in cattle breeding in general.

::

SURVEY OF THE PRESENT SITUATION IN THE EEC

Chairman: I.L. Mason

SURVEY OF THE INCIDENCE OF CALVING PROBLEMS, CALF MORTALITY, AND THEIR ECONOMIC IMPORTANCE : DAIRY AND DUAL-PURPOSE CATTLE

D. Smidt and F.W. Huth

Institute of Animal Husbandry and Animal Behaviour,
Agricultural Research Centre, Braunschweig-Volkenröde, W. Germany

ABSTRACT

This report on calving problems and calf mortality of dairy and dual-purpose breeds in the EEC relies on the following sources of information:

- scientific publications from the EEC countries,
- statistics available from ministries and organisations in the EEC countries,
- personal communications, and,
- own results.

It describes first the results obtained from observations on calving in dual-purpose breeds made at the Institute of Animal Husbandry and Animal Behaviour over many years. Based on these results, the main inter-relationships and factors influencing the occurrence of calving problems and the loss of calves are emphasised. Special results, such as the differences between farms in spite of homogeneous herds, are used to discuss the problem of objectively scoring the course of parturition.

The second section surveys the statistical data and the experimental evidence on calving problems and calf losses for the four most important dual-purpose breeds of the FRG: German Black Pied (Schwarzbunte), German Red Pied (Rotbunte), German Simmental (Fleckvieh) and German Brown (Braunvieh). The information at hand from the FRG seems, in the opinion of the reviewers, to justify a separate description of the German scene.

The third section deals with the situation in the other countries of the EEC, based on the limited information available to the reviewers. In general it can be stated that, within the EEC, comparable dairy and dual-purpose breeds experience similar rates of incidence in calving problems and calf losses. This is valid with respect to the tendency in absolute figures as well as to the relative importance of certain influential factors such

as breed, bull, calving number of cow, sex and birth weight of calf. The
more dairy-type breeds tend to be less troubled with calving problems and
with the related calf mortality than are the more beef-type dual-purpose
breeds. The systems used for data collection and recording in the EEC
countries are of differing efficiency. This is also true for the measures
taken for decreasing the occurrence of calving problems and calf losses.

The economic aspects which are discussed in the end and the still
rather sketchy framework of information prompt the reviewers to recommend a
representative EEC survey on calving problems and calf losses in dairy and
dual-purpose breeds.

INTRODUCTION

Calving problems and calf losses substantially reduce the economic returns from dairy and dual-purpose breeds. An apparently close relationship exists between the two, since the mortality rate is substantially higher for calves delivered with complications.

The economic losses are direct or indirect. Direct ones are the loss of the calf and the medical expenses arising from obstetric and postnatal care, possibly the loss of the cow and a reduced milk production. Examples of indirect economic losses are infertility, chronic disease and decreased performance of the mother, and also impaired development of the calf.

The rate of occurrence of difficult deliveries and calf mortality in dairy and dual-purpose populations depends to a greater or lesser extent on the type or breed. Other contributing factors are the localisation, the structure and the management of the farm, the age and lactation number of the cow, the bull, the duration of pregnancy and the sex of the calf - both acting primarily through the birth weight, the occurrence of twins and seasonal influences.

Due to the economic relevance of calving problems and calf mortality, the information systems employed in the EEC countries for assessing the performance of dairy and dual-purpose breeds mostly include data on the course of parturition and calf losses. Thus attempts are made to obtain sufficient information to allow a reliable determination of the rates of occurrence of these reproductive failures and for designing counter measures.

In some cases the available information is already sufficient for developing and applying appropriate measures through breeding, management of mating and animal housing with the goal of reducing calving problems and calf mortality.

This review on calving problems and calf mortality in dairy and dual-purpose cattle in the EEC is based on problem-related observations made over the course of many years at the Mariensee Institute for Animal Husbandry and Animal Behaviour (the former Max-Planck Institute) of the Agricultural Research Centre, and on publications, statistics and personal communications from the FRG and the other countries of the EEC.

We are most grateful to all colleagues who kindly responded to our request for information on this subject. Although the amount of information gathered is insufficient to warrant a representative description of the overall situation in the EEC, it should still give a general impression. The evidence provided by the information undoubtedly depends on the criteria used for the compilation and the processing of performance data and on the production structures involved.

With this background a survey of the problems arising at calving and of the mortality of calves in the EEC has to employ a rather coarse screen, such as a differentiation by breeds (populations) and regions (countries).

The report comprises:

1) the description of the results of observations on parturition performance and calf mortality in various dairy and dual-purpose breeds made over many years at the Mariensee Institute. These serve as a somewhat detailed introduction to the problems and interrelationships;

2) a review of information on calving problems and calf mortality in the FRG since the reviewers are naturally most familiar with the situation in their own country;

3) a review of the situation in the other countries of the EEC, which is limited to the information available to the reviewer.

1) Results from the Mariensee Institute

The results of this study, in which the four farms of the Institute participated, are thought to exemplify some of the basic principles and factors inherent in the problem; they will not be discussed again.

From 1965 through 1976 the calvings of the roughly 400 cows of the Institute were recorded and, with respect to the course of parturition, classified into four categories:

1) parturition proceeds without human assistance or surveillance;

2) parturition proceeds with moderate assistance by 1 - 2 persons;

3) parturition is prolonged with intensive assistance by 2 - 3 persons;

4) parturition requires veterinary attention, Caesarean section or embryotomy.

According to this classification, categories 3 and 4 are 'problem births'.

From a total of 4 332 calvings the following distribution was obtained:-

$$
\begin{array}{rll}
\text{Category 1} & : & 2\ 072 = 47.8\ \% \\
\text{"} \quad\quad 2 & : & 1\ 386 = 32.0\ \% \\
\text{"} \quad\quad 3 & : & 747 = 17.3\ \% \\
\text{"} \quad\quad 4 & : & 127 = 2.9\ \%
\end{array}
$$

The resulting rate of incidence of problem births of approximately 20% exceeds the average values for dairy and dual-purpose cattle reported in the literature. This may be due to the more accurate recording practised in the experimental farms or to other farm-specific factors.

TABLE 1 : MAJOR FACTORS INFLUENCING PARTURITION
AS FOUND IN THE MARIENSEE STUDY [1]

Parturition category	n	1 %	2 %	3 %	4 %
Total calvings	4 332	47.8	32.0	17.3	2.9
No. of lactation:					
1	1 486	28.9**	35.7*	28.9**	6.5**
2	1 045	54.3**	31.0	13.8**	0.9**
3	712	55.3**	31.8	12.1**	0.8**
4	453	59.4**	28.2	11.7**	0.7**
5-9	636	64.8**	28.0	5.3**	1.9
Sex of calf:					
male	2 198	43.5**	32.3	20.2**	4.0**
female	2 062	52.2**	31.7	14.3**	1.8**
twins	72	51.4	33.3	12.5	2.8
Birth weight of calf, kg:					
up to 30	296	57.8*	34.8	6.4*	1.0*
31 - 35	1 050	49.5	31.5	17.1	1.9
36 - 40	1 497	49.9	31.0	16.4	2.7
41 - 45	979	45.2	33.7	18.0	3.1
46 - 50	361	37.9**	31.9	25.2**	5.0*
51 and more	149	36.2*	28.9	24.2*	10.7**
Gestation length, days:					
up to 260	137	55.5	29.9	11.7	2.9
261 - 270	1 967	50.2	31.5	15.2*	3.1
271 - 280	2 058	45.8	32.6	18.9	2.7
281 - 290	170	40.0	31.8	24.1*	4.1
Weight of cow, kg:					
up to 459	570	34.4**	32.3	26.3**	7.0
460 - 510	850	39.5**	33.1	23.9**	3.5
511 - 560	1 092	46.7	35.6*	15.5	2.2
561 - 610	957	54.9**	30.3	13.3**	1.5**
611 - 660	605	57.2**	28.6	12.2**	2.0
661 and more	258	61.6**	26.8	9.3**	2.3
Breed of calf:					
German Black Pied	1 760	51.6*	30.5	16.0	1.9*
Holstein-Friesian & crosses with German Black Pied	1 309	46.0	33.9	17.0	3.1
German Red Pied	504	38.3**	26.4*	26.8**	8.5**
Angler	190	47.9	33.7	16.8	1.6
Others	569	48.8	36.9	13.2	1.1

1) The statistical significance was checked by applying the Chi-square test,
taking the averages as the expected values.

* P < 5%   ** P < 1%

Some of the factors which, as shown by this investigation, exert a major influence on the course of parturition are listed in Table 1.

The results of this analysis essentially confirm the previously described factors influencing the course of parturition in dairy and dual-purpose cattle and attributable to either the cow or the bull or the calf. They can be summarised as follows:

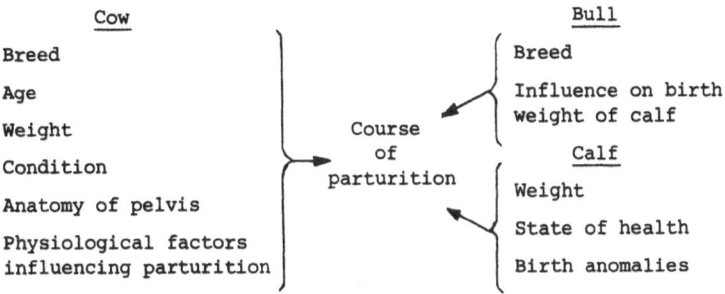

Cow
- Breed
- Age
- Weight
- Condition
- Anatomy of pelvis
- Physiological factors influencing parturition

Course of parturition

Bull
- Breed
- Influence on birth weight of calf

Calf
- Weight
- State of health
- Birth anomalies

The duration of pregnancy and sex of calf act primarily through the birth weight of the calf.

From the results of the investigation prognostic criteria for the course of parturition of the German Black Pied breed can be deduced as shown in Table 2.

TABLE 2

INFLUENCE OF BULL OR COW ON THE RATE OF PROBLEM BIRTHS WITHIN BLACK PIED BREEDS

Bull sires	Light (<35kg)	Normal (36-42kg)	Heavy calves (>45 kg)
Heifer:			
10-12 months old at first mating, retarded growth (>250 kg)	* ?	**	***
10-12 months old, sufficiently developed (~300 kg)	-	*?	**
14-18 months old, underdeveloped (~300 kg)	-	*?	**
14-18 months old, sufficiently developed (~360 kg)	-	-	*
More than 20 months, too fat (>450 kg)	-	*	**
Cow:			
Small frame, normal condition (<500 kg)	-	*	**
Small frame, too fat (550-600 kg)	-	**	***
Normal frame, normal condition (600-650 kg)	-	-	*
Normal frame, too fat (>650 kg)	-	*	**
Large frame, normal condition (650-700 kg)	-	-	-
Large frame, too fat (>700 kg)	-	-	*

* Possibly results in problem birth
** Probably results in problem birth
***Most likely to result in problem birth

It ought to be mentioned that in spite of otherwise comparable conditions in the four experimental farms the incidence of problem births in one of them considerably differed from that in the other three, which again underlines the difficulty of objectively assessing the course of parturition.

329 calves were stillborn or died after birth, which amounts to a mortality rate of 7.6%. The causes are listed in Table 3 in relation to the category of parturition.

TABLE 3

MORTALITY OF CALVES (TOTAL OF 329 CALVES = 7.6%)

Time	Cause			Parturition Category							
				1		2		3		4	
		n	%	n	%	n	%	n	%	n	%
Birth	Stillborn	143		34	10.3	32	13.1	50	15.2	16	4.9
Within 48 hrs. post partum	Death caused by problem birth	28		-	-	-	-	15	4.6	13	4.0
	Low viability	12		6	1.8	6	1.8	-	-	-	-
	Pneumonia	21		13	4.0	4	1.2	4	1.2	-	-
	Enteritis	18		6	1.8	8	2.4	4	1.2	-	-
		222	67	59		61		73		29	
From day 3 post partum	Infections, trauma, etc.	107	33	51		32		23		1	
	Total	329	100	110		93		96		30	

It is evident that an analysis of parturition problems and of calf mortality which, like this one, was restricted to the four farms of the Mariensee Institute can provide only an individual view of the overall situation in a region or country.

2) Survey of the information on calving problems and calf mortality in the FRG.

During the 1960s Dreyer (1965) and Cloppenburg (1966) systematically investigated the course of parturition of dual-purpose cattle in two German Black Pied Associations. Data were collected from test farms.

The frequency of difficult deliveries recorded amounted to 9.2% and 12.5% respectively. The causes of calving problems and the factors influencing them were analysed by the authors.

Although the surveying of representative test farms allows for a detailed insight into the existing interrelationships, this approach is apparently too demanding for a continuous monitoring of the development of the situation.

The actual information is therefore primarily based on the application of the criteria mentioned to the compilation and the processing of performance data from cattle breeding associations.

Evaluations of this kind are available to the reviewer for the German Black Pied (Schwarzbunte), the German Red Pied (Rotbunte), the German Simmental (Fleckvieh) and the German Brown (Braunvieh).

The available data do not yet cover the whole area, but they cover several large regions and can therefore be regarded as a more or less representative picture of the situation in the FRG.

Since the classifications used for evaluating the course of parturition do not exactly agree, only the frequency of difficult calvings for the various breeds are stated in Tables 4 and 5.

TABLE 4

CALVING PROBLEMS IN DUAL-PURPOSE BREEDS IN THE FEDERAL REPUBLIC OF GERMANY

Breed		Cows		Heifers		Source
		No. of Calvings	% Problem Births	No. of Calvings	% Problem Births	
German Black	a)	479 585	1.4	179 196	3.8	Vogt-Rohlf 1976
Pied	b)	4 879	0.6	1 483	3.8	Dreyer 1976
German Red	a)	10 658	1.9	4 913	4.2	Vogt-Rohlf 1976
Pied	b)	32 629	1.5	10 259	5.6	Dreyer 1976

The figures in Tables 4 and 5 confirm the well known fact that in our dual-purpose breeds calving problems occur more frequently in heifers than in cows and that male calves contribute a higher proportion than female calves. They also indicate that the frequency of calving problems in the various population ranges between 1.5 and 5%, tending to somewhat lower values for Black Pied and German Brown as compared with Simmental and Red Pied. The already mentioned differentiation by breed is thus also apparent here.

It must be emphasised again that these values in toto are lower than those found in test farms and in the analysis of single farms. This may be due to differences in the subjective classification of the calving or to deficiencies in the flow of information.

TABLE 5

COMPARISON OF BIRTHS WITH VETERINARIAN ASSISTANCE IN DIFFERENT DUAL-PURPOSE
BREEDS

Breed	Cows		Heifers	
	♂	♀	♂	♀
German Black Pied				
Calvings	14 246	13 436	4 774	4 406
Vet. assistance %	1.5	0.8	4.6	2.2
German Red Pied				
Calvings	5 017	4 815	1 465	1 308
Vet. assistance %	2.3	1.3	7.0	2.9
German Simmental				
Calvings	166 912	158 145	62 217	58 974
Vet. assistance %	3.2	1.6	8.2	2.9
German Brown				
Calvings	63 351	61 351	21 322	20 669
Vet. assistance %	2.3	1.1	6.5	2.0

Source: Annual report, 1975, of the Landeskuratorium der
Erzeugerringe für tierische Veredlung in Bayern e. V.

According to Dreyer (1965) the mortality of German Black
Pied (Ostfriesland) calves up to 14 days of age amounts to 5.49%
of 4 781 calves. 0.5% of these were abortions, 1.71% were stillborn, and 3.28% died within 14 days after birth.

Cloppenburg (1966) found a stillbirth rate for the German
Black Pied of 3.3%; a further 1.6% of the calves died within
24 hours after birth.

Out of 93 000 calves of the German Simmental, German Brown
and German Yellow breeds on 850 Bavarian farms, Walser (1975)
reported a mortality rate of 5.7% up to the age of 6 weeks.

TABLE 6

CALF LOSSES IN DUAL-PURPOSE BREEDS IN THE FEDERAL REPUBLIC OF GERMANY

Breed		Cows		Heifers		Sources
		Number of calvings	% calf losses	Number of calvings	% calf losses	
German Black	a)	479 585	2.0	179 196	4.7	Vogt-Rohlf, 1976
Pied	b)	4 879	2.0	1 483	5.6	Dreyer, 1976
	c)	376 301	3.7	146 318	7.5	Landeskontroll-verband Schles-wig-Holstein e.V., 1976
	d)	-	-	68 590	6.8	Milchkontroll-verband, Westfalen-Lippe, 1977
German Red	a)	10 658	2.2	4 913	7.5	Vogt-Rohlf, 1976
Pied	b)	32 629	3.5	10 259	8.0	Dreyer, 1976
	c)	164 451	4.2	68 109	8.1	Landeskontroll-verband Schles-wig-Holstein e.V., 1976
	d)	-	-	70 260	8.7	Milchkontroll-verband, Westfalen-Lippe, 1977

c) Data from 1973 - 75

Kräusslich (1972) reported 2.4% stillborn calves from heifers and 1.1% from cows of the German Simmental breed.

Tables 6 and 7 summarise the results from recent statistics on calf mortality and stillborn calves. The overall losses (premature births, stillbirths and calves which died within the first weeks of life) range from 2 to 6% for cows and heifers.

According to Walser (1975), 50% and more of the losses result from calving problems. Dreyer (1965) found that 24% of the German Black Pied calves from difficult deliveries were lost, which is about 5 times the normal mortality rate.

TABLE 7

COMPARISON OF BREEDS IN FREQUENCY OF STILLBIRTHS

	Cows		Heifers	
	♂	♀	♂	♀
German Black Pied				
Calvings	14 246	13 436	4 774	4 406
Stillbirths %	2.0	1.3	4.9	2.2
German Red Pied				
Calvings	5 017	4 815	1 465	1 308
Stillbirths %	2.9	1.3	5.8	2.5
German Simmental				
Calvings	166 912	158 145	62 217	58 974
Stillbirths %	2.4	1.4	5.4	1.9
German Brown				
Calvings	63 351	61 351	21 322	20 669
Stillbirths %	2.4	1.5	4.4	1.6

Source: Annual report, 1975, of the Landeskuratorium der Erzeugerringe
für tierische Veredlung in Bayern e.V.

TABLE 8

COMPARISON OF AI AND NATURAL MATING IN RELATION TO FREQUENCY OF STILLBIRTHS
IN VARIOUS BREEDS

	German Black Pied				German Red Pied			
	Cows		Heifers		Cows		Heifers	
	AI	Mating	AI	Mating	AI	Mating	AI	Mating
Calvings	17 455	10 227	4 060	5 120	6 303	3 529	1 000	1 773
Stillbirths %	1.1	2.7	2.7	4.3	1.7	2.7	2.9	5.0
	German Simmental				German Brown			
Calvings	258 822	66 235	87 987	33 204	80 031	44 671	23 793	18 198
Stillbirths %	1.4	4.0	3.1	4.8	1.4	2.8	2.8	3.2

Source: Annual report, 1975, of the Landeskuratorium der Erzeugerringe
für tierische Veredlung in Bayern e.V.

It is therefore quite understandable that cow/heifer, sex, breed, etc. tend to have the same effect on calf mortality as on problem births.

Kräusslich and Gottschalk (1975) made the interesting observation that the stillborn rate from bulls used for natural mating is higher than from artificial insemination bulls, which is probably a consequence of the more intense selection for artificial insemination (Table 8).

According to recent statistics, roughly two-thirds of the total loss of calves in dual-purpose breeds are on account of stillbirths. It should be checked if the total loss figures listed are accurate.

The section of the report dealing with the situation in the FRG focusses on the four most important dual-purpose breeds. Isolated information on local breeds by Angler and, in small numbers, Jersey, are not considered.

The recent statistics, in general, indicate that calving problems occur, on average, in a little more than 5% of the calvings and that the rate of calf mortality is considerably lower than the 10% rule-of-thumb figure.

The problems tend to be somewhat more prominent in meat-type dual-purpose breeds than in the dairy-type.

Between farms, the situation can vary considerably due to effects of region, herd structure, season, hygiene, etc.

3. Survey of the situation in the EEC, outside the FRG

The information at hand from the member countries is, to a varying degree, dependent on the local structure of cattle breeding, related to dairy and dual purpose cattle.

In the countries with a high proportion of beef cattle and crosses between dairy breeds and beef sires, their specific

problems are of prime interest, but these are outside the scope
of this review.

In the following, a few selected data will be presented,
which, as already mentioned, cannot be understood as representative of this population in their country of origin.

For three of the dairy and dual-purpose breeds (RDM, SDM,
Jersey) in Denmark, the rates of difficult deliveries from
heifers and cows have been recorded (Adler and Meding, 1968).
Table 9 shows that the differences between the breeds are sub-
stantial. The rate of calving problems clearly decreases from
RDM to SDM to Jersey.

TABLE 9

PERCENTAGES OF DIFFICULT PARTURITION, VETERINARY ASSISTANCE AND RETAINED
AFTERBIRTH FOR SINGLE BIRTHS.

Breed	Sex	Difficult calving %		Veterinary assist-ance %		Retained afterbirth %	
		1st cal-ving	Later calving	1st cal-ving	Later calving	1st cal-ving	Later calving
RDM	♂	30.7	10.0	16.2	6.8	13.2	11.4
	♀	18.5	5.8	9.8	4.2	10.0	11.6
SDM	♂	14.9	6.5	7.8	3.9	9.2	13.4
	♀	9.3	2.5	6.5	2.2	8.1	10.7
Jersey	♂	3.3	2.1	2.0	1.4	3.4	5.5
	♀	4.1	1.1	2.8	0.7	3.3	5.0

The stillborn rate from the same survey is given in Table
10. The relations are in the same order as for the calving problems, which again points to an interdependence of the two
phenomena.

The figures from official statistics for 1973/74 on the
stillbirth rate and calf mortality are shown in Table 11. The
results are, with respect to the ranking order of breeds, comparable to those obtained by Adler and Meding but show smaller
absolute differences between the breeds.

TABLE 10

THE FREQUENCY OF STILLBORN BULL AND HEIFER CALVES AT PARTURITION WITH ONE
CALF IN DENMARK

Breed	Sex	First calving		Later calving	
		Number	% dead	Number	% dead
RDM	♂	1 864	12.8	4 887	2.6
	♀	1 867	9.7	4 709	1.8
SDM	♂	441	6.6	1 198	1.4
	♀	454	4.2	1 231	1.7
Jersey	♂	249	2.8	582	1.7
	♀	220	4.1	570	1.1

TABLE 11

RATE OF STILLBIRTHS IN VARIOUS DANISH DAIRY AND DUAL-PURPOSE CATTLE

Breed	Number of calvings	% stillbirths	
		Cows	Heifers
RDM	167 017	3.1	11.3
SDM	314 314	2.6	6.9
Jersey	70529	3.7	6.2

Source: Landbrugsministeriets Produktivitetsudvalg
        Husdyrbrugsudvalget (1973/74)

In the Netherlands data on the incidence of problem births
for Dutch Friesian and for Red-and-White are shown in Table 12.

The figures from the 1960s are higher than those presently
obtained; because of the practice of artificial insemination successful measures have been taken to decrease the rate of
problem births (De Kruif, 1976).

The previously discussed interrelationship between age of
cow and sex, and birth weight of the calf have also been found
in investigations on Dutch dual-purpose breeds.

Numerous papers and statistics of Dutch authors deal with
the incidence of stillbirths and perinatal mortality.

TABLE 12

CALVING PROBLEMS IN DUTCH CATTLE BREEDS

Breed	Age	No. of calvings	% calving difficulties	Author
Dutch Friesian	Cows	1 462	5.0	Politiek (1965)
	Heifers	668	18.0	
	Cows + Heifers	1 391	10.8	Grommers et al. (1965)
Dutch Red-and-White (MRY)	Cows + Heifers	761	27.7	Grommers et al. (1965)
Dutch Red-and-White (MRY)	Heifers		19.0	Remmen (1976)

The percentage of stillborn calves, including those which died within 24 h post partum, amounts in the Dutch Friesian breed to approximately 3% for cows and 10 - 12% for heifers. The percentage for Dutch Red-and-White heifers is of the order of 15 - 20% (AI Annual Report, 1974).

Remmen (1976) quotes for the Dutch Friesian stillbirth rates of 8.6% for heifers and 2.5% for cows. The corresponding figures recorded in MRY cattle are 15.4% for heifers and 3.8% for cows. Pronounced differences between the breeds are noted; the ranking order is identical with those observed for similar breeds in the FRG.

TABLE 13

LITERATURE REVIEW OF PERINATAL MORTALITY IN DUTCH FRIESIAN CATTLE (REMMEN, 1976)

No. of births	% stillborn calves from heifers	% stillborn calves from cows	% stillborn calves from heifers & cows	Period	Reference
36 685			3.7	1950-61	Jaarverslagen Federatie voor Vereini-gingen voor KIDO VW te Noordbroek
13 013			5.7		Dabash (1962)
668	12			1/1 to 15/4/63	Politiek (1963)
1 243		1.8		1/1 to 15/4/63	Politiek (1963)
	6.5	1.2		1959/59	Stegenga (1964)
1 391			6.5	1959/62	Grommers c.s. (1965)
10 465			5.2 x)	1969	Remmen c.s. (1973)
15 945			5.5 x)	1970	Remmen c.s. (1973)
	8.5			1968	Jaarverslag K.I. 1968
		2.0		1968	" 1968
6 121	8.3			1969	" 1969
15 911		2.3		1969	" 1969
4 785	7.7			1970	" 1970
9 044		2.0		1970	" 1970
4 889	8.6			1971	" 1971
6 706		2.5		1971	" 1971
1 733	8.6			1972	" 1972
15 519		2.2		1972	" 1972
1 975	8.2			1973	" 1973
12 791		2.6		1973	" 1973
4 921	7.8			1974	" 1974
13 614		2.3		1974	" 1974

x) Calves born live after Caesarean are not included.

Tables including references are taken from Remmen (1976): 'A study of the feasibility of preventing perinatal mortality in calves.'

TABLE 14

LITERATURE REVIEW OF PERINATAL MORTALITY IN MEUSE - RHINE - YSSEL CATTLE
(REMMEN, 1976)

No. of births	% stillborn calves from heifers	% stillborn calves from cows	% stillborn calves from heifers & cows	Period	Reference
109 329			6.1	1953/61	Van Dieten (1963)
27 607	12.5				
81 722		3.9			
307 645			4.3	1962/71	Jaarverslagen K.I."De Kempen" te Oerle
70 411	11.1				
236 732		3.0			
14 392			6.9	1959/61	Jaarverslagen K.I."Peelland" te Lierop
			6.0	1962	"
4 492			7.8	1963	"
6 632	14.0	4.3		1964/67	"
1 465	15.8			1968/71	"
247		8.9		1971	"
761			7.1		Grommers c.s. (1965)
4 198	13.5			1959/62	K.I.Perspectief vereiniging "'t Land van Cuijk"
4 932	15.0			1971	Jaarverslag K.I. "Midden Brabant" 1971
5 981		4.5		1971	"
10 683			6.4 x)	1969	Remmen c.s. (1973)
22 686			6.5 x)	1970	"
	11.1	3.1		1968	Jaarverslag ...K.I. 1968
5 822	10.1			1969	" 1969
1 807		2.7		1969	" 1969
5 771	12.2			1970	" 1970
1 214		3.4		1970	" 1970
4 564	15.4			1971	" 1971
4 820		3.8		1971	" 1971
3 625	15.6			1972	" 1972
2 001		3.6		1972	" 1972
1 656	16.0			1973	" 1973
4 363		4.0		1973	" 1973
3 287	11.7	5.2		1974	" 1974

x) Calves born live after Caesarean are not included

Tables including references are taken from Remmen (1976): 'A study of the feasibility of preventing perinatal mortality in calves.

Data from Belgium on calving problems are summarised in Table 15.

TABLE 15

CALVING PROBLEMS IN VARIOUS MILK AND DUAL-PURPOSE CATTLE BREEDS IN BELGIUM

Breed	Cows		Heifers		Remarks
	No. of calvings	% diffi-cult calvings	No. of calvings	% diffi-cult calvings	
Friesian	71 362	2.03	13 304	14.24	
West Flemish Red	6 377	9.30	1 205	19.75	
East Flemish Red Pied	72 251	6.60	15 924	21.09	Data from nine provinces
MRY	119 980	5.02	23 749	14.49	
Belgian Blue-White	99 793	17.60	17 659	42.62	
East Flemish Red Pied	24 114	5.75	5 740	19.30	Average 1967 - 74

Source: Bouters (1977), Brone (1977)

Statistics on problem births include cases of embryotomy and Caesarean sections. Table 16 gives a survey of the frequency of Caesareans showing clearly that in the Belgian Blue-White breed Caesareans have to be performed in an unbelievably high number of animals due to the high percentage of double muscling in this breed.

Data are also available on calf mortality in the East Flemish breed. The multiannual average of stillborn calves is 1.8% for cows and 3.5% for heifers. The proportion of calves which died post partum is 2.2% for cows and 4% for heifers.

TABLE 16

INCIDENCE OF CAESAREAN SECTIONS IN DIFFERENT CATTLE BREEDS IN BELGIUM IN 1974

Breed	% Caesarean sections in 1974	
	Cows	Heifers
Belgian Friesian	0.45	2.57
East Flemish Red Pied	1.63	8.41
MRY	1.80	9.76
West Flemish Red	4.55	12.73
Belgian Blue-White	17.65	41.89

Source: Brone, personal communication (1977)

From the United Kingdom the incidence of problem births in an experiment on purebred and crossbred Friesians, Ayrshires and Jerseys is available and the results are summarised in Table 17. These results show that also in dairy breed crosses the sire exerts a pronounced influence on the course of parturition.

TABLE 17

INCIDENCE OF CALVING DIFFICULTIES IN DAIRY BREEDS IN THE UK

Breed of cow	No. of animals	% calving difficulties	Breed of sire of calf	No. of animals	% calving difficulties
Friesian	174	18.3	Friesian	70	40.0
			Ayrshire	24	12.5
			Jersey	80	1.3
Ayrshire	150	11.3	Friesian	35	25.7
			Ayrshire	82	8.5
			Jersey	33	3.0
Jersey	134	8.2	Friesian	69	15.9
			Ayrshire	23	0.0
			Jersey	42	0.0
1st parity	261	19.9			
2nd parity	197	4.1			
Overall	458	13.1			

Source: Monteiro, (1969)

A bulletin of the Ministry of Agriculture, Fisheries and Food gives a problem-births rate of 12% for 1 485 calvings of Friesian heifers. The data show the dependence of calving problems on age of heifer. Heifers less than 21 months old at first calving gave 27.3% difficult calvings, it decreased to 9.2% for heifers 30 to 33 months old, rising again to 23.8% when first parturition occurred after an age of 39 months. These results are in agreement with those of Dreyer (1965).

A study of Pointer et al. (1975) arrived at a figure of 13.3% for serious calving problems from 1 156 Friesian heifers.

The abovementioned study of the Ministry of Agriculture, Fisheries and Food on Friesian heifers, quotes a calf mortality rate of 13% from a total of 1 482 calvings. The mortality rate is dependent on the heifer's age at first parturition (less than 21 months: 18.2%; 27 - 30 months: 7.7%; more than 39 months: 38.1%). The mortality rate is also influenced by the season; 12.4 to 14.6% in winter compared with 8.5 to 9.3% in summer.

Table 18 contains some information on the occurrence of problem births in dairy breeds in Ireland. These figures are similar to those of the other EEC countries.

TABLE 18

INCIDENCE OF CALVING DIFFICULTY IN DAIRY BREEDS IN IRELAND

	No. of calvings	% calving problems			Source
		Cows	Heifers	Cows and heifers	
247 spring-calving dairy herds	3 116	4.2	7.7	4.7*	Cunningham et al. (1976)
Friesian	2 690	2.04	-	-	AI Surveys 1974, 1975
Friesian	356	-	-	-	Survey: North Western Cattle Breeding Soc. Ltd., 1973

* Friesian sires: 4.9% : Hereford sires: 3.4%

In France the following data were given by Ménissier (1977). In the Friesian breed, male calves give 7.36% problem births, including Caesareans, while the figure for females is only 5.28%. The corresponding values for crosses between Friesians and Holsteins are 9.04 and 3.90%, for the Normande breed 6.71 and 3.34%, and for the Maine-Anjou breed 22.7 and 11.1%. The well known differences between heifers and cows with respect to problem births have also been observed.

Calf mortalities are: for all Black-and-Whites (including French Friesian and Holstein) 5.1%, for the Normande 3.4%, for the Brune des Alpes 5.1% and for the Brown Swiss 4.7%. This amounts to an overall mortality rate of 4.9%, of which 3.1% are due to stillbirths.

The reviewers had no access to data from EEC countries not mentioned.

CONCLUSION

It must be re-emphasised here that the reviewers cannot provide a representative view of the overall situation in the EEC countries.

However, it does appear that the factors influencing the incidence of problem births and calf mortality in dairy and dual-purpose cattle breeds are similar in all EEC countries. It is further evident that considerable differences exist between the various breeds. The dual-purpose breeds are more prone to calving and rearing problems than the dairy breeds and in the more beef-type varieties they occur with a higher frequency than in the dairy-type dual-purpose breeds. The factors which exert an influence on the incidence of problem births and calf mortality, like breed, sire, calving number, sex, birth weight of calf, season, differ for the various breeds in their absolute values, but not in their ranking order.

It is not the task of this review to discuss the influence

of sire in crossbreeding programmes in dairy and dual-purpose breeds. But it should be mentioned that considerably more data are available on this than on pure dual-purpose breeds. The figures on problem births and calf mortality for individual dairy and dual-purpose breeds change drastically when bulls of beef-type dual-purpose breeds or pure beef breeds are used as compared with the use of pure dairy or dairy-type dual-purpose breeds.

The survival rate of calves as an important economic factor is, according to Kräusslich and Gottschalk (1975), considerably influenced by the rate of perinatal mortality. Here again, a close relationship of problem births and perinatal mortality is evident. The correlation coefficient calculated by Remmen (1976) amounts to 0.86.

The economic losses due to problem births and calf mortalities are composed of direct and indirect ones. Direct ones are the loss of the calf and the medical expenses arising from obstetrical and postnatal care, eventually the loss of the cow and a reduced milk production. Examples of indirect economic losses are infertility, chronic disease and a decreased performance of the mother and also impaired development of the calf.

A quantitative estimation of the economic losses can be deduced from the approximate incidence rates of 2 - 10% for problem births and calf mortality in dairy and dual-purpose breeds.

According to Kräusslich and Gottschalk (1975), a reduction of the mortality rate by 1% means a gain of at least DM 7.00 per living calf. This could, based on the number of artificially inseminated cows, result in an annual gain of 30 million DM in the FRG.

Although this example cannot be generalised without reservations, it is certainly indicative of the order of magnitude of the economic losses experienced in all EEC countries by the still-too-high rate of calf mortality.

REFERENCES

Adler, H.C. and Meding, J.H. 1968 Incidence of stillborn calves and calving difficulties in Red Danish, Black Pied Danish and Jersey cattle. Abstract in Animal Breeding Abstracts 37; 342.

Bouters, R. 1977 Personal communication.

Brone, E. 1977 Personal communication.

Cloppenburg, R. 1966 Geburtsverlauf bei Nachkommen von schwarzbunten Bullen einer westfälischen Besamungsstation. Dissertation Landw. Fakultät Göttingen.

Cunningham, E.P., Shannon, M., Fallen, T.J. and O'Byrne, T.M. 1976 A survey of reproduction, calving and culling of cows in Irish dairy herds. Irish Journal of Agricultural Research 15; 177-183.

De Kruif, A. 1976 Fertility control by means of clinical evaluation on reproduction functions in cattle. EAAP Zürich.

Dreyer, D. 1965 Geburtsverlauf und Kälberverluste, untersucht an Nachkommen ostfriesischer Besamungsbullen in Testbetrieben. Dissertation, Landw. Fakultät Göttingen. 158 pp.

Dreyer, D. 1976 Auswertung über Geburtsverlauf und Kälberverlust der Landesanstalt für Schweinemastleistungsprüfung und Zuchtwertschätzung Neumüchle.

Grommers, F.J., Brands, A.F.A. and Schoenmakers, A. 1965 Mortaliteit van kalveren bij de partus van Nederlandse runderen. Tijdschrift voor Diergeneeskunde 90; 231-244.

Kräusslich, H. 1972 Das Deutsche Fleckvieh. Der Tierzüchter 24; 538-542.

Kräusslich, H. 1975 Die Erfassung der Kälbersterblichkeit und des Geburtsverlaufs in der Besamungszucht. Der Tierzüchter 1; 8-10.

Kräusslich, H. and Gottschalk, A. 1975 Die Erfassung der Kälbersterblichkeit und des Geburtsverlaufs in der Besamungszucht. Der Tierzüchter 27; 8-10.

Landbrugsministeriets Produktivitetsudvalg Husdyrbrugsudvalget 1973/74 Kaelvningsstatistik III. Meddelelse Nr. 10.

Landeskontrollverband Schleswig-Holstein e.V. 1976 Nachkommenprüfung Bullen-Kälberverluste.

Landeskuratorium der Erzeugerringe für tierische Veredlung in Bayern e.V. Jahresbericht 1975.

Ménissier, F. 1977 Personal communication.

Milchkontrollverband Westfalen-Lippe. (Unpublished communication).

Ministry of Agriculture, Fisheries and Food (England and Wales). An investigation into the incidence of dystokia in Friesian heifers.

Monteiro, L.S. 1969 The relative size of calf and dam and the frequency of calving difficulties. Animal Production, 11; 293-306.

Pointer, C.G., Fullbrook, F.J. and Stewart, D.L. 1975 An investigation into the incidence of dystokia in Friesian heifers. EAAP Warsaw.

Politiek, R.D. 1965 Die zuchttechnische Bedeutung einer vollständigen Geburtenregistrierung der Kälber im Rahmen einer KB-Organisation. Der Tierzüchter 17; 457-459.

Remmen, J.W.A. 1976 A study of the feasibility of preventing perinatal mortality in calves. Dissertation, University of Utrecht.

Van Dieten, S.W.J. 1963 Mortaliteit van kalveren bij de partus à terme van M.R.IJ runderen. Dissertation, Faculteit der Diergeneeskunde, Rijksuniversiteit, Utrecht.

Vogt-Rohlf, O. 1976 Statistische Kälbermerkmale. Rechenzentrum zur Förderung der Landwirtschaft in Niedersachsen.

Walser, U. 1976. Kälberverluste unter besonderes Berucksichtigung von Geburtsschwierigkeiten. Der Tierzüchter 9; 10-13.

PRESENT SITUATION OF CALVING PROBLEMS IN THE EEC: INCIDENCE OF CALVING DIFFICULTIES AND EARLY CALF MORTALITY IN BEEF HERDS

F . Ménissier and J.L. Foulley

Station de Génétique Quantitative et Appliquée
Centre'National de Recherches Zootechniques - INRA
78350 Jouy en Josas, France

ABSTRACT

One fifth of the cow population in the EEC consists of suckling cows mainly located in the British Isles and in southern Europe (France and Italy) where they exploit the least productive areas. The populations are rather diverse and include hardy breeds, beef breeds, dual-purpose breeds and cross-bred cows. On the basis of scarce and often heterogeneous statistical data, we describe, by country and regions, the development of calving problems in these populations. The improvement of muscle growth potential realised by beef crossing followed by utilisation of the crossbred cows and grading up as well as by selection, leads in most cases to increase in birth weight of the calves and in size and muscularity of the dams. The result of this is a general rise in calving difficulties and consequently in their repercussions (cost, mortality, subfertility, etc). Such a development is contrary to the present goal of the suckler cow herds which is primarily that of reducing the costs related to herd management.

Most cases of calving difficulty observed in beef populations result from an incompatibility between the size of the calf and the pelvic opening of the dam. Modifications in size and muscularity have repercussions on these two components which may be different depending on whether they are of genetic or environmental origin. These two components act on calving through a threshold effect. The birth weight threshold represents one of the present limits to increase in the growth potential of calves. In the short term most of the interventions tend to reduce the birth weight of the calves so as not to exceed the threshold, but they increase at the same time the growth potential of the calves. We are thus tending towards a larger imbalance between the birth weight of the calves and their growth potential. If this development continues, calving difficulties may be replaced by other problems such as reduction of calf viability. This is the reason why priority should be given to a larger interdisciplinary discussion about research on calving problems.

The present situation of calving problems in the suckling cow herds of western Europe varies between countries and regions. A determination of the magnitude and structure of the European cattle population is necessary for understanding the incidence and development of these problems.

1.   SIZE AND CHARACTERISTICS OF SUCKLER COW HERDS IN THE EEC

The European cattle stock (Table 1) is very large - nearly
80 million head, i.e., over 31 million breeding cows distributed
in more than 40 populations (Cunningham, 1976). In contrast to
the other great cattle stocks of the world, four-fifths of the
cows are dairy cows among which the Friesian cow populations
represent about half. The other dairy cow populations are composed of milked dual-purpose breeds (22 populations) where only
a few breeds dominate (Simmental, Normande, Brown Swiss, Red-and-
White MRY). Among these dairy cows, beef crossing is still little
developed, but with large variations between populations and
between countries (Cunningham, 1976). Accordingly, an essential
part of the beef production in the EEC depends on the dairy cow
herds.

However, although representing only one-fifth of the cows,
the suckler cow herds are still of great importance for European
beef production. Primarily in absolute value, a stock of 6
million cows is far from being negligible. In addition, it may
be predicted, without committing great errors for the coming
years, that the size of the suckler cow herd will increase
either by extension of present herds from specialised beef breed
populations or by reconversion of milking into suckler cows due
to a greater specialisation of dairy production (Anonymous, 1972;
Ménissier et al., 1976; Bibé et al., 1977). Furthermore, the
suckling cow herds are exploiting more and more the areas of low
potential and productivity where intensification of milk or
cereal production is difficult (Bibé et al., 1977; Vissac, 1976).
Their production systems are therefore more extensive with low
cost inputs : minimum animal care, better adjustment of winter
feeding to minimum requirements etc.

Thus, the European suckling cow populations are rather diverse in their breeding type, management systems and localisation.
As a first approach, these populations can be grouped according
to the type of female which they exploit (Ménissier, 1976). We
distinguish between the three following groups:

TABLE 1

EEC: CHARACTERISTICS AND STRUCTURE OF THE SUCKLER COW POPULATIONS

Country	Cattle populations (a)				Structure of suckler herds (b)						Crossbred cows: beef x		
	Number (1000s)				Specialised beef breeds			Hardy breeds		Dual-purpose breeds part-ially used as suckler cows	milk	hardy	beef
	Cattle	Cows (1)	Suck-ling (2)	%Suck-ling (2/1)	% of (2)	British	Conti-nental	% of (2)	Hardy breeds				
United King-dom	14 696	5 307	1 801	34 37 b	5	He(50) SD(25) AA(17) = 92	Ch(3.5) Lm(0.6) = 4	3	WB(23) Gl(15) Sx(10) De(8) LR(5) = 61	Sh(5), Sm(2.5)	+	+	+
Ire-land	6 408	2 070	684	33	⩽5	?	?	?	= 61	?	+	+	+
France	23 896	10 174	2 496	25	52	−	Ch(892) Lm(301) BA(95) = 1 288	5	Au(58) Ga(30) others (25) = 113	Sa(150),Sm(293) MA(143),BS(140)	+	+	
Italy	8 407	3 795	744	20	66	−	Mg(258) Cn(146) Ro(49) Mr(35) = 488	4	Podolian of south (25) Sr(1.5) = 27	Pd(268),Mo(81) BS(688)	+	+	

TABLE Continued

TABLE 1 (Cont)

	Cattle populations (a)				Structure of suckler herds (b)						Crossbred cows:	
	Number (1000s)				Specialised beef breeds			Hardy breeds		Dual-purpose breeds part-ially used as suckler cows	beef x	
Country	Cattle	Cows (1)	Suck-ling (2)	%Suck-ling (2/1)	% of (2)	British	Conti-nental	% of (2)	Hardy breeds		milk	hardy beef
Belgium	2 896	1 084	66	6	?	?		?		BB(480)	+	
Luxem-bourg	208	78	6	8								
Denmark	2 956	1 203	49 87 b	4	44	He(9) AA(1) = 10	Ch(11) Lm(0.4) = 11	?		?	+	
Germany	14 364	5 639	153	3	?	AA(?)	Ch(?)	?		GY(154)?	+	
Nether-lands	4 668	2 171	0	0	-			-		-		
TOTAL	78 499	31 524	5 999	19		He,SD, AA	Ch,Lm, BA,Cn, Mg	(Various)		(Various)	+	+

a) In December 1973, according to Cunningham (1976) and the statistical office of the European Communities (Euro-stat, 1974)

b) According to Cunningham (1976) (EEC), Baker (1976) (UK), Duplan and Bougler (1976) (F), Bettini and Nardone, (1976) (I), Hanset (1976)(B), Larsen and Sejrsen (1975) (DK).

The figures give the number of cows ('000) and the percentage of the different type among suckling cows.

He = Hereford, AA = Aberdeen-Angus, SD = South Devon, WB = Welsh Black, Gl = Galloway, Sx = Sussex, De = Devon, LR = Lincoln Red, Sh = Shorthorn, Sm = Simmental, Sa = Salers, MA = Maine-Anjou, BS = Brown Swiss, Ga = Gascon, Au = Aubrac, Ch = Charolais, Lm = Limousin, BA = Blonde d'Aquitaine, Mg = Marchigiana, Cn = Chianina, Ro = Romagnola, Mr = Maremmana, Sr = Sarda, Pd = Piemontese, Mo = Modicana, BB = Belgian Blue-White GY = Gelb-vieh

c) In 1976, according to Liboriussen (1977) personal communication

a) <u>Specialised beef breed populations,</u> mostly in regions
with natural pastures of good quality: these breeds are
either beef breeds of the British type (small and fat) or
beef breeds of the Continental type (large and lean) of
French origin (former draft breeds) or of Italian origin
(former hardy or draft breeds).

b) <u>Hardy breed populations</u>, managed in marginal areas
(especially in the mountains) and formerly used for multiple
purposes (milk, meat, draft): these populations, with varied
characteristics, are still numerous but generally not very
large. They are being reconverted into suckling cows by
crossing with specialised beef breeds, but they are still
exploiting their initial environment.

c) <u>Crossbred female populations</u>, generally managed in an
environment comparable to that of their original maternal
populations. These animals are derived from beef crossing
between beef bulls and populations of either dairy or dual-
purpose breeds(Fl dairy x beef) or hardy breeds (Fl hardy
x beef) or beef breeds (Fl beef x beef). This female stock
might also include cows from dual-purpose breeds which are
no longer milked and are becoming suckler cows.

Most of these suckler cows are located in southern Europe
(France and Italy) and in the British Isles (Ireland and the
United Kingdom), where they represent more than 20% of the
breeding cows (Table 1). In France and in Italy more than half
of the suckler cow population is composed of specialised beef
breeds of the Continental type, which are almost exclusively
kept in purebreeding; the remaining suckling cow population of
southern Europe is composed of cows of hardy breeds, dual-purpose
breeds (part of Salers, Maine-Anjou in France, Piemontese and
Brown Swiss in Italy) and especially of crossbreed cows (dairy
x beef and hardy x beef) generally mated in crossbreeding. In
the British Isles, the population of purebred British beef breeds
represents less than 5% of the suckler cows, ie, scarcely more
than the hardy breeds; at present, most suckler cows are crossbred
dairy x beef (Hereford or Angus x Friesian), hardy x beef, or
beef x beef (Continental type x British type).

In these various populations of suckler cows, statistics about parturition problems, chiefly about calf mortality, are scarce and lack uniformity as regards the definition of criteria observed and populations involved. It is generally difficult to obtain information about crossbred cow populations, especially in extensive systems where most of the cows are not recorded or not even checked. Starting from the most available but still not exhaustive information, we have limited our analysis to a comparison of the main populations of suckler cows by country and large regions.

2. CALVING PROBLEMS IN NORTHERN EUROPE (Germany, Belgium, Luxembourg, Netherlands, Scandinavia)

The cattle populations in these countries are dominated by dairy cows of specialised and dual-purpose breeds, mostly kept in purebreeding but in which a trend towards crossbreeding can be noticed (Netherlands excepted). The number of suckling cows, although still small (8% of the cows) is increasing, particularly in Denmark and in Sweden where the increase has been rather fast (+7 000 and + 5 000 cows respectively, Lindhé, 1976); according to the literature there are twice as many suckling cows in Denmark now as in 1973 (87 000 cows in June, 1976; Liboriussen, 1977, personal communication). Some of these suckling females are of specialised beef breeds and are in particular used to produce the sires required for crossbreeding with dairy cows; their selection is directed towards improvement of calving conditions in crossbreeding, maintaining the growth potential at the same time (Lindhé, 1974, 1976 and 1976$_a$). These breeds are both of the British type, introduced many years ago into Scandinavia (especially Hereford) and Germany (especially Angus), and, more recently, of Continental type (first Charolais, then Limousin and Italian breeds). However, most of the suckler cows are still crossbred dairy x beef cows derived from crossing with all these beef breeds.

2.1. Scandinavia

In Scandinavia (Table 2), use of small-sized beef bulls

TABLE 2

SCANDINAVIA: INCIDENCE OF CALVING DIFFICULTIES AND EARLY CALF MORTALITY IN THE MAIN CATTLE POPULATIONS

Type of mating	Denmark Number of calvings	% Calvings Diff	Very diff	% early mortality	Ref.	Sweden Number of calvings	% Diff. calvings	% Still-births	Ref.	Norway Number of calvings	% Calvings Diff	Very diff	% Still-births	Ref.
Charolais (Ch)	3 382	14	9	4	(1)	3 623	10.9	7.1	(2)					
						-	10	-	(4)					
Hereford (He)						7 973	4.8	5.3	(3)					
						-	3	-	(4)					
Ch} x ♀ He} (back cross)						-	7	-	(4)					
						-	4	-	(4)					
Ch} x ♀ He} (beef x milk, dual-purp)						-	7	-	(4)					
						-	3	-	(4)					
Ch	114	62	10	8	(5)	-	4	-	(4)	567	11.0	3.7	2.8	(10)
						2 856	*13.2	4.2	(7)	982	6.0	2.3	3.8	(10)
						1 023	*15.1	4.7	(9)					
Lm x ♂ (milk, dual-purpose)	113	31	5	4	(5)	-	3	-	(4)	985	4.0	1.9	3.1	(10)
He	134	28	3	3	(5)	2 968	*5.8	2.8	(8)	-	-	-	-	-
						636	*7.2	3.1	(9)					
AA	-	-	-	-	-	59	(O.H)	(3.5H)	(9)	-	-	-	-	-
Milk, dual-purpose breeds	1 484	(15H)	(5H)	(8H)	(6)	8 059	*8.7	3.7	(9)	28 910	-	-	2.4	(11)

Lm = Limousin, AA = Angus, H = Heifers, (*) = adjusted for 30% of heifers and 70% of cows.

TABLE Continued

TABLE 2

SCANDINAVIA: INCIDENCE OF CALVING DIFFICULTIES AND EARLY CALF MORTALITY IN
THE MAIN CATTLE POPULATIONS (Cont)

REFERENCES

1) Danish Charolais Breeders Association (Liboriussen, 1977, person. comm.).

2) Field recording for 1971, 1973 and 1975 (Lindhé, 1976).

3) Field recording for 1971, 1973 and 1975 (Lindhé, 1976$_a$).

4) Progeny testing on farm on 5 000 calves from 1968 to 1971 (Kalm et al.,
1974).

5) Experimental results of crossbreeding on dairy (SDM) and dual-purpose
breeds (RDM), (Bech Andersen et al., 1976).

6) Results of progeny testing stations; heifers calving at 30 months
(Hansen, 1975).

7) Progeny testing on farms (Lindhé, 1976).

8) Progeny testing on farms (Lindhé, 1976$_a$).

9) Results from artificial insemination centre from 1967 to 1970 (Philipsson, 1977).

10) Progeny testing from 1972 to 1974 (Gravir, 1977, personal communication).

11) Milk recording in 1970 (Auran, 1972).

(Hereford and Limousin) on dairy cows does not increase calving difficulties and calf mortality (less than 7% and 4% respectively). However, Charolais bulls increase the mean frequency of dystocic calvings (10 to 15%), but with large differences between bulls; early calf mortality is only slightly increased. In the suckling cows of these beef breeds, essentially managed in purebreeding, the frequencies of calving problems are comparable with those observed previously with the same breeds in beef crossing: purebred or crossbred Hereford cows calve as easily as do dairy cows, whereas Charolais cows exhibit substantially greater calving difficulties (about twice as many dystocic calvings). However, this frequency of dystocic calvings is associated with a smaller increase of early calf mortality.

2.2. Germany

In Germany (Table 3), the very few statistical data available seem to indicate that, among milking cows, the large-sized dual-purpose breeds exhibit the highest frequency of difficult calvings related to the excessive size of their calves; this is even more marked in heifers. Here again, early calf mortality is almost not affected. These dairy cows of dual-purpose breeds, managed as suckler cows, have as many calving problems as the beef breed populations, 4 to 7% dystocic calvings and 3 to 8% early calf mortality. The crossbred cows from these dual-purpose populations (F1 Charolais x large-sized dual-purpose breeds) might lead to as many calving problems as large-sized beef breeds. Among the suckler cows of beef breeds, use of Angus cows (purebred or crossbred) substantially reduces the frequency of parturition difficulties (less than 10% of assisted calvings), but it does not lead to the disappearance of the early calf mortality. But, as observed in the previous countries, use of Charolais cows tends to increase the frequency of dystocic calvings (6 - 10%) without excessive increase in early calf mortality.

2.3. Belgium

In Belgium (Table 4), it can be observed once again that the large-sized and heavy-muscled dual-purpose breeds exhibit definitely more dystocic calvings than the specialised dairy

TABLE 3

GERMANY: INCIDENCE OF CALVING DIFFICULTIES AND EARLY CALF MORTALITY IN THE MAIN CATTLE POPULATIONS

Breeding types	Number tested bulls (a)		% very diffi- cult calvings (b)		% Still births		Pro- portion among suckling cows	% diffi- cult calvings	% early mort- ality
	H	C	H	C	H	C	H+C	H+C	H+C
Angus + crossbred							56%	2.2	2.1
Charolais + crossbred							25%	10.2	4.4
F1 (Angus x dual-purpose breeds									
Vorderwälder	8	12	1.8	0.4	0.4	1.3			
Gelbvieh (Yellow)							19%	4.6	3.0
Rotbunte (Red Pied)	14	18	1.7	0.5	2.5	0.9			
Fleckvieh (Simmental)	226	384	2.5	1.2	1.5	1.0			
German Brown	36	56	2.0	0.4	1.9	0.8			
Friesian	79	90	1.6	0.4	1.6	0.7			
Source and type of data	←——— MILKING COWS———→ In pure breeding (official/ recording). Official statistics from Baden/Württemberg						←——— SUCKLING COWS———→ In crossbreeding. Recording in 109 farms from 6 areas in Germany (about 4 500 cows		
Reference:	Schlöte (1977, personal communication)						Fedeler et al. (1973)		

a) H = heifers; C = cows,   b) Only caused by heavy calves

TABLE Continued

TABLE 3 (Cont)

GERMANY: INCIDENCE OF CALVING DIFFICULTIES AND EARLY CALF MORTALITY IN THE
MAIN CATTLE POPULATIONS

Breeding types	Number of calvings	% Calvings Assisted	Very diff.	% early mort-ality	Number of calvings	% Calvings Assisted	Very diff.	% early mort-ality
	H+C	H+C	H+C	H+C	H+C	H+C	H+C	H+C
Angus +crossbred	60 447	8.3 8.5	0.0 2.9	6.7 1.8	101 –	39.7	3.0	8.2
Charolais + crossbred	66 155	65.2 43.2	6.1 2.6	3.0 5.2	– 63	54.1	4.9	5.4
Fl (Angus x dual-purpose breeds	89	65.1	12.2	5.6	–			
Vorderwälder								
Gelbvieh (Yellow)					99	62.5	4.0	6.3
Rotbunte (Red Pied)					82	69.4	7.1	7.0
Fleckvieh (Simmental)								
German Brown								
Friesian					89	66.0	5.5	3.5
Source and type of data	SUCKLING COWS Farms in Westphalia				SUCKLING COWS Pure breeding (except for first calving). Experimental herd.			
Reference:	Haas (1972)				Kanning and Langholz (1975)			

a) H = Heifers; C = cows.   b) Only caused by heavy calves

breeds, chiefly in heifers. The extreme case is that of the
Belgian Blue-White breed, which, according to regions, has reached
frequencies of dystocic calvings incompatible with most production
systems (suckling and milking). Such a rise in calving problems
is connected with a search for heavy-muscled animals in order
to produce a maximum of double-muscled calves. Figure 1 perfectly
illustrates this trend.

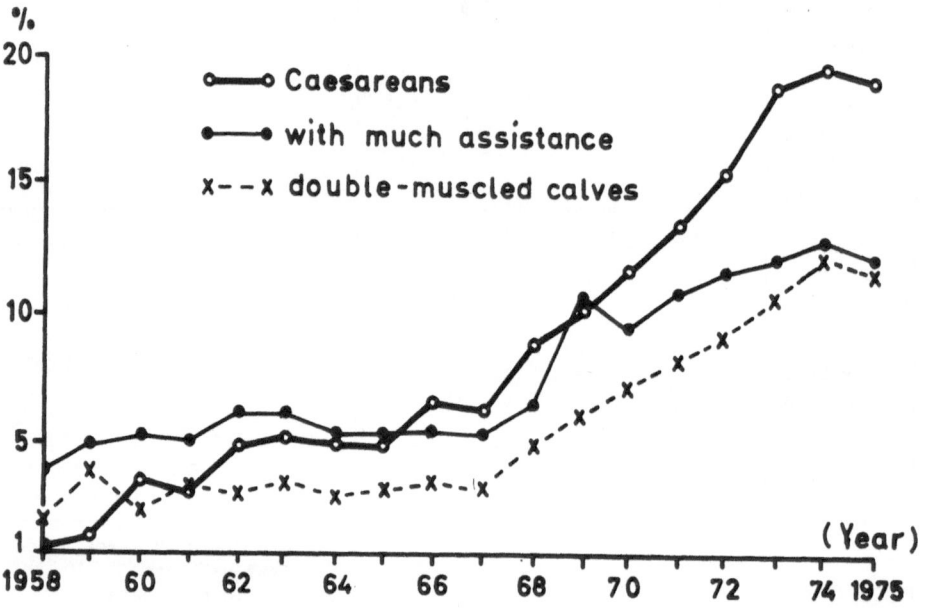

Fig. 1   Increase of Frequencies of calving difficulties and of double-muscled
         calves in Belgian Blue-White breed in the area of Namur.

Source: Report: AI Centre of Ciney, Belgium (Dimitropoulos, 1975).

As shown elsewhere (Fagot, 1965; Dérivaux et al., 1965;
Vissac et al., 1973; Ménissier, 1974), the double-muscled trait
increases the size and compactness of the calves and at the same
time leads to reduction of the pelvic opening and disturbances
of the parturition behaviour in the dam (even for those that do
not show an extreme muscular hypertrophy). In France, in the
case of a specialised beef line selected for this double-muscled
trait (Ménissier, 1974; Bibé et al., 1977), the rate of calving

TABLE 4

BELGIUM: INCIDENCE OF CALVING DIFFICULTIES AND EARLY CALF MORTALITY IN THE MAIN CATTLE POPULATIONS

Breed	Type of production (a)	1965(b) Number of calvings	1965(b) % dystocic calvings	1972(c) % of calvings: Difficult H	Difficult C	Caesarean H	Caesarean C	1974(d) Number of calvings H+C	% of calvings Difficult H+C	Caesarean H	Caesarean C
Friesian	M	77 878	1.7	5.0	1.7	1.6	0.3	–	–	–	–
Belgian MRY	M/DP	102 763	3.3	–	–	–	–	–	–	–	–
East Flemish Red Pied	DP	102 399	1.6	11.6	6.6	5.8	1.9	–	–	–	–
West Flemish Red	DP	37 856	6.9	–	–	–	–	–	–	–	–
Belgian Blue-White	DP										
Limburg		14 860	3.0	–	–	–	–	–	–	–	–
Luxembourg		8 346	10.0	–	–	–	–	–	–	–	–
Namur		44 786	9.4	–	–	–	–	13 817	32.6	30.4	15.6
Liège		–		42.4	20.1	36.5	14.7	–	–	41.9	17.7

(a) M = milk; DP = dual-purpose; H = heifers; C = cows

(b) Data from artificial insemination (Van de Plassche et al., 1965)

(c) Data from artificial insemination (Hanset, 1974)

(d) Data from artificial insemination at Namur (Dimitropoulos, 1975) and at Liège (Hanset, 1976)

difficulties is as high as in the Belgian Blue-White breed (20 to 50% Caesarean sections: Table 5). The similar situation of the dual-purpose Piemontese breed in Italy, will probably result in comparable calving problems.

TABLE 5

CALVING CONDITIONS OF DOUBLE-MUSCLED CHAROLAIS AND BLONDE D'AQUITAINE COWS IN PUREBREEDING AND CROSSBREEDING WITH DOUBLE-MUSCLED CHAROLAIS AND BLONDE D'AQUITAINE BULLS

Breed of cow	Charolais	Blonde d'Aquitaine	Total
Number of calvings (a)	76	106	182
Frequency of calvings:	(b)	(b)	
Without assistance	14%(55)	37%(67)	27%
Little assistance	13%(37)	35%(28)	26%
Difficult	18%(5)	9%(5)	13%
Caesarean	54%(3)	19%(0)	34%

(a) 5 - 6 years of mating

(b) Mean frequency, in purebreeding, in the recorded herds in France (1973 - 1974) for normal cows

Source: INRA, 1976 (unpublished data)

Thus, in northern Europe, where milking cows are dominant, purebred or crossbred suckler cows from small-sized beef breeds (essentially British type) calve at least as easily as the milking cows kept in pure- or crossbreeding, and often show fewer cases of early calf mortality. On the other hand, suckling cow populations derived from large-sized beef breeds (Charolais, for instance) or large-sized dual-purpose breeds (Simmental, for instance) cause about twice as many dystocic calvings without leading to a similar increase of the early calf mortality. These calving problems are even more severe in heifers which often calve earlier in these regions (before 30 months - Lindhé, 1976 and 1976$_a$; Kögel and Kräusslich, 1975) than in the traditional regions of suckling cows (Allen, 1975; Jarrige, 1975). Finally, frequencies of difficult calvings in suckling herds of beef breeds remain comparable with those of the other European regions.

3. CALVING PROBLEMS IN THE BRITISH ISLES (Ireland and the United
   Kingdom)

The Meat and Livestock Commission (MLC) (Great Britain) has
established statistical data about the incidence of calving
problems in the British stock of suckler cows which represents
about half of the suckler cow population of the EEC. However,
these data mainly concern pedigree herds (purebreeding) while
the largest part of this population is made up of crossbred cows
(Table 1); for the latter the establishment of statistics on
calving problems has been started more recently (Allen, 1977).
In Ireland, analysis of the data collected by Al stations
(Crowley, 1965; More O'Ferrall and Cunningham, 1977; Teehan,
1977) or field recording (Cunningham et al., 1976) confirm
rather well the observations made by the MLC.

In commercial herds (Table 6), about 5% of calvings require
forced extraction. The variations between production systems
are large. The frequency of dystocic calvings is two or three
times higher (7 to 10% v 2 to 3%) in herds using large-sized
sires (Charolais, Simmental, Limousin, South Devon) than in those
using traditional British beef sires (Hereford or Angus). Likewise, dystocic calvings are much more frequent in large-sized
cows such as crossbred Charolais and Hereford x Friesian (11% and
8% respectively) than in crossbred Angus or Hereford x beef cows
(4%) and especially the typical Blue-Grey cows (1%). Part of
these differences is due to the management system of these suckler cow populations. As a matter of fact, the large-sized
maternal breeding types are preferentially sired by paternal
breeds with a high growth potential (exotic breeds and South Devon)
and managed in the lowlands where calvings are proportionately
more difficult (6.9% of dystocias and 2.6% of early calf mortality)
than in the uplands (3.6% and 0.6% respectively) or the hills
(3.2% and 0.7% respectively) (Barnes and Kilkenny, 1976). Thus,
intensive use of suckler cows with a poorer calving ability brings
about a large increase in calving difficulties as compared with
a more extensive use of suckler cows with a higher calving ability.

TABLE 6

GREAT BRITAIN: INCIDENCE OF CALVING DIFFICULTIES IN COMMERCIAL HERDS OF SUCKLER COWS (a)

Breeding type of Cows	Number of calvings	% of dystocic calvings						Frequency of recorded mating for 100 cows				Localisations (Frequency of calvings)			Wt. of cows (kg)
		Total	Sire breed (c)					Angus	Hereford	Red South Devon	Exotic breeds	Lowland	Upland	Hill	
			Angus	Hereford breeds	Red breeds	South Devon breeds	Exotic breeds								
Beef breed x Angus	360	3.9	(d)	O	(d)		2	7	31	6	57	41	29	30	413
Beef breed x Hereford	1 268	3.9	3	10	1	14	3	17	37	23	24	43	24	33	485
Blue-Grey (b)	834	1.1	0	1	1		2	13	38	20	30	48	37	15	450
Charolais x beef breed	267	11.2	(d)	(d)			14	2	15	17	67	73	19	8	628
Angus x Friesian	61	4.9	(d)			(d)	(d)		44	15	41	23	54	23	419
Hereford x Friesian	2 018	7.9	5	6	8	8	9	4	28	35	34	72	18	10	472
Total (e)	4 858 (8 765)	5.6 —	2.5 (1.3)	3.2 (2.1)	7.0 (2.1)	10.1 —	6.9 (5.8)	—	—	—	—				Esti- mate

a) Study on field recording, with 128 commercial herds and mature cows (less than 5% of heifers). Barnes and Kilkenny, (1976)

b) Blue-Grey = crossbred between White Shorthorn and Galloway.

c) 'Red breeds' = Devon, Lincoln Red, Sussex: 'Exotic' = Charolais, Simmental, Limousin

d) Less than 50 recorded calvings

e) Similar results from Ireland (results from artifical insemination in 1976 on milking and suckling cows - Teehan, 1977)

In pedigree herds (Table 7), the mean frequency of calving problems seems lower (3.5%) than in the commercial herds. As in the previous situation, part of the differences between populations is due to their management system and contributes to increase these differences. Although there are large differences between breeds in frequency of dystocic calvings, the rates of surgical intervention and early calf mortality are generally low except in heifers. The large-sized breeds producing calves exceeding 40 kg (chiefly Charolais and Simmental) cause the highest percentage of dystocic calvings: about 10% in cows and from 10 to 20% in heifers, with an early calf mortality of 1 to 4%. This difference is still more obvious in heifers in which 4% are calved by Caesarean section and the early calf mortality reaches 6%. The small-sized British breeds produce 30 to 40 kg calves and induce definitely less dystocic calvings, but there are differences between breeds. The Aberdeen-Angus produces the least calving problems whereas the Hereford, with heavier calves (+4 to +7 kg), shows a higher percentage of dystocic calvings (+2 to +3%) and even a higher early calf mortality rate (+1%). The results from the other local breeds are situated between these two breeds but they are closer to the Hereford. It is also interesting to notice (Table 7) that the differences between large-sized Continental beef breeds and British beef breeds are substantially higher as compared with those recorded in beef crossing on dairy cows. Calving difficulties are about two to three times more frequent in suckler cows than in dairy cows whereas the early calf mortality tends to be lower.

Thus, in the suckler cow populations of the British Isles, the passage from small-sized breeds with a high calving ability to populations of larger-sized breeds (Continental beef breeds, dual-purpose breeds and crossbred dairy x beef cows) leads to an increase in the weight of the calves and to a substantial rise in the frequency of dystocic calvings (2 to 3 times more for cows and heifers). This incidence is generally increased by a more intensive management of these populations of suckler cows as compared with the local breeds. The increase in the early calf mortality was relatively less.

TABLE 7

GREAT BRITAIN: CALVING DIFFICULTIES AND EARLY CALF MORTALITY IN PEDIGREE HERDS OF SUCKLER COWS IN COMPARISON WITH MILKING HERDS

Sire breeds	Pedigree Herds (a)						
	Number of calvings		% dystocic calvings		% dead before 48 h		Birth weight (kg)
	C	H	C	H	C	H	C
Charolais	913	157	11.3	25.5	0.6	1.3	42.4 ⎫
Simmental	541	219	11.8	18.7	0.7	0.9	41.1 ⎬
South Devon	700	141	5.6	11.3	2.1	4.3	41.2 ⎭
Limousin	-	-	-	-	-	-	-
Lincoln Red	1 420	299	1.3	3.3	2.2	4.7 ⎫	
Devon	805	148	3.1	3.4	1.2	2.0 ⎬	37.6
Sussex	1 606	293	1.5	2.3	2.2	4.0 ⎭	
Welsh Black	1 700	321	1.2	3.1	0.7	1.9	
Hereford	3 749	718	3.7	10.4	0.4	1.1	36.6
Angus	1 012	218	0.7	2.3	0.0	0.0	29.3
Total	12 446	2 514	3.5	8.3	1.0	2.1	-

Sire breeds	Pedigree and commercial herds (b)							
	% dystocic calvings			Surgical calvings		% dead before 48 h		Birth weight (kg)
	C (b)	C (c)	H (c)	C (c)	H (c)	C (c)	H (c)	C
Charolais		9.0		1.2		4.6		43.2
Simmental	9.1	8.9	14.2	1.2	3.8	4.4	6.3	42.1
South Devon	7.0	8.7		0.9		4.2		41.4
Limousin	-	7.4		0.7		3.8		39.5
Lincoln Red		6.7	9.7	0.3	2.0	3.2	4.7	39.3
Devon	2.8	6.4		0.3		3.6		38.5
Sussex		4.5		0.2		2.1		36.6
Welsh Black	-	-	-	-	-	-	-	-
Hereford	3.6	4.0	6.5	0.3	0.6	2.0	2.9	34.3
Angus	1.4	2.4	4.1	0.1	0.0	1.6	0.0	30.4
Total	-	-	-	-	-	-	-	-

TABLE Continued

TABLE 7 (Cont)

GREAT BRITAIN: CALVING DIFFICULTIES AND EARLY CALF MORTALITY IN PEDIGREE
HERDS OF SUCKLER COWS IN COMPARISON WITH MILKING HERDS

Sire breeds	Crossbreeding in milking herds (d)					
	Number of calvings		% dystocic calvings		% dead before 48 h	Birth weight (kg)
	C	H (e)	C	H (e)	C	C (e)
Charolais	4 656	123	3.2	5.7	3.9	46.1
Simmental	4 499	362	3.5	8.8	4.5	43.9
South Devon	–	–	–	–	–	–
Limousin	972	388	2.4	8.2	3.3	41.6
Lincoln Red	–	–	–	–	–	–
Devon	–	–	–	–	–	–
Sussex	–	–	–	–	–	–
Welsh Black	–	–	–	–	–	–
Hereford	15 728	1 913	1.1	2.7	2.8	40.8
Angus	–	3 619	–	1.7	(5.8)	–
Total	21 850 1 595 (Friesian in purebreeding)		2.5	5.7	–	40.4

a) From field recording in 799 pedigree herds (Allen, 1977; Barnes and
   Kilkenny, 1976).
b) From field recording in 928 pedigree and commercial herds (Barnes and
   Kilkenny, 1976).
c) From field recording in about 800 pedigree and commercial herds (Kilkenny
   and Stollard, 1976).
d) Pooled results of some reports of progeny testing from Milk Marketing
   Board (O'Connor, 1977).
e) From results of progeny testing by Milk Marketing Board (Frapel, 1975).

4. CALVING PROBLEMS IN SOUTHERN EUROPE (Italy and France)

4.1. Italy

In spite of the presence of large populations of suckler cows in Italy, we have not found any overall statistical data on the incidence of calving problems in Italy, either in the large-sized breeds (Chianina, Marchigiana, Romagnola) or in the Piemontese breed, or in the hardy breeds.

The Piemontese breed (milking cows) is more and more used and selected as a beef breed for terminal crossing on dairy cows (Bettini and Nardone, 1976), especially in introducing the double-muscle trait. According to the latest results of the testing of sires on the slaughter value of their purebred progeny (Anonymous, 1976), 5.4% of the 7 829 registered calvings (86% multiparous cows) are made by forced extraction and concern calves weighing 15 kg more than the average (43 kg). When bred to Friesian cows, comparable results are found (4.4% of very difficult calvings for calves weighing 12 kg more than the average birth weight of 44.3 kg). This suggests that the increase in the muscularity of Piemontese bulls, as in the Belgian Blue-White breed, is accompanied by an increase in calving problems.

As regards the populations of hardy breeds or of crossbred hardy x beef cows, the first experimental results obtained in Sardinia (extensive production system - Casu et al., 1975) give already some indications (Tables 8A and B). In old cows of hardy breeds (Table 8A), commercial crossing with beef bulls (even with the Charolais breed) does not cause any calving problems; less than 5% of the calvings required assistance which in all cases was small and early calf mortality was not common. Only cows of dual-purpose breeds or crossbred dual-purpose cows showed some parturition difficulties. The first information on the use of crossbred hardy x beef cows (Table 8B) shows that the frequency of calving difficulties in these populations of suckling cows might increase as compared with that of the hardy breed populations. This is all the more true as these cows are in crossbreeding with bulls of heavy-muscled breeds.

TABLE 8

ITALY: CALVING DIFFICULTIES AND EARLY CALF MORTALITY IN EXTENSIVELY MANAGED SUCKLER HERD IN THE HILLS OF SARDINIA

A: MATURE COWS (7 to 12 years old) from 1964 to 1970. (Casu et al., 1975).

Sire breed	Charolais		Piemontese (a)		Limousin		Mean			Sample (b)			
Dam breed	Dif.	Dead	Dif.	Dead	Dif.	Dead	No.	Dif.	Dead	Weights of cows	calves at birth (c)	PO	BW/PO
Sarda	4.9	0.6	0	3.1	0	3.4	407	2.2	2.2	279	33.0	261	0.12
Modicana	0	0	0.9	1.7	0	3.6	270	0.4	1.4	386	38.1	319	0.12
Brown Swiss	12.5	1.3	7.7	1.3	0	0	187	9.6	1.1	380	38.6	258	0.16
Brown x Modic.	5.0	7.0	7.4	0	0	0	111	5.4	2.6	-	-	-	-
Average:	5.1%	1.3%	2.4%	1.9%	0%	2.8%	975	3.2%	1.8%	-	-	-	-

Dif: % of assisted calvings (including minimal assistance).
Dead: % of calf mortality before one day old.

PO: Pelvic opening.
BW: Birth weight.
a)   Double-muscled sires
b)   Observations on a sample of cows mated with the three sire breeds.
c)   Heifers.

TABLE Continued

TABLE 8 (Cont)

ITALY: CALVING DIFFICULTIES AND EARLY CALF MORTALITY IN EXTENSIVELY MANAGED SUCKLER HERD IN THE HILLS OF SARDINIA

B: HEIFERS AND COWS (from 1973 to 1975 (a). Casu et al., 1975).

Dam breed	% difficult calvings						Charolais		Piemontese (b)	
	Heifers			Cows						
	No.	Dif.	Caes.	No.	Dif.	Caes.	No.	Dif.%	No.	Dif.%
Sarda (Sr)	34	12	0	38	3	0	37	11	35	3
Modicana (Mo)	25	8	0	21	10	5	36	6	20	10
Brown Swiss (BS)	17	35	6	(2)	(100)	(O)	12	42	7	43
Charolais x Sr	18	11	6	13	8	0	18	6	13	15
x Mo	14	36	0	7	29	14	12	17	9	56
Piemontese x Sa	11	55	9	10	20	10	13	38	8	38
x Mo	18	33	11	25	12	4	26	19	17	24
Charolais x BS	12	58	8	8	38	25	11	45	9	56
Average	149	26	4	124	13	5	155	19	118	21

Dif: % of assisted calvings (including minimal assistance).
Caes: % of surgical calvings.

a) In crossbreeding with Charolais and Piemontese sires.
b) Double-muscled sires.

In Italy, improvement of the productivity of suckling cows in the local populations due to use of beef breeds with a high muscle growth potential, is most certainly accompanied by an increase in the frequency of calving problems.

4.2. France

Because of the importance of suckler cows and in particular the part played by beef breeds, parturition difficulties have been examined with more interest for field recording and for the programmes aimed at improving the productivity of the herds. At the national level, calving difficulties have been codified in 5 categories, (1 = without assistance; 2 = little assistance; 3 = much assistance or forced extraction; 4 = Caesarean section; 5 = embryotomy), based on the degree of assistance independent of the cause or origin of the resistance to foetus expulsion. However, in this system of performance recording, the early calf mortality is generally not well recorded due to a lack of accuracy concerning the criteria to be observed. With the data of the field recording of the last 4 years (1973 - 76), we obtained the frequency of calving difficulties in the main populations of French suckler cows: beef breeds, hardy and dual-purpose breeds, even if a fraction of the latter is entirely or partly milked. This information concerns both commercial and pedigree herds. We have used only purebreeding data because crossbreeding is either little developed (beef breed populations) or very developed (hardy breed populations) but with few recorded herds. We have completed these results with some data of different surveys (technico-economic management of farms and progeny testing). As in Great Britain, the results presented by breed include production-system differences since these populations are exploited in different areas with different farming traditions.

In all these populations of suckler cows, calvings are commonest at the beginning of spring, especially in beef breeds (Charolais) and hardy breeds. Accordingly, the first calvings generally take place at the age of about 3 years (Jarrige, 1975). It is only in breeds with a less marked breeding season that the age at first calving falls to 32 to 33 months (for instance

Limousin) and exceptionally 2 years. Another characteristic of
these suckler cows is that embryotomies have been almost replaced
by Caesarean sections. For instance, about 15 years ago, embryotomies represented about 75% of the surgical interventions in
the Charolais breed (2.2% out of 3.0% - Auriol et al., 1961)
whereas today less than 7% of the interventions are embryotomies
(0.2% out of 3.2%). Within each population of suckler cows, the
incidence of calving problems is rather variable.

4.2.1. Hardy breed populations (Table 9)

These breeds are mountain breeds now utilised as suckler
cows (Gascon in the Pyrenees, Aubrac in the south of the Massif
Central) or still as milking cows (Salers in the Massif Central).
In purebreeding, their calving ability is excellent: the frequency
of dystocic calvings is very low and even lower than 1% in the
populations of the Massif Central; cases of surgical interventions
are almost nonexistent. Even in the heifers, 98 to 99% of calvings take place easily without much assistance. This high calving
ability has led to a large development of crossbreeding (often
exceeding 50%) with beef breeds and especially with Charolais
bulls. Generally applied to adult cows, beef crossing slightly
increases the frequency of dystocic calvings without, however,
exceeding 2% in the populations of the Massif Central (1.3%, 2.1%
and 0.9% difficult calvings, respectively in Gascon, Salers and
Aubrac, for 1 800 recorded calvings). In these herds, early calf
mortality does not reach 3% and represents about one-third of
the overall pre-weaning mortality (Liénard and Legendre, 1974)
In first calving and especially early first calving (2 years),
commercial crossing greatly increases the frequency of dystocic
calvings even when using a sire breed such as Blonde d'Aquitaine
(Bibé et al., 1973 and 1976). In this situation, the early calf
mortality is substantially increased. Because of the extension
of commercial crossing with beef breeds, the utilisation of crossbred hardy x beef suckling cows is more and more current in these
areas. The few observations made in the Massif Central tend to
show that their calving ability is rather high (Matray, 1973).
On the other hand, their more intensive management in crossbreeding, with first calving at 2 years, leads to much more frequent
calving problems (dystocic calvings and early calf mortality)

TABLE 9

FRANCE: CALVING CONDITIONS AND BIRTH WEIGHT IN PUREBRED HARDY BREEDS - Source: Field recording from 1973 to 1976*

Calving conditions	Gascon					Salers					Aubrac				
	Birth weight (kg)	Frequency(%) all	Frequency(%) calving parity 1	2	≥3	Birth weight (kg)	Frequency(%) all	Frequency(%) calving parity 1	2	≥3	Birth weight (kg)	Frequency(%) all	Frequency(%) calving parity 1	2	≥3
1. Without assistance	38.3	61.7	45.6	55.1	55.4	38.3	87.1	82.6	95.8	91.9	33.1	73.7	51.3	58.4	57.1
2. With little assistance	36.1	35.9	51.8	43.0	42.1	37.8	12.0	16.1	3.7	7.8	33.3	25.7	47.5	41.6	41.6
3. With much assistance	46.5	2.3	1.7	0.9	2.6	35.3	0.5	1.0	0.5	0.4	35.3	0.4	1.3	0	1.0
4. Caesarean	40.5	0.2	0.8	0.9	0	39.9	0.4	0.3	0	0	31.0	0.2	0	0	0.3
5. Embryotomy	-	0	0	0	0	-	0	0	0	0	-	0	0	0	0
Not specified	37.8	(501)	(45)	(41)	(102)	37.1	(829)	(75)	(46)	(156)	32.9	(319)	(9)	(10)	(37)
Total (1+2+3+4+5) =	37.7	1 021	118	107	233	38.2	2 636	311	216	849	33.1	1 093	80	77	296
Easy (1+2)	37.5	97.6	97.5	98.1	97.4	38.3	99.1	98.7	99.5	99.6	33.1	99.5	98.8	100	98.6
Much assistance (3)	46.5	2.3	1.7	0.9	2.6	35.3	0.5	1.0	0.5	0.4	35.3	0.4	1.3	0	1.0
Surgical Intervention (4+5)	40.5	0.2	0.8	0.9	0	39.9	0.4	0.3	0	0	31.0	0.2	0	0	0.3
Difficult (3+4+5) =	46.0	2.4	2.5	1.8	2.6	37.3	1.0	1.3	0.5	0	38.8	0.6	1.3	0	1.4

* Unpublished data

than hardy breed populations (Bibé et al., 1973 and 1976). This
is all the more true as heifers from heavy-muscled breeds and
sires especially selected for terminal crossing, are used.

4.2.2. Dual-purpose breed populations (Table 10)

These are populations of large-sized breeds managed in
eastern France (chiefly Simmental) as milking cows and in southern
(chiefly Brown Swiss) and western France (Maine-Anjou) as milking
or suckling cows. Crossbreeding with beef breeds is widely
practised except perhaps in the east on the Simmental population.
As shown elsewhere (Foulley et al., 1975), the frequency of dystocic calvings is much higher in these populations than in hardy
breed populations - about 5% in Simmental and Brown Swiss, and
even 10% in Maine-Anjou. In Simmental and Brown Swiss populations,
where the calf birth weight ranges around 43 kg, the percentage
of surgical interventions is generally lower than 1.0%; in heifers,
dystocic calvings are becoming more frequent and exceed 10%. In
the Maine-Anjou breed, of larger size than the previous ones and
producing heavier calves (49.3 kg), the frequency of dystocic
calvings is extremely high (9%) leading to about 5% surgical
interventions in the overall population (10% in 3-year-old heifers).
Even in adult cows, the proportion of calvings by surgical intervention still ranges from 2 to 3%. These frequencies of difficult
calvings, which are higher than in most of the French beef breeds,
are confirmed by the observations made during progeny testing of
the Maine-Anjou bulls (50.4 kg calf birth weight and 4.9% surgical
calvings - Union Maine-Anjou, 1977) or pedigree herds (49.4 kg
and 3.9% respectively - UPRA Maine-Anjou, 1977). Even if double-
muscled calves occur in this breed (probably less than 5%), this
trait is not specially selected in the breeding animals as in
the Belgian Blue-White breed. This poor calving ability that we
have confirmed and anlaysed (Ménissier et al., 1974), does not
prevent crossbreeding with Charolais bulls in the field, which
increases the calving difficulties even more; about half of the
Maine-Anjou cows are used in beef crossing and generally with the
Charolais. We do not know the early calf mortality rate in this
population, in which the highest proportion of twin births is
recorded (Ménissier and Frebling, 1974).

TABLE 10

FRANCE: CALVING CONDITIONS AND BIRTH WEIGHT IN PUREBRED DUAL-PURPOSE BREEDS - Source: Field recording from 1973 to 1976. Unpublished data

Calving conditions	Maine-Anjou					Simmental(a)					Brown Swiss				
	Birth weight (kg)	Frequency(%) all %	calving parity 1	2	≥3	Birth weight (kg)	Frequency(%) all %	calving parity 1	2	≥3	Birth weight (kg)	Frequency(%) all %	calving parity 1	2	≥3
1. Without assistance	47.3	46.2	26.3	44.8	56.4	43.3	51.1	35.7	54.5	58.1	43.2	52.9	38.7	52.9	57.1
2. With little assistance	50.0	44.6	56.1	46.4	38.3	43.7	43.4	53.3	40.7	38.4	42.4	41.8	50.0	40.2	38.2
3. With much assistance	55.3	4.4	7.3	4.2	2.9	46.9	4.7	9.3	4.6	3.3	45.0	4.8	11.3	6.9	3.9
4. Caesarean	57.1	4.7	10.2	4.5	2.3	45.7	0.7	1.7	0.2	0.3	43.0	0.5	0	0	0.5
5. Embryotomy	60.2	0.1	0.2	0.1	0.1	-	0	0	0	0	58.0	0.1	0	0	0.3
Not specified	49.7	(524)	(84)	(97)	(163)	42.3	(554)	(29)	(19)	(36)	41.3	(70)	(1)	(9)	(16)
Total (1+2+3+4+5) =	49.3	21306	4 155	3 527	8 964	43.7	3 671	591	545	1 386	43.0	879	106	102	387
Easy (1+2)	48.6	90.8	82.4	91.2	94.7	43.5	94.6	89.0	95.2	96.5	42.9	94.7	88.7	93.1	95.4
Much assistance (3)	55.3	4.4	7.3	4.2	2.8	46.9	4.7	9.3	4.6	3.3	45.0	4.8	11.3	6.9	3.9
Surgical Intervention (4+5)	57.2	4.8	10.4	4.6	2.5	45.7	0.7	1.7	0.2	0.3	46.0	0.6	0	0	0.8
Difficult (3+4+5)	56.3	9.2	17.7	8.7	5.6	46.7	5.4	11.0	4.8	3.5	45.1	5.4	11.3	6.9	4.7

a) Eastern Red-and-White

TABLE 11

FRANCE: INCIDENCE OF EARLY CALF MORTALITY (before 48 hours) IN SOME POPULATIONS OF SUCKLING COWS

Suckler cow populations	Heifers			Cows			Heifers + cows		
	No. (a)	0-48 h %	Overall % (b)	No. (a)	0-48 h %	Overall % (b)	No. (a)	0-48 h %	Overall % (b)
Salers: (50-60% cross-breeding with Charolais sire)	549	3.2	14.1	2 335	2.7	9.2	2 996	2.9	10.1
Aubrac: (>90% cross-breeding with Charolais sire)	342	4.8	16.3	1 865	2.3	8.7	2 207	2.7	10.1
Charolais: (in pure-breeding)									
Nièvre	587	7.2	11.0	2 336	5.2	8.1	2 953	5.6	8.7
Saône et Loire	386	7.3	14.1	1 389	3.7	6.2	1 775	4.4	7.9
Allier	226	11.7	17.0	781	6.8	11.7	1 007	7.9	12.9
Aisne	422	12.8	20.1	1 437	7.3	13.3	1 859	8.5	14.8
Total	1 621	9.4	15.0	5 973	5.6	9.4	7 594	6.4	10.6
Limousin: (in pure-breeding and ranching conditions)									
Corrèze	229	4.8	9.1	1 158	1.5	4.7	1 411	2.2	5.4
Hte Vienne	195	4.4	8.3	987	2.1	3.6	1 182	2.5	4.4
Total	424	4.6	8.7	2 145	1.8	4.2	2 769	2.3	4.9

a) Number of breeding cows in the recorded herds.

b) % calves born which died between birth and weaning

Source: Liénard and Legendre (1974)

Besides these populations of dual-purpose breeds, crossbred
dairy x beef cows are becoming more and more numerous especially
in western France and also in southern France. Calving problems
are not very well known in this type of population. Progeny
testing results of Charolais bulls in eastern and western France
(Table 12) show that the frequencies of dystocic calvings and
of surgical interventions are similar to or slightly higher than
those in dairy populations (Friesian, Montbéliard and Normande).
Their rate of calving difficulty is close to that of Charolais
cows, whereas it remains lower than that of Maine-Anjou cows.
A better knowledge of these populations of crossbred dairy x beef
cows would contribute to improving their efficiency.

4.2.3. Specialised beef breed populations (Table 13)

As these breeds are preferentially kept in purebreeding, the
number of observations available to estimate calving problems is
higher in these suckling cows. There are very large differences,
mainly of genetic origin, between the populations of French beef
breeds. The Limousin breed, with a calf mean weight of 37.3 kg,
exhibits the highest calving ability: only 0.4% calvings by surgical intervention and about 2% difficult or dystocic calvings.
In heifers, where the first calving is a little earlier (32 to
33 months - Jarrige, 1975), there are less than 6% difficult
calvings. Thus, the calving problems are comparable with those of
the small-sized British breeds and perhaps even less than those of
the Hereford. The Blonde d'Aquitaine population producing
heavier calves (44.4 kg) exhibits already a slightly higher proportion of calving difficulties (5.6% difficult calvings) and
more particularly in the heifers (9.8%). However, the frequency
of surgical interventions remains lower than 1% whatever the age
of the females. The calving ability of Blonde d'Aquitaine cows
should be comparable to that of the middle-sized British beef
breeds (South Devon). In the Charolais population, despite a
mean calf birth weight similar to that of the Blonde d'Aquitaine
(43.6 kg), the frequency of dystocia is substantially higher
(8.4%) and the incidence of surgical calvings is also higher (3.1%).
The difficulties are greatest in the heifers - 19% difficult
calvings among which half are made by surgical intervention;

TABLE 12

FRANCE: CALVING DIFFICULTIES IN DAIRY, DUAL-PURPOSE AND BEEF x DAIRY CROSSBRED COWS IN BEEF CROSSING WITH CHAROLAIS BULLS (a)

Dam breeding type	Western France (c)			Eastern France (d)		
		% calvings			% calvings	
	No.	Difficult	Caesarean	No.	Difficult	Caesarean
Friesian	–	–	–	1 886 (283)	7.1 (7.5)	0.2 (0.0)
Montbéliard (Simmental dairy type)	–	–	–	3 978 (628)	5.8 (6.4)	0.4 (0.6)
Normande	2 765 (467) (b)	5.5 (4.8)	1.3 (1.5)	–	–	–
Crossbred: Dairy x beef	–	–	–	1 863 (112)	7.6 (5.1)	0.9 (0.0)
Charolais x Normande	830 (120)	7.1 (5.7)	2.0 (1.8)	–	–	–
Maine-Anjou	523 (91)	14.0 (18.4)	4.7 (5.2)	–	–	–
Charolais	– (117)	– (7.8)	– (2.7)	– (1 152)	– (5.3)	– (1.9)

a) Results of field progeny testing of Charolais bulls selected in crossbreeding

b) The results in brackets are obtained from bulls selected also in purebreeding.

c) Selection unit: UCAIA-VENDEE et INTERSELECTION (1975 and 76) – 28 bulls selected in crossbreeding, 5 out of them also in purebreeding.

d) Selection unit: UNION-CENTRE-EST (1974, 75 and 76) – 60 bulls selected in crossbreeding, 11 out of them also in purebreeding.

Source: ITEB reports; Fabrègue, (1977).

TABLE 13

FRANCE: CALVING CONDITIONS AND BIRTH WEIGHT IN PURE BEEF BREEDS

Calving condition	Charolais					Blonde d'Aquitaine					Limousin				
	Birth weight (kg)	Frequency (%)				Birth weight (kg)	Frequency (%)				Birth weight (kg)	Frequency (%)			
		all	calving parity:				all	calving parity:				all	calving parity:		
			1	2	≥3			1	2	≥3			1	2	≥3
1 = Without assistance	42.6	54.4%	34.0	54.0	62.0	43.7	66.7%	59.7	67.1	69.9	37.2	89.4%	76.9	90.8	96.2
2 = With little assistance	43.9	37.2%	47.5	38.7	33.5	45.1	27.7%	30.6	27.9	25.2	37.8	8.5%	17.5	7.5	6.2
3 = With much assistance	47.5	5.3%	10.3	4.7	3.3	48.8	5.2%	9.1	4.7	4.6	40.6	1.7%	4.9	1.5	0.9
4 = Caesarean	49.4	3.0%	7.9	2.6	1.3	49.3	0.3%	0.7	0.2	0.3	38.2	0.3%	0.6	0.3	0.3
5 = Embryotomy	50.2	0.2%	0.4	0.1	0.1	53.4	0.04%	0.04	0.005	0.03	43.9	0.03%	0.07	0.04	0.02
Not specified	43.1	(13 434)	(1367)	(996)	(2961)	44.5	(2896)	(193)	(151)	(383)	37.0	(12 241)	(1826)	(1639)	(6909)
Total (1+2+3+4+5) =	43.6	203 865	31765	24 417	75 275	44.4	26 050	2851	2221	6403	37.3	78 081	12 647	10 599	42 985
Easy (1+2)	43.1	91.6%	81.5	92.7	95.5	44.1	94.4%	90.3	95.1	95.1	37.3	97.9%	94.4	98.2	98.8
Much assistance	47.5	5.3%	10.3	4.7	3.3	48.8	5.2%	9.1	4.7	4.6	40.6	1.7%	4.9	1.5	0.9
Surgical intervention (4+5)	49.4	3.1%	8.5	2.7	1.4	49.8	0.4%	0.7	0.2	0.3	38.7	0.4%	0.7	0.3	0.3
Difficult (3+4+5)	48.2	8.4%	18.5	7.3	4.6	48.9	5.6%	9.8	5.0	4.9	40.3	2.1%	5.6	1.8	1.2

Source: Field recording from 1973 to 1976; unpublished data.

part of the problem still remains at second calving. This high
incidence of calving difficulties, which has been known for a
long time, (Auriol et al., 1961; Bourrin and Marcher, 1965), is
confirmed in the commercial herds (Texier and Legendre, 1976).

The early calf mortality rate in these beef breed populations
(Table 11), seems primarily to be related to calving difficulties
although there are large differences between areas. It is 1.5 to
2 times higher in heifers than in cows; it reaches 4% to 9% in
the Charolais production area where about two-thirds of the cases
are due to parturition difficulties (Texier and Legendre, 1976;
Liénard and Legendre, 1974) whereas it only represents 2 to 3%
in the Limousin production area.

In these beef breed populations, early calving at 2 years,
mainly occurring experimentally or accidentally, increases all
these calving problems (Table 14). Despite a certain choice of
bulls to sire the heifers, the frequencies of different calvings
and surgical interventions are 2 or 3 times higher than in 3-year
old heifers (ie, 4 to 7 times higher than in adults). Here again
the early calf mortality increases proportionally less, partly
due to a more efficient surveillance at calving. Until now, the
incidence of these calving problems has been the main barrier to
the realisation of early calving in suckler cow herds (Liénard,
1975) and more particularly in the Charolais breed (Legendre,
1974; Carrère and Liénard, 1976).

Thus, in the French populations of suckler cows, the
incidence of calving problems is very different from one
population to another. Almost non-existent in hardy breeds
used in purebreeding, calving difficulties progressively increase
as the growth potential of the calves increases and reach high
frequencies in large-sized breeds (such as the Maine-Anjou)
or heavy-muscled breeds (such as the Charolais). In these
breeds with poor calving ability, the early calf mortality
rate is often higher because of the parturition difficulties.

TABLE 14

FRANCE: INCIDENCE OF CALVING PROBLEMS IN PUREBRED BEEF BREEDS FOR THE FIRST CALVING AT 2 YEARS OLD IN COMPARISON WITH OTHER CALVINGS

	Charolais			Blonde d'Aquitaine			Limousin		
	1st calving at 2 years	1st calving at 3 years	Other calvings	1st calving at 2 years	1st calving at 3 years	Other calvings	1st calving at 2 years	1st calving at 3 years	Other calvings
	(a)	(b)	(b)	(a)	(b)	(b)	(a)	(a)	(b)
Number: calvings	951	31 765	162 100	229	2 851	23 199	284	12 647	65 448
years	8	4	4	3	4	4	2	4	4
sires	75	-	-	18	-	-	20	-	-
Age at calving (months)	25.0	(3 years)	(≥4y)	25.5	(3 years)	(≥4 y)	27.4	30 - 36	(≥ 4 years)
Weight after calving (kg)	495	-	-	499	-	-	464	-	-
% of Caesareans	20.1	8.5	2.2	13.1	0.7	0.3	1.5	0.7	0.3
% of very difficult calvings	46.5	18.5	6.6	21.7	9.8	5.1	11.6	5.6	1.4
Calf birth weight (kg)	37.4	42.0	43.9	39.4	40.9	44.8	35.1	35.6	37.6
Gestation length (days)	283.7	-	-	291.5	-	-	287.1	-	-
Calf mortality: birth - 48 h(%)	11.9	(c) 9.4	(c) 5.6	7.7	-	-	10.6	(c) 4.6	(c) 1.8
48 h - weaning (%)	6.7	5.6	3.8	4.5	-	-	5.8	4.1	2.4
Total (%)	18.7	15.0	9.4	12.5	-	-	16.4	8.7	4.2

a) Performance from progeny testing stations for breeding qualities. (Unpublished data)

b) Field recording from 1973 to 1976, except for mortality.

c) Field results of many herds over several years, according to Liénard and Legendre, (1974).

5. CALVING PROBLEMS ARE THE CONSEQUENCE OF AND THE LIMITING
    FACTOR TO THE IMPROVEMENT OF SUCKLER COW PRODUCTIVITY

Despite scarcity and disparity of the information collected,
it can be concluded that the diversity of the populations used
as suckler cows and of their management systems has led to rather
different situations concerning the frequency of calving problems:

In the populations of hardy breeds (Italy, France and Great
Britain), frequencies of calving difficulties (1 to 2%) and of
early mortality (1 to 2%) are very low, even in the heifers (2 to
3% and 1 to 5% respectively). The calving difficulties in these
populations are little increased by the extensive development of
beef crossing, in spite of higher calf birth weight. This is
even more marked if the heifers, as in most systems, are kept
in purebreeding as replacement stock or at least are not used in
crossbreeding with heavy-muscled breeds and early first calving.
The crossbred hardy x beef cows which are more and more exploited
in these areas, but generally in more intensive systems, present
more risks of calving difficulties, chiefly if they are sired by
bulls especially selected for terminal crossing.

In the populations of dual-purpose breeds,often managed
more intensively than the previous ones and with variable crossing rates, the frequency of calving difficulties is much higher
in purebreeding (5 to 9% in cows and 10 to 18% in 3-year-old
heifers). As the dual-purpose breeds have evolved towards either
largesized animals (Maine-Anjou and Simmental, for instance) or
heavy-muscled ones notably because of the double-muscled trait
(Piemontese and Belgian Blue-White breeds), the frequencies of
calvings by surgical interventions are now so high that they have
almost reached or overstepped the threshold compatible with most
management systems. The abilities of crossbred dairy x beef
cows often used in the same areas have not been well established.

In the populations of specialised beef breeds, the frequencies
of dystocic calvings are very different from one population to
another, ranging from low frequencies (1 to 3%), comparable with
those of hardy breeds, in the small-sized breeds (Hereford, Angus,

Limousin), to higher frequencies (6 to 12%), comparable with
those of dual-purpose breeds, in the large-sized breeds (in particular Charolais). The frequency of early calf mortality increases
in these populations when the number of dystocic calvings gets
higher.

In addition to the problem of frequency, calving difficulties
have also important and complex repercussions on the productivity
of the selection cows. Primarily, calving difficulties increase
the production costs of the calves because of the interventions
required and risks of mortality of the calf and even of the dam.
It is obvious that in the case of the dystocic calvings, the survival rate of the calves at birth is substantially reduced; in
the Charolais breed for example (Table 15), it can be seen that
the early calf mortality is 2 to 3 times higher in calves delivered
by Caesareans or forced extraction than in those delivered normally. As compared with forced extraction, Caesarean operations
improve the survival of calves immediately after the birth, but
they do not reduce the mortality rate at the moment of parturition;
furthermore, the postnatal mortality seems to be a little higher
in calves delivered by Caesarean sections. The cost of the intervention at parturition is rather variable according to the mode
of extraction of the foetus (Ménissier, 1974_a). Apart from the
supervision of calvings which is difficult to quantify in medium-
sized farms, the relative incidence of intervention costs is
rather small in beef herds as compared with the commercial value
of the calves, even in the case of veterinary assistance or
surgical intervention. The extreme cases are those of the double-
muscled calves where the cost of a Caesarean section represents
less than one-third of the calf value whereas in specialised dairy
breeds, the cost of intervention often exceeds the calf value.
On the other hand, the increase in the mortality of calves due
to difficult parturition has a much greater repercussion. We
have estimated (Foulley et al., 1976) that an increase of about
1% in the number of Caesareans in a Charolais herd increases less
than 1% the weaned calf cost, but 1% reduction of the calf crop
increases this cost by 2 - 4%. These relative incidences are
closely related and they depend directly on the production system.

This explains the variation in the acceptance of calving problems by the breeders.

TABLE 15

VARIATION OF CALF MORTALITY ACCORDING TO DIFFICULTIES AT CALVING IN THE CHAROLAIS BREED (a)

Condition of birth	No. of calves born	Calf mortality (%) (b)						Overall
		Mortality before 2 days old			Mortality after 2 days old			
		At birth	0-48 h	Up to 2 days	2 days - 1 month	1-4 months	2 days - 4 months	
Caesarean	167	14.4	2.4	16.8% (5.3)	6.6	3.6	10.2% (5.3)	27.0 (10.7)
Forced extraction	228	13.2	6.6	19.8% (16.6)	5.3	0.9	6.2% (6.8)	26.0 (23.4)
Little assistance	297	3.0	1.7	4.7% (3.3)	4.7	0.7	5.4% (2.3)	10.1 (5.6)
Without assistance	130	3.1	0.8	3.9% (2.6)	5.4	0.8	6.2% (2.1)	10.1 (4.8)

a) Station results of heifers calving at two years in purebreeding (unpublished data). The results of commercial Charolais herds are given in brackets (2 318 calvings from heifers and cows according to Texier and Legendre, 1976).

b) Dead calves as % of calves born

Secondly, calving difficulties have repercussions on the subsequent performance of the suckler cows. In the case of calvings by Caesarean operations in beef heifers, we observed a reduction of their milk production (by about one-third) especially during the first month of suckling (Bonnet, 1973 - unpublished data). This incidence is not to be neglected in the case of suckling cows with a poor milk yield and managed extensively with calves having a high growth potential. The most important repercussion is that exerted on the subsequent fertility of cows with calving problems (Table 16). Caesarean operations considerably reduce (by 20 to 30%) the fertility of the cows and forced extractions also lead to substantially lower fertility. This incidence seems primarily to depend on a prolongation of post partum anoestrus, but also is due to a lower fertility at the first oestrus, both leading to a lengthening of the calving interval (Dreyer and Smidt, 1966; Hanset, 1966; Brinks et al., 1973; Laster et al., 1973; Vissac et al., 1964). This necessarily results in higher culling rates in the heifers or young cows and this is prejudicial to herd management (administration and sel- ection) except in the systems where the slaughter value is very high (double-muscled or heavy-muscled heifers).

TABLE 16

EFFECT OF CALVING CONDITIONS ON POST PARTUM FERTILITY OF BEEF BREED HEIFERS

Calving conditions	Charolais heifers after 1st calving at 2 yrs old				Beef breed cows			
	No.	% pregnant			1st calving at 2 years old		1st or 2nd calving at 3 years old	
		to 1st AI	Overall	(a)	No.	% pregnant	No.	% pregnant (a)
Caesarean	94	35%	61%	(76)	31	61%	21	38%
Difficult	146	49%	80%	(79)	69	75%	61	78%
Easy	238	63%	91%	(87)	38	92%	59	81%
Reference	Progeny testing ITEB station. (Legendre (Ménissier, 1974a) 1974)				Crossbreeding experiment between beef breeds. INRA (Ménissier, 1974)			

a) AI for 60 - 70 days starting 30 - 40 days after calving

Finally, the present development of the beef cow populations of the EEC towards production of calves with a high muscle-growth potential leads to an increase in the calf birth weight as well as in the size and muscle development of their dams. Such a development, primarily realised by commercial crossing with specialised beef breeds, then by utilisation of the crossbred cows and even by grading up or by specific selection, induces more and more frequent parturition problems. This phenomenon is all the more important if the calving ability of the initial maternal population is already small in purebreeding and as the management of the herds becomes more intensive. Furthermore, in order to exploit the improvement in the growth potential, intensive tech- niques are used more and more to reduce the incidence of calving difficulties and especially their repercussions on early calf mortality (increased surveillance, more systematic interventions,

more frequent use of Caesarean sections, etc). Independently
of their efficiency, these interventions lead to rupture of the
natural "dam-calf-environment" equilibrium established formerly
in different production systems and often by natural selection.
Conversely, the improvement of the productivity of the suckler
cow herd is achieved through a reduction of production costs and
in particular of the labour part of these costs (herds of larger
size, minimum care, calvings more grouped within the breeding
season, etc). Consequently, the main goal in suckling cow herds
is to reduce the frequency of calving problems as compared with
the present situation. Because of the discrepancy between this
objective and the development observed during the last few years,
calving problems represent one of the factors limiting the
improvement of beef herd productivity.

Before defining more accurately for each field how to improve
this situation, it seemed to us to be useful to determine the
causes of the increasing calving difficulties in beef cattle.

6. CALVING PROBLEMS RESULTING FROM A MORPHOLOGICAL INCOMPAT-
IBILITY BETWEEN DAM AND FOETUS

Calving difficulties being more frequent in heifers and
especially in those calving early (2 years), our analysis of
calving problems has been mainly made in these populations of
suckler females (Abdallah, 1971 and 1971$_a$; Couteaudier et al.,
1971; Ménissier, 1974$_a$ and 1975$_a$; Ménissier et al., 1975$_a$). The
principal conclusions are the following:

6.1. Calving difficulties might be due to an abnormality in the
presentation or position of the foetus at the moment of its
expulsion. However, these cases are comparatively rare (Auriol
et al., 1961; Bourrin and Marcher, 1965) and do not always lead
to serious dystocia. In heifers of beef breeds (Table 17) normal
and unknown presentations represent about 95% of the parturitions;
the other cases mainly represent posterior presentations requiring
some assistance. The cases of uterine torsion requiring much
assistance or a surgical intervention are rare (0% to 2%). Most
certainly these small percentages calling for surgical

interventions are always included in calving statistics, also for those breeds producing light calves and having only rare calving difficulties. Furthermore, it seems to us that abnormal presentations tend to be more frequent with large-sized breeds which have already shown calving difficulties. Certainly twin births, in spite of the small size of the calves, often require assistance (generally slight) because of the abnormal presentation or position of the calves.

6.2. In most situations, the major difficulties at calving are due to the excessive size of the calf as compared with that of the dam, ie, a morphological incompatibility between dam and foetus (Ménissier, 1974_a and 1975_a).

TABLE 17

FREQUENCY OF ABNORMAL PRESENTATION AND POSITION OF FOETUS AT BIRTH IN FRENCH BEEF COWS

Reference	No. of calvings	% dystocic calvings	Type of presentation (%)		
			Normal (or unknown)	Posterior and other abnormals	Torsion of uterus
1st calving at 2 years old:Charolais (a)	829	48.1	96.0	3.8	0.2
Crossbreeding between French beef breeds (b) *1st calvings at 2 years old	⎰149 ⎱231	38.7 10.5	95.3 97.9	2.7 1.7	2.0 0.4
*Other calvings	432	14.4	93.4	5.5	1.1

a) In progeny testing station (Ménissier et al., 1975_b)
b) In French crossbreeding experiment with beef breeds (unpublished data)

This incompatibility leads to a certain resistance of the foetus to the expulsion which, by increasing order of difficulties, appears first between the head and girdles (scapular and then pelvic) of the foetus and the vulva and pelvic canals (posterior

and then anterior) of the dam (Ménissier et al., 1974 and
unpublished data). The cases of extreme difficulties requiring
surgical intervention are those due to disproportion between
shoulders or rump of the foetus and the anterior pelvic canal
or pelvic opening of the dam. On the basis of a more statistical
approach (analysis of correlations) of the variations in frequency of parturition difficulties, we have made a hierarchical
classification of the different causes of resistance to the
expulsion of the foetus and their interrelationships (Abdallah,
1971 and 1971$_a$; Ménissier,1974$_a$ and 1975$_a$: Ménissier et al.,
1975$_a$).

6.2.1. The size of the calf is most certainly the factor highest
correlated with parturition difficulties. Birth weight explains
around 50% of the variability in the frequency of difficult
calvings (r = +0.6 to 0.8). When the frequency of dystocic
calvings decreases, as in the case of multiparous cows, this
correlation rapidly decreases, thus confirming the absence of a
relationship between difficult calvings and size of calves in
older dams (Couteaudier et al., 1971). The calf birth weight
varies much from one population to another. This variability
is either of environmental origin (maternal feed restriction,
induced parturition, twin births, etc) or of genetic origin
(crossbreeding and selection) and in the latter case it is
closely related to the calf growth potential. The morphology
of the calf expressed by its dimensions (length and width) at a
fixed birth weight, also represents a source of variation that
might cause differences in the frequency of dystocic calvings
(Abdallah, 1971$_a$; Ménissier, 1974$_a$). However, variations in
morphology independent of calf size (constant birth weight) are
generally small compared with those of birth weight, except perhaps in the long-bodied breeds of southern Europe (Blonde d'Aquitaine and Chianina) (Ménissier et al., 1974 and 1978). At all
events, these differences in morphology can only compensate for
a small part of the excessive calf birth weight in very large
breeds (Ménissier et al., 1978).

6.2.2. Pelvic opening of the dams (measured on the live animal)
(Ménissier and Vissac, 1971) represents the second cause of
variation in the frequency of calving difficulties and explains
more than 10% of this trait (r = -0.2 to -0.4). As for birth
weight, this relationship is closer in situations where dystocic
calvings are more frequent (heifers and first calvings at 2 years).
The pelvic opening of a cow is obviously positively related to
her size: large cattle have larger pelvic openings than small
ones; those of heifers are smaller than those of mature cows.
However, there are variations in the pelvic opening which are
independent of the size of the animal (at constant weight for
instance, - Abdallah, 1971_a; Bibé et al., 1976). These relative
variations are both of environmental origin (age and calving
parity, for instance) and of genetic origin (breeding type in
particular). From an environmental point of view, the relative
growth of the pelvic opening seems to be at least equal to the
increase in body size and even superior in the case of variation
with the age of the dams between 2 and 6 years (Ménissier et
al., 1978); as a matter of fact, the pelvic opening or the
pelvis is a late maturing region. Conversely, from a genetic
point of view, increase in size of cattle (and hence their own
potential) would induce only a small relative increase in their
pelvic opening (Taylor et al., 1975; Ménissier et al., 1978).
Likewise, the muscle development of cattle generally tends to
reduce the pelvic opening relative to body weight (Abdallah
et al., 1971; Vissac et al., 1973; Ménissier, 1974_a and 1975_a;
Ménissier et al., 1974). Thus, improvement of muscle growth
potential (genetic) in beef cattle is accompanied by a reduction
of the pelvic opening relative to their size, whereas increase
in weight due to better feeding conditions tends to have the
opposite effect.

6.2.3. Frequency of calving difficulty is also affected by
another maternal character, ie, the preparation for parturition
or the maternal behaviour. Although estimated subjectively
(vulva and udder congestion, relaxation of sacro-sciatic ligaments), the variation in maternal behaviour explains around 10%
(r = -0.2 to -0.5) of the variation in calving difficulty. As

regards this criterion, we have also observed genetic differences between (Abdallah et al., 1971[a]; Couteaudier et al., 1971)
and within breeds (Ménissier et al., 1975[a]). These variations
have not been interpreted especially from an endocrinological
point of view.

6.3. In fact, calving difficulties are the result of an interaction between various maternal factors and the size at the moment of delivery. The relationships between the frequency of
dystocic calvings and these different traits are thus more complex and the correlations only reflect general trends. The
incidence of calf birth weight or pelvic opening of dam on the
frequency of dystocic calvings is non-linear: the effect of each
of these factors on calving difficulties depends on the mean
level of the other factor. Thus, the action of each of these
traits correspond to 'threshold effects' depending on the other
factors (Ménissier, 1974[a] and 1975[a]). This is, for instance,
the well known case (Belíc and Ménissier, 1968; Ménissier, 1975[a];
Philipsson, 1977) of the threshold effect of birth weight in a
definite cow population, i.e., with a fixed mean pelvic opening
(Figure 2). A rise in birth weight only increases the frequency
of calving difficulties beyond a threshold. For a definite
population, this threshold is determined by the pelvic opening
of the cows (Ménissier 1974[a] and 1975[a]), i.e., for a first approx-
imation by their breeding type and their parity. If in such a
population we obtain a variation in the mean weight of calves
by genetic techniques or environmental ones, the repercussions
on calving difficulties then only depend on the position of calf
mean birth weight relative to this threshold. Consequently, it
appears to us to be very important when characterising and analysing calving ability (i.e., incidence of calving difficulties in
a population or production system) to specify not only the mean
parameters of calf birth weight and frequency of dystocic
calvings, but also the position of this threshold. For example,
in the Charolais breed (Figure 2), we know that calving ability
is limited by a calf birth weight of 35 kg in 2-year-old heifers,
of 40 to 45 kg in 3-year-old heifers and beyond 55 kg in older
cows aged 4 years and more, whereas mean weights of calves are

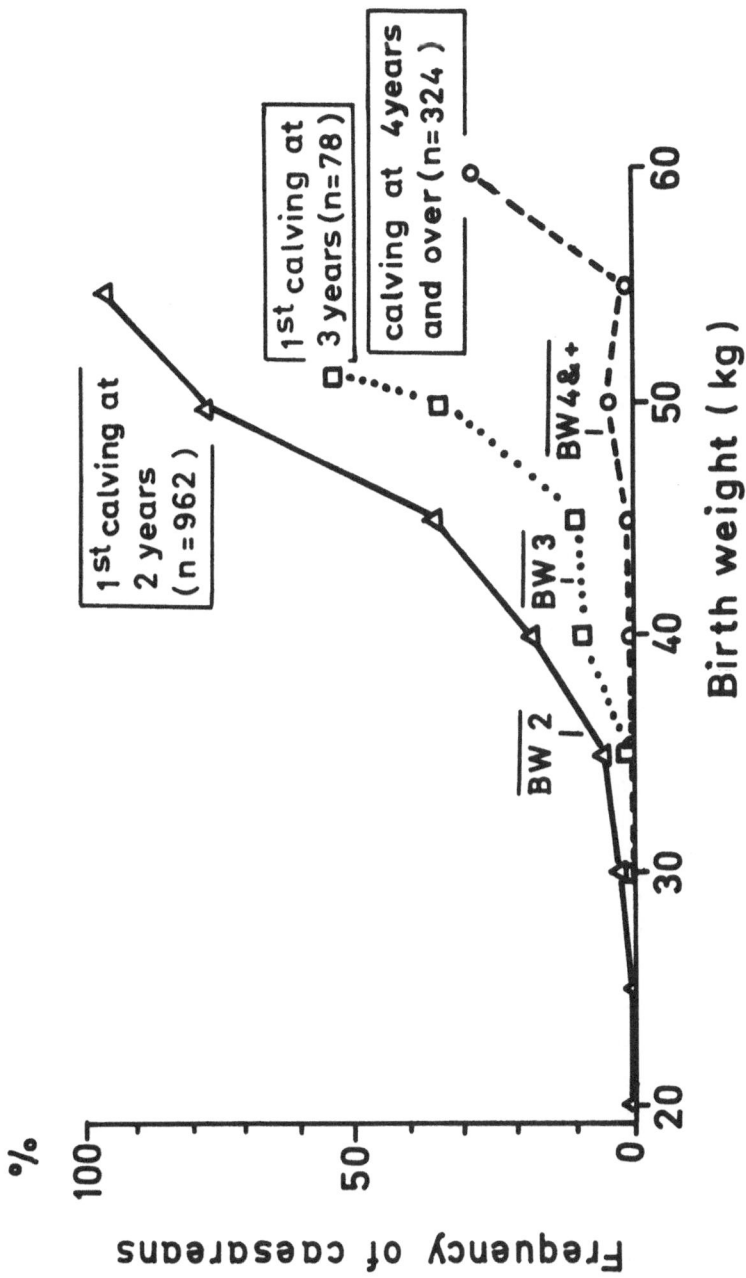

Fig. 2. Threshold effect of birth weight on the frequency of calving difficulties according to age of dams (Charolais in purebreeding).

generally located around 37.4 kg, 42.0 kg and 43.9 kg, respectively for these females. At the present time we are attempting
to define the thresholds for breeding females of the other French
beef breeds (Figure 3). According to the findings of Stollard
and Kilkenny (1976) concerning multiparous cows of British breeds,
the thresholds would be located around 40 kg for the small breeds
(Angus and Hereford) and around 45 kg for the large ones (Charolais, Simmental and also South Devon). The threshold of adult
Friesian cows probably ranges around 42 to 45 kg (Foulley and
Ménissier, 1976; Philipsson, 1977; Ménissier et al., 1978). More
generally a determination of the optimum birth weight of calves
in a population would be easier if the probability of dystocic
calvings in relation to birth weight were previously established
for this population.

Furthermore, the birth weight, indicator of uterine foetal
growth, depends on the growth potential of the calf and on the
incidence of uterine environment or uterine maternal effect.
The environmental uterine effect might be of genetic or non-
genetic origin. From a genetic point of view, there are not only
the already mentioned relations between calf size and weight, and
pelvic opening of the dam, but also relations between direct
effects and uterine maternal effects on the calf birth weight
(Philipsson, 1976; Foulley and Ménissier, 1978). Any change in
muscle growth of calves by genetic means will therefore affect
both the direct and the maternal effects. Although the maternal
effects are not well known in beef breeds (Foulley and Ménissier,
1978), it is certain that increase in size results in larger direct
and maternal effects on calf birth weight. The optimum birth
weight sought by genetic means should therefore be close to the
threshold weight for calving difficulties otherwise the postnatal
growth potential would be too much reduced. The non-genetic
interventions will influence chiefly maternal effects. However,
the goal of such interventions (both genetic and environmental)
should not be to reduce the calf birth weight to the maximum
without taking into account their growth potential. Indeed, a
large reduction of the calf weight by environmental means (feed
restriction, twin births, induced parturition), at the same time

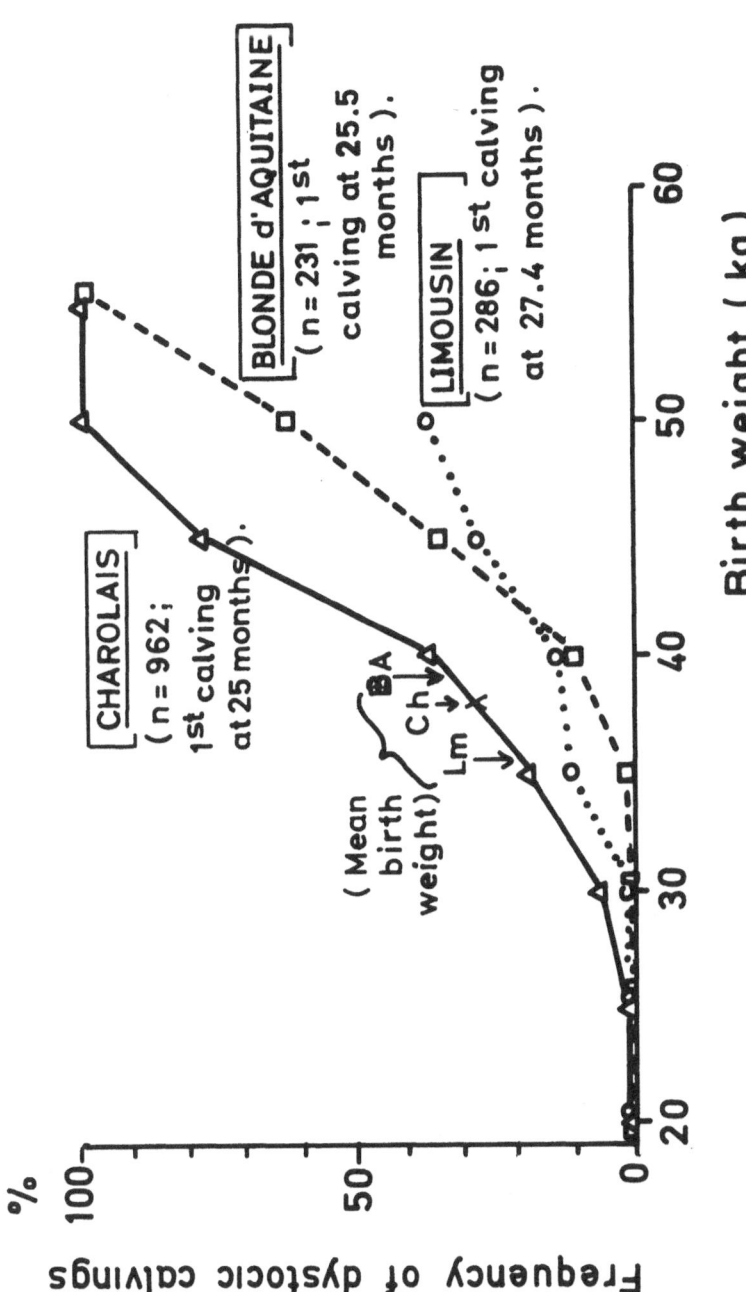

Fig. 3. Threshold effect of birth weight on the frequency of calving difficulties according to breed of dams (1st calving at 2 years old in purebreeding).

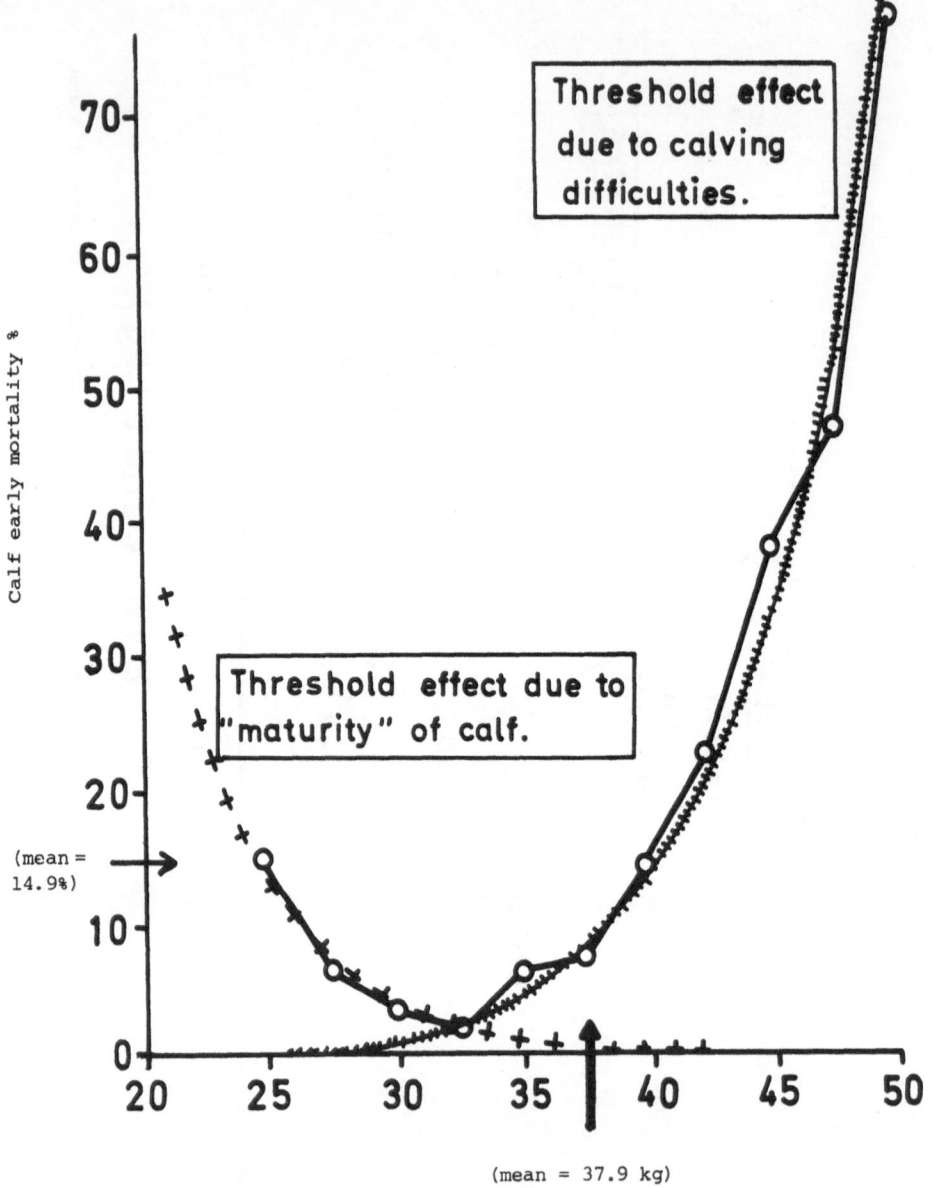

Fig. 4. Relationship between early calf mortality and calf birth weight
(Interpretation of phenomenon according to data of Remmen, 1975 =
2782 calves of MRY heifers).

as increasing the calf's growth potential, might lead to smaller but less mature calves (actual birth weight/potential birth weight) and consequently of lower viability. Thus according to the arguments on early mortality developed by Bradford, (1972, 1974) in sheep, it can be imagined that in addition to the threshold effect on high birth weight concerning calving difficulties, another threshold effect will appear on low birth weight connected with the lack of maturity of the calf (Ménissier, 1975$_b$). The case reported in Figure 4 (MRY breed) may be an illustration of this. Any change in the uterine growth parameters of the foetus should be made with caution.

7. CONCLUSION

Improvement of suckling cow productivity will unavoidably lead to search for maximum muscle growth potential of the calves (foetal period included). The upper limit of growth potential is determined by the birth weight threshold beyond which the rate of calving difficulties exceeds that tolerable for the production system in the population considered. In the long term it is possible, by selection, to modify this limit or at least the repercussions on calving difficulties due to increase in the postnatal growth potential. In the short term, most solutions will tend to reduce the birth weight of calves with a too high growth potential. A certain number of techniques or results will be reported on this topic during the present seminar. We are certainly tending towards a greater and greater conflict in beef cattle populations, between, on the one hand the increasing muscle growth potential of the calf and, on the other, the necessity of practising more and more efficient intervention to reduce the uterine growth of these calves. Consequently, in addition to the economic aspect of the system, we ultimately run the risk of replacing parturition difficulties by other problems such as reduction of calf viability which is perhaps one of the most important.

Such an antagonism would be reduced and even prevented if we accept from today to develop more interdisciplinary discussion rather than perfecting separately in each discipline techniques which are probably efficient in the short term, but whose repercussions are not known.

REFERENCES

Abdallah, O.Y. 1971 Variations génétiques de l'aptitude au vêlage et de ses
    composantes: revue bibliographique. Bulletin technique du Département
    de Génétique animale, Institut National de la Recherche Agronomique,
    France, No. 13, 180 pp.

Abdallah, O.Y. 1971$_a$ Variations génétiques de l'aptitude au vêlage et de ses
    composantes. Thèse de doctorat es-Sciences naturelles, Paris, 330 pp.

Abdallah, O.Y., Ménissier, F. and Vissac, B. 1971 Liaison entre la musculature
    des races et leur aptitude morphologique au vêlage: résultats prélimin-
    aires. Xe Congrès International de Zootechnie, juillet, 1971, Versailles,
    4 pp.

Abdallah, O.Y., Hadjej, M.S. and Ménissier, F. 1971$_a$ Variations du comporte-
    ment pré et post-partum entre races rustiques, race à viande et croise-
    ment de première génération. Xe Congrès International de Zootechnie,
    juillet 1971, Versailles, 6 pp.

Allen, D.M. 1975 Age of cows at first calving in the United Kingdom. In:
    First seminar on Nutrition and Management, 'The early first calving of
    heifers and its impact on beef production'. EEC programme of research
    on beef production, June 4 - 6th 1975, Copenhagen. CEC, EUR 5545e,
    pp. 33-36.

Allen, D.M. 1977 Ease of calving in beef herds. British Cattle Breeders'
    Club, Winter Conference 1977, Cambridge, Digest, No 32, pp. 62-67.

Andersen, B. Bech, Liboriussen, T., Thysen, I., Kousgaards, K. and Buchter, L.
    1976 Crossbreeding experiment with beef and dual-purpose sire breeds
    on Danish dairy cows. Livestock Production Science, 3; 227-238.

Anonymous, 1972 Rapport entre la production de lait et la production de
    viande de boeuf en Europe. Etude (no. WS/C 8978/C) de l'Organisation
    des Nations Unies pour l'Alimentation et l'Agriculture, (FAO), Rome,
    170 pp.

Anonymous, 1976 Prove de progenie sulla razza bovina Piemontese. La razza
    bovina Piemontese. 8, 83-192.

Auran, T. 1972 Factors affecting the frequency of stillbirths in Norwegian
    cattle. Acta Agriculturae Scandinavica, 22, 178-182.

Auriol, P., Dumont, B.L., Lefebvre, J. and Duplan, J.M. 1961 Caractéristiques
    générales des vaches charolaises et croissance de leurs produits.
    France-Elevage Charolais, 3, 1-28.

Baker, H.K. 1976 Beef breeds for beef herds in the United Kingdom. In: First Seminar on Genetics, 'Optimisation of cattle breeding schemes'. EEC programme of research on beef production, November 26-28th 1975, Dublin, CEC, EUR 5490e, pp. 173-185.

Barnes, B. and Kilkenny, J.B. 1976 Calving survey in pedigree and commercial beef herds. Meat and Livestock Commission (Report on summer scholarship) October 1976, 21 pp.

Belic, M. and Ménissier, F. 1968 Etude du quelques facteurs influençant les difficultés de vêlage en croisement industriel. Annales de Zootechnie, 17; 107-142.

Bettini, T.M. and Nardone, A. 1976 Status and application of breeding schemes in Italy. In:First Seminar on Genetics, Optimisation of cattle breeding schemes'. EEC programme of research on beef production, November 26-28th 1975, Dublin. CEC, EUR 5490e, pp. 249-261.

Bibé, B., Frebling, J. and Ménissier, F. 1973 Schéma d'utilisation des races rustiques en croisement avec des races à viande. Fédération Européenne de Zootechnie, septembre 1973, Vienne, 30 pp.

Bibé, B., Frebling, J., Gillard, P. and Ménissier, F. 1974 Incidence de l'utilisation de taureaux culards en croisement avec des femelles de races laitières sur la production de viande de jeunes bovins. I. Croissance pondérale, consommation alimentaire et développement corporel jusqu'à l'abattage. ler Congrès Mondial de Génétique appliquée à l'Elevage, Madrid, 3; 857-866.

Bibé, B., Frebling, J., Ménissier, F. and Vissac, B. 1976 Utilisation des races rustiques en croisement avec des races à viande: exemple de la race Gasconne. Annales de Génétique et de Sélection animale, 8; 233-264.

Bibé, B., Frebling, J., Ménissier, F. and Vissac, B. 1977 Optimum cross-breeding plans. In: Second seminar on Genetics, 'Crossbreeding experiments and strategy of breed utilisation to increase beef production'. EEC programme of research on beef production, February 9-11th, 1976, Verden, CEC, EUR 5492e, pp. 34-54.

Bibé, B., Frebling, J., Ménissier, F. and Vissac, B. 1977a Double-muscled sires for terminal crossing: French experiments. In: Second seminar on Genetics: 'Crossbreeding experiments and strategy of breed utilisation to increase beef production'. EEC programme of research on beef production, February 9-11th 1976, Verden, CEC, EUR 5492e, pp. 72-96.

Burrin, R. and Marcher, R. 1965 Age moyen et nombre de vêlages moyen des vaches réformées dans les élevages Charolais de 1959 à 1963. Syndicat de Contrôle des Performances des Elevages Nivernais, Nevers, 11 pp. (ronéoté).

Bradford, G.E. 1972 The role of maternal effects in animal breeding. VII. Maternal effects in sheep. Journal of Animal Science,35; 1324-1334.

Bradford, G.E., Taylor, St C.S., Quirke, J.F. and Hart, R. 1974 An egg-transfer study of litter size, birth weight and lamb survival. Animal Production,18; 249-263.

Brinks, J.S., Olson, J.E. and Carroll, J.E. 1973 Calving difficulty and its association with subsequent productivity in Herefords. Journal of Animal Science,36; 11-17.

Carrère, G.and Liénard, G. 1976 Résultats économiques d'un groupe d'élevages de vaches Charolaises. Evolution sur 4 ans: 1971-1974. Bulletin technique du CRZV de Theix, (23), 19-34.

Casu, S., Boyazoglu, J.G., Bibé, B. and Vissac, B. 1975 Systèmes d'amélioration génétique de la production de viande bovine dans les pays méditerranéens: les recherches sardes. Bulletin technique du Département de Génétique animale, Institut National de la Recherche agronomique, France, No. 22, 50 pp.

Couteaudier, J.F., Regis, R. and Ménissier, F. 1971 Possibilités de sélection de l'aptitude au vêlage en race Charolaise. Xème Congrès International de Zootechnie, 17-19 juillet, Paris-Versailles, 10 pp.

Crowley, J.P. 1965 The effect of Charolais bulls on calving performance. Irish Journal of Agricultural Research, 4, 205-213.

Cunningham, E.P. 1976 The structure of cattle populations in the EEC. In: First seminar on Genetics, 'Optimisation of cattle breeding schemes'. EEC programme of research on beef production, November 26-28th, Dublin. CEC, EUR 5490e, pp. 13-30.

Cunningham, E.P., Shannon, M., Fallen, J.J. and O'Byrne, T.M. 1976 A survey of reproduction, calving and culling of cows in Irish dairy herds. Irish Journal of Agricultural Research, 15; 177-183.

Derivaux, J., Fagot, V. and Huet, R. 1964 Le problème des dystocies. Essais de pelvimetrie. Annales de médecine vétérinaire, 108; 335-365.

Dimitropoulos, E. 1975 Rapport d'activité du Centre, secteur insémination artificielle 1975. Centre d'I.A. de Ciney, province de Namur (Belgium) 87 pp.

Dreyer, D. and Smidt, D. 1966 Welche Bedeutung haben Schwergeburten und Kälberverluste für die züchterische Beurteilung von Besamungsbullen. Tierzüchter, 18; 528-529.

Duplan, J.M. and Bougler, J. 1976 Current status, results and trends of AI cattle breeding in France. In: First seminar on Genetics, 'Optimisation of cattle breeding schemes'. EEC programme of research on beef production, November 26-28th 1975, Dublin. CEC, EUR 5490e, pp. 287-307.

Eurostat, 1974 Annuaire de statistique agricole. Office statistique des Communautés Européennes, 273 pp.

Fabrègue, P. 1977 Resultats établis à partir du contrôle de descendance des taureaux Charolais; ITEB (données non publiées - communication personelle).

Fagot, V. 1965 Dystocies et pelvimètrie. Zootechnia, 14; 60-64.

Fedeler, H.J., Haring, F., Langholz, H.J. and Pabst, W. 1973 Die Fleischrinderhaltung in der Bundesrepublik Deutschland. Züchtungskunde, 45 31-44.

Foulley, J.L., Ménissier, F., Gaillard, J. and Nebrada, A.M. 1975 Aptitudes maternelles dès races laitières, mixtes, rustiques et à viande pour la production de veaux de boucherie par croisement industriel. Livestock Production Science, 2; 39-49.

Foulley, J.L. and Ménissier, F. 1976 A few reflexions on the genetic improvement of the calving ability of the Limousine breed. Technical meeting of the International Limousin Council, Oklahoma City (USA), September 1975; 13 pp.

Foulley, J.L., Ménissier, F. and Vissac, B. 1976 Calving ability in French beef breeds and its genetic improvement. Beef Improvement Federation Meetings, Kansas City, May, 1976; 68 pp.

Foulley, J.L. and Ménissier, F. 1978 Selection for calving ability in French beef breeds. In: Seminar on 'Calving problems and early viability of the calf'. EEC programme of research on beef production, May 4-6th, Freising, 20 pp.

Frappel, J.P. 1975 Reprints of overhead slides used for lecture at British Cattle Veterinary Association Meeting, Langford, Bristol, 15th April 1975; 4 p.

Haas, B. 1972 Die Fleischrinderzucht und haltung im Rheinland und in Westfalen sowie Wege zu ihrer Verbesserung. Thesis, Bonn.

Hansen, M. 1975 Calving performance in Danish breeds of dual-purpose cattle. Livestock Production Science, 2; 51-58.

Hanset, R. 1966 Le problème de l'hypertrophie musculaire ou caractère 'culard' dans la race bovine de Moyenne et Haute Belgique. European Association of Animal Production, 9th International Congress of Animal Production, Edinburgh, 21 pp.

Hanset, R. 1974 La selection en race bovine Blanc-Bleu-Belge: Objectifs -
    programmes. Rapport non publiée, 15 pp.(Communication personelle).

Hanset, R. 1976 Cattle selection in Belgium. In: First seminar on Genetics,
    'Optimisation of cattle breeding schemes'. EEC programme of research
    on beef production, November 26-28th 1975, Dublin, CEC, EUR 5490e, pp.263-272

Jarrige, R. 1975 Age of cows at first calving in France. In: First seminar
    on Nutrition and Management: 'The early calving of heifers and its
    impact on beef production'. EEC programme of research on beef pro-
    duction, June 4-6th 1975, Copenhagen. CEC, EUR 5445e, pp. 10-13.

Kalm, E., Pabst, W., Langholz, H.J. and Lindhé, B. 1974 Purebred versus
    crossbred calves from Hereford and Charolais bulls in Sweden. Proceed-
    ings of Working Symposium on Breed Evaluation and Crossbreeding
    Experiments, 15-21st September, 1974, Zeist, pp.189-198.

Kanning, K. and Langholz, H.J. 1975 Mutterkuhhaltung mit Rassen unter-
    schiedlicher Nutzungsrichtung (Mimeo) 11 pp.

Kilkenny, J.B. and Stollard, R.J. 1976 Calf birth weights in beef breeding
    herds and the relationships between birth weight and calf mortality and
    calving difficulties. 62nd meeting of the British Society of Animal
    Production, Harrogate, March 1976. Animal Production, $\underline{22}$, 159-160.

Kögel, S. and Kräusslich, H. 1975 Current situation and potential increase
    in beef and veal production from female cattle in the Federal Republic
    of Germany. In: First seminar on Nutrition and Management: 'The early
    calving of heifers and its impact on beef production'. EEC programme
    of research on beef production, June 4-6th, 1975, Copenhagen, CEC,EUR 5545e
    pp. 14-22.

Larsen, J. Brolund and Sejrsen, K.R. 1975 The Danish cattle population and
    beef production. In: First seminar on Nutrition and Management, 'The
    early calving of heifers and its impact on beef production'. EEC pro-
    gramme of research on beef production, June 4-6th, 1975, Copenhagen,
    CEC, EUR 5545e, pp.2-9.

Laster, D.B., Glimp, H.A., Cundiff, L.V. and Gregory, K.E. 1973 Factors
    affecting dystocia and the effects of dystocia on subsequent repro-
    duction in beef cattle. Journal of Animal Science, $\underline{36}$, 695-705.

Legendre, J. 1974 Le vélage à deux ans des génisses Charolaises. Supplément
    au Bulletin technique du CRZV de Theix (numéro spécial). Octobre 1974
    249-261.

Liénard, G. 1975 Economic aspects of early calving in suckling herds. In:
    First seminar on Nutrition and Management, 'The early calving of heifers

and impact on beef production'. EEC programme of research on beef production, June 4-6th, 1975, Copenhagen, EUR 5545e pp. 177-190.

Liénard, G. and Legendre, J. 1974 Productivité en veaux des troupeaux de vaches allaitantes (méthode d'analyse et résultat). Supplément au Bulletin technique du CRZV de Theix (numéro spécial), octobre 1974, 47-70.

Lindhé, B. 1974 Improvement in beef-breeding by selection. 1st World Congress on Genetics applied to Livestock Production. Madrid, vol 1, 655-669.

Lindhé, B. 1976 On the possibility of standardising the performance and progeny testing of bulls throughout the member countries in order to achieve comparative results. 7th World Hereford Conference, July 1976, Banff, Canada, 14 pp.

Lindhé, B. 1976a Report of selection procedures of Charolais cattle in Sweden, Personal communication.

Matray, M. 1973 Comparaison des qualités d'élevage des femelles de race pure et croisées charolais, Charolais, (29) 61-67.

Ménissier, F. and Vissac, B. 1971 Possibilités d'amélioration des conditions de vêlage par sélection. I. Technique de mesure de l'ouverture pelvienne des bovins. Annales de Génétique et de Sélection animale, 3, 207-214.

Ménissier, F. 1974 Hypertrophie musculaire d'origine génétique chez les bovins: description, transmission et emploi pour l'amélioration de la production de viande. 1er Congrès mondial de Génétique appliquée à l'Elevage, Madrid 1974, vol 1, (Séances plènieres), 85-107.

Ménissier, F. 1974a L'aptitude au vêlage des races à viande françaises: l'origine des difficultés de vêlage et leur amélioration génétique. Supplément au Bulletin technique du CRZV de Theix (numéro spécial), octobre 1974, 139-170. Also in 'Optimum breeding plans for beef cattle'. Bulletin technique du Departément de Génétique animale, Institut National de la Recherche Agronomique, France, numéro 21, 57-102 (in English).

Ménissier, F. and Frebling, J. 1974 Aptitude à la gémellité des races à viande françaises: observations en élevage et constitution d'un troupeau de sélection. Fédération Européenne de Zootechnie, Août 1974, Copenhagen 17 pp.

Ménissier, F., Bibé, B. and Perreau, B. 1974 Possibilités d'amélioration des conditions de vêlage par sélection. II. Aptitude au vêlage de trois races à viande françaises. Annales de Génétique et de Sélection animale, 6, 69-90.

Ménissier, F. 1975[a] Genetic aspects related to use of beef breeds. In: First seminar on Nutrition and Management, 'The early calving of heifers and its impact on beef production'. EEC programme of research on beef production, June 4-6th, Copenhagen, CEC, EUR 5545e, pp. 81-122.

Ménissier, F. 1975[b] L'amélioration génétique de la mortalité périnatale chez les bovins: Recherches et perspectives. Rapport présenté à la Commission des Recherches Bovines de l'INRA, 22nd octobre 1975 au CRZV de Theix, 15 pp. (not published).

Ménissier, F., Vissac, B. and Frebling, J. 1975 Optimum production plans for milk and meat: specialised beef herds and beef crossing in dairy herds. In: 'Optimum breeding plans for beef cattle'. Bulletin du Departément de Génétique animale, INRA, France, (numéro 21), 2-56.

Ménissier, F., Foulley, J.L. and Gogue, J. 1975[a] Estimation des paramètres génétiques et phénotypiques en race Charolaise, pour les génisses vèlant à 2 ans en station de contrôle des qualités d'élevage. (Not published).

Ménissier, F. 1976 Comments on optimisation of cattle breeding schemes: beef breeds for suckler herds. In: First seminar on Genetics, 'Optimisation of cattle breeding schemes'. EEC programme of research on beef production. November 26-28th 1975, Dublin, EUR 5490e pp. 125-151.

Ménissier, F., Sapa, J., Gogue, J., Bonaiti, B. and Frebling, J. 1978 Comparison of the main European cattle breeds used in industrial crossing on French Friesian dairy cows: preliminary results on calving difficulties. In: Seminar on 'Calving problems and early viability of the calf'. EEC programme of research on beef production, May 4-6th, 1977, Freising, 12 pp.

More O'Ferrall, G.J. and Cunningham, E.P. 1977 Review of Irish experiments on beef breeds and crosses. In: Second seminar on Genetics, 'Crossbreeding experiments and strategy of breed utilisation to increase beef production'. EEC programme of research on beef production, February 9-11th, 1976, Verden, CEC, EUR 5492e, pp. 277-296.

O'Connor, L.K. 1977 Calving surveys of beef crossing on Friesian dairy cows (reports on progeny testing of bulls for AI). (Unpublished data, personal communication).

Philipsson, J. 1976 Studies on calving difficulty, stillbirth and associated factors in Swedish cattle breeds. III. Genetic parameters. Acta Agriculturae Scandinavica, 26, 211-220.

Philipsson, J. 1977 Studies on calving difficulty, stillbirth and associated factors in Swedish cattle breeds. VI. Effects of crossbreeding. Acta Agriculturae Scandinavica, 27, 58-64.

Remmen, J.W.A. 1975 Calving performance and calf mortality. 26ème Réunion annuelle, Fédération Européenne de Zootechnie, juin 1975, Warsaw, 15 pp.

Taylor, St.C.S., Monteiro, L.S. and Perreau, B. 1975 Possibility of reducing calving difficulties by selection. III. A note on pelvic size in relation to body weight of cattle. Annales de Génétiques et de Sélection animale, 7, 49-57.

Teehan, T.J. 1977 Report on calving surveys carried out by Irish AI Stations April 1977, Department of Agriculture 10 pp. (Unpublished data - personal communication).

Texier, C. and Legendre, J. 1976 L'exploitation des troupeaux Charolais commerciax du Nivernais. Analyse des résultats de reproduction des troupeaux de vaches allaitantes. Charolais (43), 6-19.

Union Maine-Anjou, 1977 Résultats de testage en ferme des taureaux Maine-Anjou. (Unpublished data - personal communication).

UPRA Maine-Anjou 1977 Résultats partiels sur les veaux déclarés à la naissance en 1976. (Unpublished data - personal communication).

Van de Plassche, M., Bouters, R., Spincemailles, J. and Herman, J. 1968 Wird beim Rind die Trächtigkeit durch eine vorausgegangene Schnittenbindung beeinträchtigt? Zuchthygiene, 3, 62-69.

Vissac, B. 1976 Optimierung europäischer Rinderzuchtplanung. Gedanken und Ansätze aus französischer Sicht. Der Tierzüchter, (11), 486-488.

Vissac, B., Ménissier, F. and Perreau, B. 1973 Etude caractère culard. VII Croissance et musculature des femelles. Déséquilibre morphologique au vêlage. Annales de Génétique et de Sélection animale, 5, 23-38.

Vissac, B., Perreau, B., Mauléon, P. and Ménissier, F. 1974 Etude caractère culard. IX. Fertilité des femelles et aptitude maternelle. Annales de Génétique et de Sélection animale, 6, 35-48.

DISCUSSION

F. Pirchner (West Germany)

Dr. Menissier, you mentioned that size has only a small
influence. Could you quote a rough figure; is it 5% or less?

F. Ménissier (France)

Less, as seen from our studies in Maine-Anjou, Charolais
and Limousin. However, in southern Europe with breeds like
Blonde d'Aquitaine or Chianina more variation between breeds
may be encountered.

F. Pirchner

Did you say that muscularity or pelvic size has an influence
on calving problems?

F. Ménissier

We have two problems, that of size and that of muscularity
and both act in the same direction.

T. Liboriussen (Denmark)

You mentioned that Blonde d'Aquitaine have a slightly higher
birth weight than Charolais; is this due to a higher frequency
of heavy muscling in Blonde d'Aquitaine?

F. Ménissier

The situation in Blonde d'Aquitaine is very surprising
concerning the high birth weight. I think the rapidly increasing
birth weight can be explained by an efficient selection for
size and muscularity; many recordings are from cows bred by
these selected sires. Nevertheless the incidence of calving
difficulties is less than in Charolais. But it also should be
considered that the dam in the Blonde d'Aquitaine breed is
of a different type; it is a very large sized cow, possibly
changing very rapidly in the future.

A. Osinga (The Netherlands)

You said that both pelvic opening and birth weight are not

linear but threshold effects. My question is, does this also apply to gestation length, which is itself correlated with birth weight? With extended gestation the percentage of dystocia is increased.

J.L. Foulley *(France)*

I think there are different factors to be considered, for instance differences between breeds. The relation between weight and gestation length is not the same for all breeds. For instance the Limousin breed which has a relatively low birth weight has a long gestation length of about 287 days, while the Charolais with a gestation length of about 284 days has a higher birth weight (3 - 4 kg more). Within breeds, differences in the relationship between birth weight and gestation length have to be considered according to the use of the breed. If the sire breed is considered, a relatively high relation to the weight of calf and to calving difficulties is obvious. On the contrary, for dam breeds practically no genetic relation between gestation length and birth weight was shown by French data.

F. Pirchner

Two thresholds were mentioned, the low and the high birth weight. There is a straight decline of viability on both sides. This is a very nice case for selection for the intermediate.

F. Ménissier

This is quite a nice approach. For example in sheep the threshold effect is greater on the side of lack of maturity while in cattle the effect of high birth weight is more important.

GENETIC FACTORS AND BREEDING FOR CALVING PERFORMANCE. PART 1

Chairman: R.D. Politiek

SELECTION FOR DOUBLE-MUSCLING AND CALVING PROBLEMS

R. Hanset and M. Jandrain

Faculté de Médecine Vétérinaire (U. Lg)
Rue des Vétérinaires, 45
B-1070 Brussels, Belgium.

ABSTRACT

The Belgian meat market overemphasises any increase in muscular development. This is the reason why double muscling has been favoured by breeders of the Blue Belgian breed. This selection has succeeded in producing super-meaty animals of very high value for the butcher but has increased calving difficulties to a large extent.

Data concerning this problem are analysed and the following points are studied: annual increase in difficult calvings, influence of the sire, the type and birth weight of the calf, the age of the mother. Data on crosses between Blue-White bulls and Black-and-White cows are also given. The problem of selection against calving difficulties in the Blue-White breed is discussed.

1. INTRODUCTION

Selection in Belgian Blue-White breed directed towards muscular hypertrophy has led in less than 20 years to a complete change in the morphology of the cattle bred by the top breeders.

Originally, the Blue Belgian was a dual-purpose breed. Like the majority of dual-purpose breeds, this one stood at a crossroads: it had to specialise either as a dairy breed or as a beef breed.

The structure of the farms in Central and Upper Belgium, the high prices awarded to double-muscled cattle, the high rate of genetic gain in this respect, all these conditions have pushed the breed in the direction of a super-meaty breed.

To understand this move and the acceptance of its draw-backs, it must be borne in mind how high are the prices paid for double-muscled cattle on the Belgian market, in comparison with those for conventional cattle (Table 1).

TABLE 1

PRICES (IN B.F.) OF DOUBLE-MUSCLED AND CONVENTIONAL CATTLE IN BELGIUM.

Double-muscled		Conventional	Difference
Calf, 2 weeks old	20 000	5 000	+15 000
Heifers (per kg live weight)	115	60	+55 per kg
Bulls " " " "	105	60	+45 per kg
Cows " " " "	80	50	+30 per kg
35 B.F. = 1 U.S. dollar			

These differences are not fully explained by the degree of hypertrophy of the muscles. If we consider the carcass as a whole, the degree of hypertrophy is about 20 to 30%. But it is highest for the muscles of the rump and the shoulder. This hypertrophy is linked with an important reduction in the connective tissue which explains the tenderness of the meat. As a result

several cuts are promoted to the rank of prime cuts. Furthermore, it is well known that the carcass of the double-muscled
animal has less fat, lighter bones and a better dressing percentage. From the butcher's point of view it represents almost
an ideal.

While the top breeders have achieved complete fixation of
the double-muscle character in bulls and cows and if all AI bulls
are of that type, on the other hand, on commercial farms, cows
of the conventional type are still numerous. So, the cow population bred by AI bulls has a mixed composition.

2. CALVING PROBLEMS
2.1. <u>General situation</u>

The increase in the calving difficulties has run parallel to
the increase in the frequency of double-muscled calves. When the
selection for muscular hypertrophy started, Caesarean section
was already a usual practice in Belgium. At the beginning, some
fears were expressed regarding breeding operations relying upon
surgery. But now, what is considered as amazing outside Belgium,
has become common practice.

The cost of the Caesarean section (2 200 B.F.) is rather low
in comparison with the price of the double-muscled calf, of the
fattened bull or heifer, and of the cow if she is not too old.

In Table 2, the incidence of Caesareans is given for three
Belgian breeds: Black-and-White, Red-and-White, Blue-White.

TABLE 2

THE INCIDENCE (IN %) OF CAESAREANS IN THREE BREEDS (PROVINCE OF LIEGE).

YEAR	Black-and-White		Red-and-White		Blue-White	
	Heifers	Cows	Heifers	Cows	Heifers	Cows
1960	0.5	0.03	1.3	0.1	8.5	2.3
1965	1.56	0.26	5.7	0.72	26.08	5.26
1970	2.31	0.52	10.06	2.61	37.71	11.91
1975	2.57	0.45	9.76	1.80	48.02	25.70

In the Blue-White breed, the annual increase has been 2.3 % for heifers and 1.76 % for cows (period 1965-75).

The 'within sire' rate of increase, during the period 1971-75, has been 1.58 % per annum for the cows (ten bulls: range 0 - 3.34 %). This shows that the increase is mainly due to the change in the genetic structure of the cow population because of the accumulation of genes for double-muscling, with, as a result, an ever increasing frequency of double-muscled calves and of double-muscled cows, with narrower pelvic opening and less ability for calving.

2.2. Factors affecting calving

2.2.1. Birth weight

Among the factors leading to calving difficulties, the weight of the calf at birth is of great importance and the double-muscle character is known for its influence on the birth weight. (Table 3).

TABLE 3

DOUBLE-MUSCLING, BIRTH WEIGHT (IN KG) AND CAESAREANS

Birth weight	Double-muscling	Conventional	Difference
Experim. data (two sexes)	46.7 (n = 12)	38.9 (n = 9)	+7.8
Field data (males only)	51.7 (n = 48)	43.6 (n = 103)	+8.1
% Caesareans			
Experim. data	58.3 (n = 12)	11.1 (n = 9)	
Field data	81.2 (n = 48)	13.6 (n = 103)	

Two kinds of data are given in Table 3. The 'experimental data' have been obtained from a back-cross between F_1 cows (Black-and-White cows x double-muscled bull) and their father; the 'field data' refer to bull calves entered in 1976 for progeny-testing in the station of Ciney.

For these two groups of data, the effect of the double-muscle character amounts to 8 kg, and influences tremendously the proportion of calvings by Caesarean.

2.2.2. Conformation of the calf

The conformation of the double-muscled calf, by itself has a detrimental effect on calving (Table 4).

TABLE 4

PERCENTAGE OF CAESAREANS, AT CONSTANT BIRTH WEIGHT

	Birthweight classes		
Type of calf	40 kg	45 kg	50 kg
Double-muscled	60% (n = 5)	75% (n = 12)	100% (n = 32)
Conventional	2.5% (n = 39)	14.3% (n = 35)	42.9% (n = 14)

2.2.3. Age of the mother

That first calvers have more calving problems is a well known fact and it is true in any breed (Tables 2, 5, 6).

TABLE 5

AGE OF THE MOTHER, TYPE OF CALF (MALE CALVES) AND PERCENTAGE OF CAESAREANS

	Type of calf	
No. of calving	Double-muscled	Conventional
1	100% (n = 7)	50% (n = 10)
2	73.7% (n = 19)	11.5% (n = 26)
3	83.3% (n = 6)	9.5% (n = 21)
4	90.0% (n = 10)	14.3% (n = 14)
5 and more	55.7% (n = 66)	6.2% (n = 32)
	Source: Data Testing Station, Ciney, 1976 (male calves)	

TABLE 6
AGE OF THE MOTHER, SEX OF THE CALF (DOUBLE-MUSCLED CALVES) AND PERCENTAGE OF CAESAREANS.

Age of the mother	Sex			
	Male		Female	
Heifers	93.9% (n = 49)		81.0% (n = 42)	
Cows	64.0% (n = 222)		35.5% (n = 200)	
	Source: Hanset (1977)			

2.2.4. Conformation (double muscling) and pelvic opening

In the double-muscled animal the bone measurements are reduced and in particular the measurements of the pelvis. This characteristic of the double-muscled female is well known to the obstetrician (Derivaux et al., 1964).

Concerning this problem, we have data on bulls of the two types, slaughtered at the uniform age of one year (Table 7).

TABLE 7

MEASUREMENTS (ON THE CARCASS) OF THE PELVIC OPENING, ADJUSTED FOR LIVE WEIGHT FOR CONVENTIONAL AND DOUBLE-MUSCLED BULLS SLAUGHTERED AT 12 MONTHS OF AGE.

	Adjusted means (for live weight)		
Number	Double-muscled 34	Conventional 89	Significance
Pelvic height (cm)	15.92	16.67	P < 1 %
1/2 Pelvic width (cm)	5.69	6.05	P < 1 %
Source: Data Testing Station, Ciney.			

We do not know to what extent selection of bulls on their own pelvic opening would be effective in increasing the pelvic opening of their daughters and in improving the calving ability.

2.2.5. The effect of the sire

If we compare AI bulls for their percentage of Caesareans, significant differences are observed and these differences show some degree of repeatability.

The heritability of Caesarean incidence has been estimated by the method of Robertson and Lerner (1949), for all or none traits.

The phenotypic expression of the character by the calf is 0 or 1, according as the calf is born by Caesarean or not, while the genotypic value is the probability of normal birth for that particular calf.

Heritability estimates are given in Tables 8 and 9 for

TABLE 8

HERITABILITY OF CAESAREAN INCIDENCE (BLUE-WHITE, PROVINCE OF LIEGE).

YEAR	h^2 (heifers)	$SE(h^2)$	n	%	h^2 (cows)	$SE(h^2)$	n	%	N
1970	0.030	± 0.028	95.7	0.377	0.023	± 0.013	435.3	0.120	15
1971	0.029	± 0.020	91.1	0.371	0.037	± 0.013	380.7	0.140	28
1972	0.048	± 0.038	55.5	0.367	0.063	± 0.025	262.1	0.148	21
1973	0.158	± 0.070	50.9	0.414	0.047	± 0.020	226.7	0.167	23
1974	0.214	± 0.078	59.4	0.419	0.069	± 0.025	248.8	0.177	25
1975	0.075	± 0.048	62.6	0.491	0.102	± 0.041	244.2	0.258	18

h^2 = heritability; $SE(h^2)$ = standard error of the estimate; n = progeny-group size;

% = percentage of Caesareans; N = number of sires.

TABLE 9

HERITABILITY OF CAESAREAN INCIDENCE (BLUE-WHITE, PROVINCE OF NAMUR).

YEAR	h^2 (heifers)	$SE(h^2)$	n	%	h^2 (cows)	$SE(h^2)$	n	%	N
1971	0.033	± 0.016	158.1	0.249	0.054	± 0.017	496.9	0.104	29
1972	0.055	± 0.022	163.9	0.282	0.043	± 0.014	460.2	0.124	28
1973	0.037	± 0.016	168.8	0.308	0.029	± 0.010	487.0	0.133	32
1974	0.000	±	126.2	0.303	0.029	± 0.011	335.3	0.156	29
1975	0.010	± 0.011	133.2	0.304	0.035	± 0.012	352.5	0.157	30

h^2 = heritability; $SE(h^2)$ = standard error of estimate; n = progeny-group size;

% = percentage of Caesareans; N = number of sires.

two AI studies and separately for the matings with heifers and cows.

The standard deviations of the true progeny-group means are given in Table 10. This standard deviation is the square root of the sire component of variance; this estimate is of the order of 3 to 6%.

TABLE 10

STANDARD DEVIATION (σ_s) OF THE TRUE PROGENY-GROUP MEANS (SQUARE ROOT OF THE SIRE COMPONENT OF VARIANCE, σ^2_s).

YEAR	Province of Liège		Province of Namur	
	Heifers	Cows	Heifers	Cows
1970	0.042	0.025		
1971	0.041	0.034	0.039	0.035
1972	0.052	0.045	0.053	0.034
1973	0.098	0.040	0.044	0.029
1974	0.114	0.050	0.000	0.031
1975	0.068	0.070	0.024	0.034
Average	0.069	0.044	0.032	0.033

The repeatability of a bull's progeny test (percentage of Caesareans) has been studied by:

a) the correlation between the progeny test on heifers, and the progeny test on cows (Table 11);

b) the intra-class correlation of the deviations from stud-year average for bulls used five years in succession (Table 12).

A repeatability of 0.6 for consecutive progeny tests, each test being based on 250 progeny, corresponds to a heritability of 2.4%, knowing that the correlation between two tests based on n offspring each is equal to $\frac{n}{n + k}$, where $k = \frac{4 - h^2}{h^2}$.

Such a repeatability is equivalent to that of successive progeny tests for milk yield each based on 30 daughters ($h^2 = 0.20$.)

The lower repeatability in the case of the heifers, has probably several reasons:

a) A smaller number of progeny

b) The adjustment of the attitude of the farmer, once the first results of a bull are known; if a young bull is recognised as suitable for heifers, the next year its test will be less favourable and inversely if its first test was not good. The same applies to the cows, at least to some extent.

TABLE 11

CORRELATION BETWEEN THE PROGENY TEST ON HEIFERS AND THE PROGENY TEST ON COWS.

YEAR	Province of Liège		Province of Namur	
1970	0.315	(15)		
1971	0.504**	(28)	0.637**	(29)
1972	0.256	(21)	0.499**	(28)
1973	0.521*	(23)	0.479**	(32)
1974	0.677**	(25)	0.476**	(29)
1975	0.460	(18)	0.328	(30)
Average	0.461		0.484	
(In brackets, number of sires; significance: * = 5 % ** = 1 %).				

TABLE 12

REPEATABILITY (INTRA-CLASS CORRELATION) OF A BULL'S PROGENY TEST.

	Province of Liège		Province of Namur	
Heifers	0.44	F^6** 28	0.20	F^{14} * 60
Cows	0.58	F^9*** 40	0.64	F^{14} *** 60
Significance: * = 5%, ** = 1%, *** = 1 ^{o}/oo.				

2.2.6. Blue-White bulls in crossbreeding with Black-and-White cows.

Four AI bulls have been used in crossbreeding with Black-and-White cows as well as in pure breeding.

The results are given in Table 13.

TABLE 13

PERCENTAGES OF CAESAREANS IN CROSSBREEDING AND PURE BREEDING.

Name of the bull	Crossbreeding		Pure breeding	
	1973	1974	1973	1974
Emile	1.92 (52)	1.96 (102)	10.84 (83)	8.69 (69)
Sultan	4.16 (24)	5.88 (51)	19.3 (1010)	18.99 (1369)
Tenor	0.0 (30)	0 (86)	13.64 (579)	11.19 (1027)
Other bulls	2.32 (129)	1.89 (264)	16.31 (3298)	16.20 (3826)
In brackets, the number of calvings.				

A considerable drop in the incidence of Caesareans is observed when Blue-White bulls are mated to Black-and-White cows.
Two important things have changed when going from purebreeding
to crossbreeding. Fifty percent of the genes of the foetus are
of Black-and-White origin and the parturient is a pure Black-
and-White female. Nevertheless, it seems that there is a relation
between the performance of the bull in purebreeding and that
in crossbreeding. The bull Tenor performs well in the two cow
populations while the bull Sultan is less good in the two
instances.

3. DISCUSSION

In considering the case of the Belgian Blue-White we encounter the most extreme situation that we are aware of, concerning calving problems.

We can reduce the elements of the calving to a few
components:
　　1)　The calf component, subdivided into:
　　　　a -- the direct sire effect on the calf - the genes
　　　　received from the father:
　　　　b - the direct dam effect on the calf - the genes re-
　　　　ceived from the mother:
　　　　c - the indirect dam effect on the calf - the developing
　　　　foetus being influenced by the maternal environment.

2) The parturient component, or calving ability, this being the expression of the pelvic size, the slackening of the ligaments, the contraction of the uterus etc. This component is determined by the sire of the cow (direct sire effect on the cow) and by the dam of the cow (direct dam effect on the cow).

The steady increase in the percentage of Caesareans is due, to a large extent, to the transformation of the genetic make-up of the cow population- one could consider the situation as a kind of 'within breed' grading-up.

The change undergone by the female population is exhibited in the 'direct dam effect on the calf' and in the 'parturient component'. Nevertheless, the genotype of the bulls does not remain constant.

If we compare bulls of different ages, used in the same population and at the same time (year 1975), we find that bulls born during the period 1961-65, caused 37.2 % of Caesareans with heifers and 21.7 % with cows, while younger bulls, born during the period 1969-72, caused 51.9 % of Caesareans with heifers and 27 % with cows. But one could argue that these two groups of bulls were not mated to similar cows.

For the Blue-White breed, the crucial point is: is it possible, without altering what has been achieved regarding the muscular development, the growth rate and the adult size, to alleviate the burden of the Caesarean?

We would have to act on the two levels: the calf component and the parturient component.

Regarding the genetic structure of the cow population, the breed is not in a stable situation but in a fast changing one. We could make a selection among AI bulls on their percentage of Caesareans. We then act on the direct sire effect on the calf (the only one of interest if we have crossbreeding in mind).

If the bulls were selected on the results of their first 250 calvings and if the 50 % best were retained, the expected improvement for the future calving would be equal to 2 % (h^2 = 0.02).

Nevertheless, the proportion of double-muscled calves born annually, would go on increasing, as well as the proportion of cows with poorer calving ability. As a result of the above-mentioned selection, there would be, at best, a slowing down of the present rate of increase in the percentage of Caesareans and at the end of the grading-up process, a lower final level of Caesareans.

At the same time, a selection on the female side for calving ability (pelvic size, etc.) ought to be supplied on the bulls' mothers.

AI bulls could also be selected on the basis of the calving ability of their daughters (direct sire effect on the calving ability of the cow).

Thus, AI bulls would be selected on their direct effect on the foetus and on their direct effect on their daughters as parturient.

We do not know whether these two effects are genetically correlated or not, in the population considered.

The genetic progress could be evaluated by the comparison of the selected bulls with control bulls, in purebreeding as well as in crossbreeding. In this latter case, the genes received from the dam (direct effect of the dam on the calf) and the calving ability (the parturient component) may be assumed constant and so the direct sire effect on the calf can be distinguished from the other effects.

Is such a selection worth undertaking or will the challenge to mould double-muscled animals with a moderate level of

calving difficulties, in purebreeding and consequently in crossbreeding be accepted?

When Blue-White bulls are crossed with Black-and-White cows
the two components, the cow component and the parturient component, are modified and consequently a spectacular drop in the
incidence of Caesarean sections is observed. On the contrary,
if we used bulls from another breed on the actual Blue-White
cow population, only the calf component (direct sire effect)
would be affected, at least during the first two or three years.
A drop in the percentage of Caesareans, obtained in the first
year, would be followed by a temporary increase, until the
introduction of crossbred females in the breeding population.

REFERENCES

Dempster, E.R. and Lerner, I.M. 1950. Heritability of threshold characters
    Genetics 35, 212-236.

Derivaux, J., Fagot, V. and Huet, R. 1964. Le problème des dystocies.
    Essais de pelvimétrie. Ann. Méd. Vét. 108, 335-365.

Hanset, R. 1967. Le problème de l'hypertrophie musculaire ou caractère
    'culard' dans la race bovine de Moyenne et Haute Belgique. Ann. Méd.
    Vét. 111, 140-176.

Hanset, R. 1972. L'interference du caractère 'culard' et de la sélection
    basée sur la conformation, dans la race bovine de Moyenne et Haute.
    Belgique. Ann. Méd. Vét. 116, 27-56.

Hanset, R. and Ansay, M. 1972. Régions privilégiées d'hypertrophie
    musculaire, chez le bovin culard. Ann. Méd. Vét. 116, 17-25.

Hanset, R. 1974. Modèles de sélection en faveur d'un gène: application au
    cas du gène du 'culard' en race Blanc Bleu Belge. Ann. Méd. Vét. 118,
    507-518.

Hanset, R. 1975. L'exploitation du gène 'culard' en croisements. La
    détermination de l'importance relative des différents compartiments
    dans les schémas de croisement. Ann. Méd. Vét. 119 1-10.

Hanset, R. 1977. Crossbreeding plans with double-muscled cattle. In: EEC
    Seminar, 'Crossbreeding experiments and strategy of breed utilisation
    to increase beef production', Verden, Feb. 9-11, 1976, CEC, EUR 5492e,
    pp. 55-66.

Hanset, R., Ansay, M. and Jandrain, M. 1977. Morphology of the double-
    muscled calf. Discrimination between the double-muscled and the
    normal. In: EEC Seminar 'Crossbreeding experiments and strategy of
    breed utilisation to increase beef production', Verden, Feb 9-11, 1976,
    CEC, EUR 5492e, pp. 399-411.

Ménissier, F. 1974. Hypertrophie musculaire d'origine génétique chez les
    bovins: description, transmission, emploi pour l'amélioration de la
    production de viande. 1er Congrès Mondial de Génétique Appliquée à
    l'Elevage, Vol. 1. Séances plénières: 85-107.

Robertson, A. and Lerner, I.M. 1949. The heritability of all-or-none traits.
    Genetics. 34, 395-411.

REPRODUCTIVE PERFORMANCE IN CROSSBREEDING:
RESULTS FROM A CURRENT EXPERIMENT
IN THE FEDERAL REPUBLIC OF GERMANY

H.-J. Langholz, R.F. Diehl and W. Pabst
University of Animal Husbandry and Genetics,
University of Göttingen, Fed. Rep. of Germany

ABSTRACT

Reproductive performance following commercial crossbreeding of Charolais, German Simmental, German Yellow and German Red-and-White sires with German Friesian and partly with German Brown dams, was studied. The characters studied were 60 - 90 day non-return rate, gestation length, calving performance and peri- and postnatal losses. Altogether 2 044 crossbred and 5 676 straightbred matings were performed on 741 farms.

The only positive effect of crossbred mating was a slightly increased non-return rate mainly in the case of above average quality of insemination technique.

Gestation length was increased by up to 7 days when crossing German Friesian dams with sires of large-framed breeds (Charolais, German Simmental, German Yellow). No increase in gestation length, however, was observed when crossing breeds originating from northern Germany (German Friesian x German Red-and-White) or breeds originating from southern Germany (German Brown x German Simmental or German Yellow).

Calving problems were considerably increased by breeding large-framed bulls to German Friesian but not to German Brown cows. These problems are due to heavier birth weights of calves and to some extent are responsible for the increased rate of early mortality. The frequency of postnatal losses of crossbred Friesian calves sired by large-framed breeds, was surprisingly high, reaching 11 - 18% which is double or triple the incidence for straightbred calves. Genetic potential of crossbred calves sired by breeds not originating from the area where the calves are born and raised was apparently inferior to the genetic potential of the original straightbred Friesians.

No influence of farm management and feeding factors on reproductive performance could be detected. There were, however, distinct differences between progeny groups especially regarding dystocia and survival rate. Thus there is a need for and a fair chance of selecting sires which do not produce intolerable reproductive problems.

1. INTRODUCTION

According to genetic theory we would normally expect a better reproductive performance by using crossbreeding instead of straightbreeding. This could hold also for breeding straightbred cows to bulls of another breed. Indeed, the results available so far from crossbreeding experiments show an overall positive effect of crossmating straightbred cows.

However, from the different experiments it appears that the effect of crossbreeding depends on,

1) The degree of environmental adversity, and,
2) The degree of adaptation of the straightbreds to the environmental conditions in question.

As summarised by Koger (1973),Bos indicus x Bos taurus crossbreeding shows a greater effect under the adverse conditions of the Gulf Coast than crossbreeding among Bos taurus breeds carried out mainly under the better conditions of the mid-western states of the United States (Table 1).

TABLE 1

REPRODUCTIVE PERFORMANCE OF STRAIGHTBRED SUCKLER COWS AFTER CROSSBREEDING (% ADVANTAGE OVER STRAIGHTBRED CONTROLS). (Koger, 1973)

	Bos indicus x Bos taurus crosses		Bos taurus breed crosses	
	No. of controls*	% advantage	No. of crosses	% advantage
Calving rate	81	- 8.6	87	+ 2.3
Survival rate	83	+ 7.5	90	+ 3.6
Weaning rate	67	+ 5.1	78	+ 1.7

* Straightbred mated cows

At a low level of fertility the crossbreeding effect is more distinct as indicated by the studies of Wiltbank (1973) and Donald and Russell (1968). (Table 2).

TABLE 2

CONCEPTION RATE (CR) TO FIRST INSEMINATION BY TYPE OF MATING (Wiltbank, 1973; Donald and Russell, 1968)

	No.	CR	Heifers No.	CR	Cows No.	CR
Crossbreeding	970	63%	113	42%	97	51%
Purebreeding	447	58%	109	33%	68	41%

The great superiority of crossbred calves in survival rate under the adverse conditions of the Gulf Coast may, to some extent, be due to the fact that none of the parental breeds has been evolved in that specific environment.

With regard to the reproductive performance that can be expected from crossbreeding dairy-type dual-purpose cows with bulls of beef breeds, we should realise,

1) That we start from a high level of fertility in the dam breed, and,
2) That the dam breed has been evolved for centuries under the special environmental conditions of the production area. Thus, in general, the response to crossbreeding to be expected is of less significance for this special situation.

2. MATERIAL

The following study on reproductive performance in commercial crossbreeding for beef from dairy herds refers to the first series of crossbreeding experiments currently performed in this field in the Federal Republic of Germany. Results are based on crossing Charolais, German Simmental, German Yellow and German Red-and-White as sire breeds with German Friesian and partly with German Brown as dam breeds. They refer to 2 044 crossbred inseminations and 5 676 straightbred inseminations carried out simultaneously on 741 farms.

Reproductive performance is measured by the following characters: 60 - 90- day non-return rate, gestation length, calving performance, peri- and postnatal losses. Detailed information was collected on feeding and management structures in order to isolate possible correlations with the level of fertility and calving performance.

3. RESULTS

Considering that reproductive performance has a low heritability and that confounding of systematic environmental effects with the breed cannot be excluded in any case, the results obtained should be interpreted as tendencies rather than as true breed differences.

3.1. Non-return rate (NRR)

There is a slightly better conception rate, indicated by a 1 - 2% better non-return rate (NRR), after mating to a different sire breed. As can be seen from Table 3, this superiority is mainly expressed in connection with above-average quality of AI services (74.6% NRR for crossbreeding vs 72.2% NRR for straightbreeding). Apparently in the case of poorer quality of AI service the insufficiencies of AI management, especially of correct heat detection, conceal the genetic effect of crossbred matings.

TABLE 3

NON-RETURN RATE AFTER PURE- AND CROSSBREEDING (Diehl, 1976)

Breed combination*	n	Non-return rate %		
		Average	Technician above average	Technician below average
GF x GF	5676	69.3	72.2	65.8
Ch x GF	926	70.6	74.1	67.1
GS x GF	523	70.8	72.1	67.1
GY x GF	321	68.6	76.7	57.6
GRW x GF	274	71.0	78.5	65.1
Total average of cross-breeding	2044	70.4	74.6	65.7

* Here, and in all the following Tables: GF = German Friesian: Ch = Charolais: GS = German Simmental: GY = German Yellow: GRW = German Red-and-White

Breed differences are evident; on average there is no advantage of crossing German Yellow sires. Poorer semen quality of
some of the bulls used for crossbreeding might have reduced the
significance of crossbred insemination as well as the tendency
of farmers to select poorer breeders for crossbred inseminations.

As can be directly seen from Table 3, one of the main systematic effects on the non-return rate is the efficiency of the
inseminator, including the accuracy of his recording. The effect
of year should also be noted. The effect of farm management and
feeding structures on the non-return rate is, however, more or
less negligible. None of the factors analysed - herd size, area
of grassland available per cow, kind of roughage in the diet,
amount of concentrates per cow and amount of hay per cow - shows
a correlation with the fertility situation, as demonstrated in
Table 4. There may be a tendency for a better non-return rate
in very small herds using little concentrates and a lot of hay,
but this is more likely to be the expression of individual
attention rather than an effect of the extreme feeding ration.
The breakdown of the results into crossbred and straightbred
inseminations does dot show any differences in the effect of the
analysed environmental factors.

3.2. Service period

A lower incidence of prolonged service period following the
first insemination would indicate a lower frequency of early
embryonic mortality. A higher vitality of the crossbred embryo
should result in a lower frequency of prolonged service periods
after the first insemination. However, results listed in Table
5 do not show any difference of this kind between crossbred and
straightbred inseminations. But again, this result may be con-
fused by the fact that there is a tendency to apply crossbreeding to poorer breeders.

3.3. Gestation length

Gestation length following mating of large-framed sire breeds
with German Friesian dams is significantly prolonged by 4.2, 6.7
and 7.0 days in the case of breeding to German Simmental, German

TABLE 4

EFFECT OF HERD MANAGEMENT AND FEEDING ON NON-RETURN RATE (MODIFIED FROM DIEHL, 1976)

Herd size	NRR %	Grassland/cow	NRR %	Kind of roughage	NRR
< 10 cows	74.9	< 70 ar	73.9	Grass silage	69.7
11 - 15 cows	69.1	71 - 90 ar	67.8	Silage from sugar beet tops	72.0
16 - 20 cows	69.1	91 - 140 ar	72.6	Maize silage	73.8
21 - 35 cows	70.5	> 140 ar	72.6	Fodder beets	69.2
> 35 cows	71.9			Silage from maize and sugar beet tops	71.2
				Silage from maize, sugar beet tops and grass	70.1

Concentrates/cow/year	NRR %	Hay/cow/day	NRR %
< 500 kg	77.5	< 4 kg	68.7
501 - 750 kg	71.7	5 - 6 kg	71.7
751 - 1000 kg	68.9	7 - 9 kg	70.2
1001 - 1250 kg	69.5	> 9 kg	74.3
1251 - 1500 kg	73.0		
> 1500 kg	70.1		

TABLE 5

FREQUENCY OF RETURNING TO OESTRUS WITHIN NORMAL AND PROLONGED PERIODS FOLLOWING THE FIRST INSEMINATION

Breed combination	Number of first inseminations	% returns within: 1 - 24 days	25 - 200 days
GF x GF	6883	14.6	18.9
Ch x GF	1108	14.7	18.6
GS x GF	598	15.2	19.9
GY x GF	307	15.6	16.9
GRW x GF	357	14.0	18.8

Yellow and Charolais sires respectively (Table 6). In the case of crossing German Red-and-White with German Friesian and German Simmental, and German Yellow with German Brown cattle, no prolongation of gestation length can be observed. But there is a distinct difference between the north German lowland breeds (Friesian and Red-and-White), and the south German breeds (Simmental, Yellow and Brown cattle).

TABLE 6.

GESTATION LENGTH FOLLOWING PURE- AND CROSSBREEDING (DIEHL, 1976)

Breed combi- nation	n	Gestation length, days	Standard deviation
German Friesian dams			
GF x GF	1790	279.7	8.5
Ch x GF	541	286.7	7.8
GS x GF	290	283.9	7.6
GY x GF	131	286.4	8.8
GRW x GF	136	280.0	9.6
German Brown dams			
GB x GB	137	288.7	6.8
GS x GB	179	288.2	6.6
GY x GB	107	288.7	5.6

Sex differences were negligible and insignificant for the crossbreds from large-framed sire breeds, but they were significant for straightbred German Friesian gestations, being 1.7 days longer for males which is in agreement with El Kashab (1974). Prolonged gestation is correlated with a higher frequency of calving problems, as shown in an analysis of variance, which again is mainly due to heavier calves from longer pregnancies. However, the correlation between birth weight of calf and gestation length turns out to be not very close: it is $r_p = 0.22$ and $r_p = 0.10$ for German Brown and German Friesian dams respectively. Thus stillborn calves were carried, on average, 5.4 days longer. Moreover, attention should be drawn to the fact that postnatal losses on fattening farms - this means after a transfer at 10 days of age in northern Germany -

occurred when gestation length was increased by 2.5 days though
the calves had slightly below average birth weights and none of
them derived from a difficult birth.

We have to conclude that gestation length requires increased attention, especially when selecting bulls for commercial
crossbreeding.

3.4. Calving difficulties

Frequency of dystocia following crossbreeding dairy-type
dual-purpose cows with large-framed beef-type bulls is
significant. The results obtained in this experiment are in
good agreement with others of this kind (Philipsson, 1971;
Andersen et al., 1976) showing for German Yellow, German
Simmental and Charolais sires crossed with German Friesian
cows an incidence of 4 - 5% deliveries requiring veterinarian
assistance, which is three times as much as for straightbred
Friesians (Table 7). If the farmers' judgment is included,
frequency of dystocia increases to 9.5, 15.5 and 18.0%
respectively for the three sire breeds mentioned versus 5.5%
for straightbred Friesians. Birth weight of the crossbred
calves was increased by 3.3 kg for German Yellow crosses, by
4.6 kg for German Simmental crosses and by 6.3 kg for Charolais
crosses (Table 8).

TABLE 7

FREQUENCY OF DYSTOCIA AND PERINATAL CALF LOSSES AFTER PURE- AND CROSSBREEDING
(Diehl, 1976)

Breed combi-nation	n	Calving assistance (veterinarian help)		Perinatal losses (up to 10 days of age)	
		Number[x]	Percent	Number[xx]	Percent
GF x GF	1928	30 (5)	1.55	47 (31)	2.44
Ch x GF	574	25 (4)	4.36 **	24 (13)	4.18 **
GS x GF	300	15 (5)	5.00 **	15 (1)	5.00 **
GY x GF	137	7 (1)	5.11 **	2 (-)	1.46
GRW x GF	144	5 (1)	3.47	2 (1)	1.39

[x] ( ) number of Caesarean sections;
[xx] ( ) number of stillborn calves.

TABLE 8

BIRTH WEIGHT OF COMMERCIAL CROSSBRED CALVES FROM GERMAN FRIESIAN AND GERMAN BROWN DAMS

Breed combination	Number of calves	Birth weight, kg $\mu \times \hat{c}$	Error $s_{\hat{c}}$
		German Friesian dams	
GF x GF	138	38.6[x)]	.85
Ch x GF	160	44.9[x)]	.79
GS x GF	158	43.2[x)]	.75
GY x GF	113	41.9[x)]	.80
GRW x GF	117	40.7[x)]	.85
		German Brown dams	
GB x GB	11	44.5	1.84
GS x GB	173	46.8	.85
GY x GB	54	47.9	1.04

[x)] 7-day weight.

With respect to dystocia and birth weight, crosses between German Red-and-White and German Friesian show intermediate results as compared with those of crosses from large-framed breeds.

Crosses from German Brown dams did not reveal a higher incidence of dystocia although birth weight was increased by 2.3 kg for Simmental and by 3.4 kg for Yellow x Brown crosses.

None of the analysed farm management and feeding effects showed a significant influence either on calving performance or on survival rate. Attention, however, should be drawn to the distinct differences between the different progeny groups with regard to dystocia, extreme values ranging between 9.1 and 22.7% for Charolais sires and between 8.3 and 25% for German Simmental sires when crossed with German Friesian dams (Table 9). Thus there is a fair chance of and need for selecting bulls for commercial crossbreeding which do not have the disadvantage of giving an intolerably high frequency of calving problems.

TABLE 9

INFLUENCE OF SIRE ON DYSTOCIA AND PERINATAL CALF LOSSES (UP TO 10 DAYS OF AGE) (Diehl, 1976)

	Charolais x Friesian		Simmental x Friesian	
	Min.	Max.	Min.	Max.
Dystocia				
Size of progeny group	22	176	72	20
Dystocia	9.1%	22.7%	8.3%	25.0%
Perinatal losses				
Size of progeny group	50	74	72	24
Perinatal losses	0.0%	12.2%	1.4%	8.3%
Dystocia	16.0%	17.6%	8.3%	20.8%

3.5. Peri- and postnatal losses

The last but not least important trait of reproductive performance is the survival rate. Since, at least in northern Germany, calves for fattening are transferred to specialised fattening farms at an age of 10 days, we have to differentiate between losses up to that age and losses occurring during the rearing period on the fattening farm.

In both cases a distinct increase of losses among crossbred calves from large-framed sire breeds on German Friesian dams was observed. The losses of 4.2 and 5.0% respectively among Charolais and German Simmental x German Friesian crossbred calves of up to 10 days of age were double those for straightbred German Friesians (2.4%; Table 7). Losses of Charolais and German Simmental x German Friesian crossbred calves during rearing rose to 11.4 and 11.1% respectively and again were double those for the straightbred German Friesians (6.1%; Table 10). German Yellow x German Friesian losses climbed even up to 17.7%, whilst no increased losses could be observed for this breed combination up to 10 days of age.

An explanation of the higher losses among crossbred calves up to 10 days of age may to some extent be the higher frequency of calving problems; however, the pre- and perinatal factors cannot be held responsible for the extremely high rate of loss among crossbreds during the rearing period, especially since no relation was observed between these losses and calving problems or birth weight. Insufficient genetic potential of the crossbred calves must be assumed to be one of the main reasons.

TABLE 10

CALF LOSSES DURING THE REARING PERIOD (IN % OF PURCHASED CALVES), BY GENETIC GROUP AND SEX (Pabst, 1977 Personal communication)

Genetic group	No. of calves purchased		Percentage of losses (10 days to 3 months)	
	♂	♀	♂	♀
GF x GF	256	206	8.1	5.3
Ch x GF	186	192	10.2	13.0
GS x GF	134	154	13.0	9.1
GY x GF	55	64	18.2	17.2
GRW x GF	159	106	6.3	6.6
GB x GB	52	36	9.6	5.6
GS x GB	82	82	8.5	4.9
GY x GB	72	49	1.4	4.1
Total	996	889	9.1	8.6

With reference to the initial statements we may advance the hypothesis that the sire breeds are not well adapted to the unhygienic calf environment of northern Germany. Thus, even if hybrid vigour does occur in the crossbreds, the genetic potential of the crossbred calf from sire breeds evolved in other environments will not make up for the additive genetic potential of the straightbred German Friesians adapted to the environment of northern Germany. That German Red-and-White x German Friesian crosses do not show an increased rate of loss and that the crosses from south German sire breeds on German Brown dams even show a higher survival rate, support this hypothesis.

Another explanation, suggested by Pirchner (1976), is that there may be a special incompatibility when crossing Charolais, German Simmental and German Yellow with German Friesian. But in such a case we should have observed similar problems in commercial crossbreeding projects with related Friesian populations.

When we do not find such serious losses in Friesian cross-breds in other regions and populations we must conclude that the calf environment must be extremely disadvantageous under the north German conditions. In addition, the experimental procedure of gathering calves from 300 farms and filling up the fattening farms continuously with young stock over a period of about 4 months is likely to increase the hygienic burden significantly. Moreover, the experimental procedure of trying to fill up missing numbers immediately may have tended to lead to a more disadvantageous situation for the crossbred calves. Thus the percentage of losses of crossbred calves may be somewhat overestimated.

On the other hand, distinct genetic differences become evident when, between different crossbred progeny groups, there are clear differences in the percentage of losses up to 10 days of age, ranging from 0 to 12.2% for Charolais and from 1.9 to 8.3% for German Simmental sires crossed with German Friesian dams (Table 9).

Moreover, as is evident from Table 9, high incidence of calving difficulties does not necessarily lead to a high rate of early mortality. Thus, it is not enough to consider in selecting sires for commercial crossbreeding, only a decrease in calving problems. The rate of peri- and postnatal losses have to be included in the selection decision.

Summing up from the comprehensive German study on commercial crossbreeding, only a slight positive effect of crossbred mating on conception can be observed, whereas the effect on gestation length, calving performance and survival rate is clearly disadvantageous. Marked differences between crossbred progeny groups, however, indicate a chance of reducing these disadvantages to a large extent by consequent sire evaluation and selection. More detailed studies are needed to quantify the chance of finding such bulls.

REFERENCES

Andersen, B. Bech, Liboriussen, T., Thysen, I., Kousgaard, K. and Buchter,
   W. 1976. Crossbreeding experiment with beef and dual-purpose sire
   breeds on Danish dairy cows. Livestock Production Science 3, 227-238.

Diehl, R.F. 1976. Vergleichende Untersuchung zur Fortpflanzungsleistung
   nach Einfachkreuzung beim Rind. Diss. Göttingen, 121 pp.

Donald, H.P. and Russell, W.S. 1968. Some aspects of fertility in purebred
   and crossbred dairy cattle. Animal Production 10, 465-471.

El Kashab, S. 1974. Untersuchungen zur Trächtigkeitsdauer beim Rind.
   Diss. Göttingen, 106 pp.

Koger, M. 1973. Summary in 'Crossbreeding Beef Cattle', Series 2, pp. 434-447.

Langholz, H.-J. and Pabst, W. 1977. Stratified experiment on commercial
   crossbreeding for beef in dairy herds of the FRG. In: EEC seminar on
   'Crossbreeding experiments and strategy of breed utilisation to increase
   beef production', Verden. 9-11 Feb. 1976. CEC, EUR 5492e pp. 313-343.

Philipsson, J. 1971.Kalvingssvårigketer och dödfödslar vid korsning mellan
   Charolais och SLB. NJF's kongress, Uppsala 1971, Seksjon V. Husdyrbruk,
   122-123.

Pirchner, F. 1976. Zusammenfassende Betrachtung zum Thema 'Fruchtbarkeit
   und züchterische Aspekte', Hülsenberger Gespräche 1976.Schriftenreihe
   der Schaumann-Stiftung zur Förderung der Agrarwissenschaften, pp. 75-76.

Wiltbank, J.N. 1973. Heterotic effects influencing reproduction when
   crossing European beef breeds. In: 'Crossbreeding Beef Cattle', Series
   2, PP 143-152. University of Florida Press, Gainesville.

INFLUENCE OF SIRE BREED ON CALVING PERFORMANCE, PERINATAL MORTALITY AND GESTATION LENGTH

T. Liboriussen

National Institute of Animal Science
Department of Cattle and Sheep Experiments
Rolighedsvej 25, 1958 Copenhagen V, Denmark

ABSTRACT

The paternal influence of 15 European beef and dual-purpose breeds on calving performance, early viability, birth weight and gestation length, has been studied in a Danish crossbreeding experiment with RDM (Danish Red) and SDM (Black Pied Danish) as dam breeds. The experiment was carried out in two series, of which the second series is still in progress. There are from 45 to 130 calvings per sire breed.

The preliminary results gave the following ranking of sire breeds, according to percentage of easy calvings (unassisted or minor assistance by one person): Angus, Braunvieh, Gelbvieh, Hereford, Limousin, Danish Red-and-White (DRK), Simmental, Blonde d'Aquitaine, West Flemish Red, Chianina, Piemontese, Blue-White Belgian, South Devon, Charolais, Romagnola.

1. INTRODUCTION

The paternal influence on calving difficulty has been studied in various crossbreeding experiments (Bergstrøm, 1973;
Laster, 1974; Lindhé, 1968;LSTSC, 1973; Philipsson,1976). It
can be concluded from these experiments that the sirebreed has
influence on the ease of calving and that calving difficulties
occur mainly when large breeds are used as sirebreeds.

Several authors have dealt with the direct causes of calving
difficulty and perinatal mortality (Dreyer, 1965; Abdallah, 1971;
Hansen, 1972; Ménissier et al., 1974). It seems that birth
weight is the trait which is responsible for the major part of
the calving difficulties. Birth weight is determined by pre-
natal growth rate and gestation length. Both these traits may,
therefore, be indirectly responsible for calving difficulties.

Misplacement of the foetus and size and morphology of the
pelvic opening of the dam are other factors often mentioned as
direct causes of calving difficulties. According to Abdallah,
(1971) the morphology of the foetus (ie its dimensions relative
to its weight) is also of importance for the frequency of difficult calvings.

2. MATERIALS AND METHODS

This investigation is based on 1307 calvings from two series
of a Danish crossbreeding experiment. In the first series, the
following breeds were compared as sire breeds in crossbreeding
with RDM (Danish Red), SDM (Black Pied Danish) and Danish Jersey
cows:

Simmental,
Charolais,
DRK (Danish Red-and-White),
Romagnola,
Chianina,
Hereford,

Blonde d'Aquitaine,
Limousin.

The calves from this series were born in 1972, 1973 and
1974. The results from a calving survey of 1 006 calvings from
this series have been presented by Andersen et al. (1976).

The second series was initiated in 1975 as part of the common EEC crossbreeding programme. In this series, seven other
beef or dual-purpose breeds are being used as sire breeds for
crossbreeding with RDM and SDM cows. They are:

Angus,
Gelbvieh, (German Yellow)
Piemontese,
South Devon,
Braunvieh, (Brown Swiss)
West Flemish Red,
Blue-White Belgian.

Three of the Charolais bulls used in the first series are
also being used in the second. At present approximately 50
calves have been born by each of the sire breeds in the second
series. As this series is still in progress, the results concerning these breeds will be considered preliminary.

The data from the first two series have been pooled. Data
from 131 observations from the first series, where the Jersey
was the dam breed, have been excluded, and so have observations
on twins (35 pairs) and observations where the gestation length
was less than 210 days.

The calves were born on private farms, and the degree of
calving difficulty was recorded by the herd owners on question-
naires. According to the amount of assistance furnished at
parturition, the calvings were classified as follows:

a) Easy (no, or minor assistance only given by one person)

b)   Difficult (pulling assistance by the use of chains or rope, but without veterinary assistance).

c)   Very difficult (all calving which required veterinary assistance.

The gestation length was calculated as the number of days from last insemination to parturition.

The calves were weighed on their arrival at the test station at the age of 2 - 4 weeks. From their weight and age at arrival their birth weight has been estimated as: weight at arrival - 0.4 x days of age. Calves which were more than 28 days old before they arrived at the test station were excluded from the analyses concerning birthweight. An analysis of regression of weight on age at arrival showed no significant sire breed or dam breed effect on the coefficients of regression.

For the examination of the viability of the calves, they were divided into three 'viability classes':

a)   Calves that lived long enough to be brought to the experimental stations (average age - 16 days).

b)   Calves that died on private farms, <u>after</u> 48 hours after birth.

c)   Calves that were either stillborn or died during birth or within 48 hours after birth.

3.   RESULTS

Table 1 shows the effects of sire and dam breed and sex of calf on the frequency of easy, difficult and very difficult calvings, and on the viability of the calves. The sire breeds are listed according to percentage of easy calvings.

TABLE 1

INFLUENCE OF SIRE AND DAM BREED AND SEX OF CALF ON CALVING DIFFICULTIES AND
CALF MORTALITY

Sire breeds	n	% easy calvings	% difficult calvings	% very difficult calvings	Viability %		
					a	b	c
Angus	40	92.5	7.5	0.0	97.5	0	2.5
Braunvieh	49	69.4	30.6	0.0	93.9	6.1	0.0
Gelbvieh	44	68.2	23.7	4.6	93.2	0.0	6.8
Hereford	123	65.9	29.2	4.9	97.6	1.6	0.8
Limousin	101	59.4	34.7	5.9	96.0	0.0	4.0
DRK	121	56.2	37.2	6.6	95.0	1.7	3.3
Simmental	124	54.0	39.9	16.1	97.6	0.0	2.4
Bl.d'Aq.	78	50.7	44.1	5.2	97.4	1.3	1.3
W.Fl.Red	80	51.3	42.4	6.3	88.8	2.4	8.8
Chianina	112	42.9	46.4	10.7	95.5	1.8	2.7
Piedmont	56	41.1	51.8	7.1	85.7	3.6	10.7
B.W.Belg.	54	40.7	59.3	0.0	92.6	0.0	7.4
South Devon	55	38.2	58.2	3.6	94.6	0.0	5.4
Charolais	150	34.0	51.3	14.7	90.7	2.0	7.3
Romagnola	122	26.5	53.7	19.8	95.0	0.8	4.2
Dam breeds							
RDM	653	51.0	40.6	8.4	94.8	1.2	4.0
SDM	654	49.4	41.7	8.9	93.6	1.7	4.7
Sex							
Males	702	41.5	48.4	10.1	92.2	2.1	5.7
Females	605	60.2	32.6	7.3	96.9	0.7	2.5
Total	1307	50.2	41.1	8.7	94.3	1.5	4.2

a) Viable at the age of 16 days
b) Died at the age of 2 - 16 days
c) Died before birth, during birth or within 48 hours

Chi square tests for the distributions on calving performances and viability classes gave the following results:

Sire breed on calving performance: Chi^2 = 150.4, DF = 28, P<0.001

  "   "   "   viability   : Chi^2 =   42.7, DF = 28, P<0.05

Sex on calving performance   : Chi^2 =   46.7, DF =   2, P<0.001

Sex on viability   : Chi^2 =   13.6, DF =   2, P<0.01

For the two dam breeds, the distributions were not significantly different either for calving performance ιor for calf
mortality.

Calves which had had a difficult or very difficult birth
had reduced viability. This appears from Table 2.

TABLE 2

VIABILITY AFTER EASY, DIFFICULT AND VERY DIFFICULT CALVINGS

	Calvings						Total	
	Easy		Diffi-cult		Very diffi-cult			
	n	%	n	%	n	%	n	%
Viable at the age of 16 days	638	97.6	5ol	93.1	94	81.7	1233	94.3
Died at the age of 2 - 16 days	6	0.9	6	1.1	6	5.2	18	1.4
Died before, during or within 48 hours of birth	1o	1.5	31	5.8	15	13.4	56	4.3
Total	654	50.2	538	41.2	115	8.7	1307	1oo.o

Chi^2 = 51.2   DF = 4   P<0.001

Average gestation lengths, birth weights and the frequency
of malpresentations for sire breeds, dam breeds and sexes, are
shown in Table 3.

The influence of sire breed on gestation length and birth
weight is highly significant, P<0.001, and the frequency of malpresentation is significantly different at the 0.05 level
(Chi^2 = 25.4, 14 DE).

The effects of dam breed and sex are also highly significant
for gestation length and birth weight, and malpresentation

TABLE 3

INFLUENCE OF SIRE AND DAM BREED AND SEX OF CALF ON GESTATION LENGTH, BIRTH WEIGHT AND FREQUENCY OF MALPRESENTATIONS*

Sire breeds	Gestation length days			Birth weight kg			Malpre- * sented, %
	n	$\bar{x}$	SD	n	$\bar{x}$	SD	
Angus	36	281.2	5.0	36	34.5	5.8	2.5
Braunvieh	49	284.8	4.6	43	43.8	7.8	0.0
Gelbvieh	44	286.3	6.3	37	42.7	7.2	4.6
Hereford	123	282.7	4.8	107	40.5	6.7	6.5
Limousin	101	287.3	5.7	83	42.5	5.8	6.9
DRK	121	279.4	5.1	109	44.2	7.8	8.3
Simmental	124	285.6	5.4	114	46.3	7.3	12.9
Bl. d'Aq.	77	285.5	6.8	74	46.0	7.8	5.2
W. Fl. Red	80	281.9	7.7	70	49.4	8.6	10.0
Chianina	112	288.2	7.2	104	47.3	8.0	8.9
Piedmont.	56	287.9	6.0	46	44.3	10.3	17.9
B.W. Belg.	54	284.6	5.6	50	46.5	7.2	3.7
South Devon	55	284.4	5.7	52	41.9	6.9	3.6
Charolais	150	287.3	6.4	126	49.7	8.1	12.0
Romagnola	121	286.6	8.4	107	47.8	8.3	10.7
Dam breeds							
RDM	653	286.1	5.9	587	45.7	8.5	9.3
SDM	654	283.9	7.3	571	44.7	8.2	7.7
Sex							
Males	702	285.5	7.2	607	47.2	8.5	10.1
Females	605	284.5	6.2	551	43.0	7.7	6.6
Total	1307	285.0	6.8	1158	45.2	8.4	8.5

* Including all calvings where the calf was not correctly presented for birth.

occurred more frequently for males than for females (Chi^2 = 5.12, 2 DF, P<0.05)

4. DISCUSSION

In order to analyse the causal factors that might be responsible for the observed effects of breed and sex on the degree of calving difficulty, the material has been separated into:

1) Calvings with normal presentation of calf, and,
2) Calvings with abnormal presentation of the calf (including all kinds of misplacements).

From Table 3 it appears that especially Piemontese, Simmental and Charolais have a high frequency of malpresentations, and from Table 2 it appears that these breeds also have a relatively high frequency of very difficult calvings and a relatively high mortality. A part of this high frequency of very difficult calvings can be accounted for by malpresentations. As can be seen from Table 4, malpresented calves have a much higher frequency of very difficult calvings, and much higher mortality than normally presented calves.

Abnormal presentations occurred in 55 of 115 very difficult calvings, and 24 out of a total of 75 calves that died before, during or after birth, had been malpresented. It can therefore be concluded that in this investigation malpresentation of the foetus has been the main causal factor for 47.8% of the very difficult calvings, and for 32.0% of the peri- and neonatal mortality.

The relatively great influence that malpresentations have had on the total frequency of very difficult calvings and mortality can probably be ascribed to the fact that all cows had had at least one previous calving. If the analysis had been carried out on material which also included first calvings, one might expect that the size of the foetus, especially in relation to the size of the cow, would have more importance as a causal factor.

It is difficult to explain why the incidence of malpresentations is higher for some sire breeds than for others and the clarification of this problem calls for more research.

TABLE 4

INFLUENCE OF CALF PRESENTATION ON DEGREE OF CALVING DIFFICULTY AND VIABILITY

Number	Normal 1196	Malpresented 111	Total 1307
Calving difficulty			
Easy (%)	54.2	5.4	50.2
Difficult (%)	40.8	45.1	41.1
Very difficult (%)	5.0	49.6	8.7
Viability			
Viable at the age of 16 days (%)	95.7	78.4	94.3
Died at the age of 2 - 16 days (%)	1.0	6.3	1.5
Died before, during or within 48 hours after birth (%)	3.3	15.3	4.2

The influence of birth weight on calving difficulty was analysed by distributing all normally presented calves into nine birth weight classes and calculating the frequency of easy, difficult and very difficult calvings within each of these birth weight classes (Table 5).

TABLE 5

CALVING DIFFICULTY FOR DIFFERENT BIRTH WEIGHTS

Birth weight classes, (kg)	n	Easy calvings (%)	Difficult calvings (%)	Very difficult calvings (%)
<35.0	104	77.9	22.1	0.0
35.0 - 38.5	123	74.0	23.6	2.4
38.5 - 41.5	150	66.0	34.0	0.0
41.5 - 44.0	122	62.3	32.0	5.7
44.0 - 46.0	112	61.6	33.9	4.5
46.0 - 48.5	130	50.8	44.6	4.6
48.5 - 51.5	116	34.5	58.6	6.9
51.5 - 55.0	95	37.9	53.7	8.4
>55.0	127	31.5	58.3	10.2
Total	1079	55.4	40.0	4.6

The frequency of _easy_ calvings decreased almost linearly with increasing birth weight, and it is remarkable that even in the heaviest birth weight class (>55 kg), approximately one third of the calvings were accomplished without pulling assistance.

For birth weights below 41.5 kg very difficult calvings seldom occurred (only 3 out of 377 calvings). For the weight group between 41.5 and 51.5 kg the frequency of difficult calvings was almost constant at about 5%. For heavier birth weights the frequency of very difficult calvings again increased.

From the results shown in Table 5 the calvings can roughly be divided into three birth weight classes, according to the risk that birth weight will cause calving difficulty.

a) No calving difficulty: birth weight less than 41.5 kg.
b) Some   "   "   : birth weight between 41.5 and
                             51.5 kg.
c) Much   "   "   : birth weight higher than 51.5 kg.

Table 6 shows the distribution of calving according to the various sire breeds in these three weight classes. For those sire breeds which had more than 30 offspring in the intermediate birth weight class, the percentage of very difficult calvings among these is shown in the last column of the Table.

Charolais, West Flemish Red, Romagnola and Blue-White Belgian, especially, have given relatively large numbers of calves with high birth weights; birth weights higher than 51.5 kg occurred more frequently for male than for female calves.

Birth weight is, however, not the only factor causing calving difficulties for normally presented calves. From the last column of Table 6 it appears that within the intermediate weight group, the percentage of very difficult calvings is different for the various sire breeds, dam breeds and the two sexes. As far as the effects of sire breed and sex are concerned, breed and sex influence on the morphology of the calf is probably the causal factor.

TABLE 6

DISTRIBUTION (%) OF CALVES BY VARIOUS SIRE BREEDS ON BIRTH WEIGHT CLASSES

| Sire breed | n | Birth weight classes | | | % very difficult calvings in intermediate group |
		(<41.5 kg) %	41.5 - 51.5 kg %	>51.5 kg %	
Angus	36	88	12	0	-
Braunvieh	43	32	54	14	-
Gelbvieh	37	46	38	16	-
Hereford	103	56	40	4	4.9
Limousin	79	50	39	11	9.7
DRK	101	40	39	21	5.2
Simmental	102	27	53	20	5.6
Bl.d'Aq.	70	30	51	19	2.8
W.Fl.Red	63	16	49	35	0.0
Chianina	96	21	53	26	2.0
Piedmont	39	38	44	18	-
B.W.Belg.	50	30	46	24	-
South Devon	51	57	31	12	-
Charolais	113	17	44	39	4.7
Romagnola	96	21	52	27	20.0
Dam breed					
RDM	541	34	45	21	3.7
SDM	538	36	44	22	7.1
Sex					
Males	559	26	45	29	6.4
Females	520	44	44	12	4.4
Total	1079	35	45	20	5.4

Especially the morphology of the Romagnola crossbred calf seems to have caused greater calving difficulty than would be expected from its birth weight.

The influence of dam breed on the frequency of very difficult calvings in the intermediate weight group indicates that RDM

cows are able to give birth to heavier calves than SDM cows.
The physical background for this effect can either be the fact
that RDM cows have relatively larger pelvic openings or that the
morphology of an RDM crossbred calf is more favourable.

The high incidence of difficult calvings after crossbreeding
with South Devon, can be explained neither by frequency of malpresentations, by high birth weight nor by unfavourable morphology.

5. CONCLUSION

There are great differences between sire breeds, in their
influence on calving difficulty and viability of the crossbred
offspring. The viability of the crossbred calf is highly dependent on the degree of calving difficulty. Breeds causing much
calving difficulty also cause relatively high peri- and neonatal
mortality.

The paternal influence and the influence of the sex of the
calf on calving difficulty mainly act through the birth weight,
and to some extent also through the morphology of the calf.
Besides these two causal factors, the sire breed might also
affect calving difficulty through the paternal influence on the
frequency of malpresented calves.

REFERENCES

Abdallah, O.Y. 1971 Variations génétiques de l'aptitude au velage et de ses composantes. Thesis, Paris, 310 pp.

Andersen, B. Bech, Liboriussen, T., Thysen, I., Kousgaard, K. and Buchter, L. 1976 Crossbreeding experiment with beef and dual-purpose sire breeds on Danish dairy cows. II. Calving performance, birth weight and gestation length. Livestock Production Science 3: 227-238.

Bergstrøm, P.L. 1973 Gebruikskruising voor vleesproduktion bij rundvee. Rapport B117, Instituut voor Veeteeltkundig Onderzoek "Schoonord", Zeist, The Netherlands. 127 pp.

Dreyer, D. 1965 Geburtsverlauf und Kälberverluste, untersucht an Nachkommen ostfriesischer Besamungsbullen in Testbetrieben. Diss., Göttingen, 158 pp.

Hansen, M. 1972 Kaelvningsforløb samt relationer mellem dette og nogle kød-produktionsegenskaber hos RDM og SDM. Thesis, Copenhagen, 73 pp.

Laster, D.B. 1974 Factors affecting pelvic size and dystocia in beef cattle. Journal of Animal Science 38: 497-503.

Lindhé, B. 1968 Crossbreeding for beef with Swedish Red-and-White cattle: Part I. Performance under varying field conditions. Lantbrukshögskolans Annaler 34: 465-516.

LSTSC 1973 Limousin and Simmental Test Steering Committee. Report of the calving surveys on bulls imported to Great Britain in 1970-1971. Mimeo, 15 pp.

Ménissier, F., Bibé, B. and Frebling, J. 1974 Calving of three French beef breeds: preliminary results. Livestock Production Science 1: 217-218.

Philipsson, J. 1976 Studies on calving difficulty, stillbirth and associated factors in Swedish cattle breeds. VI. Effects of crossbreeding. Acta Agriculturae Scandinavica 26: 58-63.

BREEDING CONSIDERATIONS FOR MINIMISING DIFFICULT CALVING

R.T. Berg

Department of Animal Science, The University of Alberta,
Edmonton, Canada, T6G 2E3

ABSTRACT

Factors influencing calving difficulty in beef herds are examined using two data sources. The first set of data was from two breed populations maintained by the University of Alberta at its experimental beef breeding ranch and the second was obtained from the Canadian Charolais Association 'Conception to Consumer' progeny testing programme.

Age of dam, which was closely related to weight of dam, was an important factor in calving difficulty. Most of the problem was associated with 2 and 3-year-old heifers in the University herds. Breed differences in calving difficulty were apparent in the University herds with 27% of Hereford births requiring assistance compared with 15% of the Synthetics. Two-year-old Hereford heifers received assistance in 59% of their calvings, 3-year-olds in 23% while the comparable figures for the Synthetics were 41% and 6% respectively.

In the Charolais progeny testing programme sire/breed ranking for increasing calving difficulty was: Chianina, Limousin, Charolais. Regressing calving assistance on sire progeny average birth weight indicated from 1 to 40% increase in calving difficulty for an increase of 1 kg in birth weight of calves.

Calf weight as a percentage of cow weight seemed to show promise for explaining variation in calving difficulty. Calf weight, cow weight and calf weight as a percentage of cow weight explained 20 - 30% of the variance in calving difficulty. Possible breeding methods for alleviating calving difficulty are discussed.

1. INTRODUCTION

Beef production trends in North America indicate an emphasis
on faster growing cattle and an increased use of some of the
large European breeds for crossbreeding. Concurrently there has
been an increase in the incidence of difficult parturition. The
economic consequences can be quite important with increased labour costs for observation and assistance, increased veterinary
costs, losses of calves and cows, and delayed breeding of cows
suffering difficult births. Calving difficulty has problems
associated with its assessment; besides being a threshold character, it is subject to vagaries of the judgment of the
attendant. Degree of difficulty is also largely a subjective
judgment.

Calving difficulty is both a trait of the cow and of the calf.
A bull can thus have influence both as sire and as maternal
grandsire. A number of factors have been shown to be related to
calving difficulty including age of dam, gestation length, birth
weight and sex of calf. The heritability of the trait is reported to be low and no clear recommendation for alleviating the
problem has been put forth.

Factors influencing calving difficulty in beef herds were
examined from data collected on two breed populations maintained
by the University of Alberta and from data collected by the
Canadian Charolais Association from its progeny testing programme.
Births were scored from 0 to 5, 0 being an unassisted birth, 1
slight hand assistance and 5 the most difficult requiring Caesarean section for foetal extraction. For the purpose of this
paper the data are classified as assisted or unassisted births.
For the Charolais Association data scores of 0 and 1 were considered unassisted while for the University of Alberta data only
zero score was considered unassisted. The Charolais Association
progeny test did not include 2-year-old dams.

2. RESULTS

2.1. Breed and age of dam and sex of calf

In Table 1 are presented data from the two University of Alberta breed populations by age of dam and sex of calf. There was more calving difficulty experienced in the Hereford group. In the Synthetic group the problem was pretty much confined to 2-year-olds but in the Herefords a significant percentage of 3-year-olds also required assistance. Male calves in both breeds caused more calving difficulty than females.

TABLE 1

ASSISTED CALVINGS BY BREED AND AGE OF DAM AND BY SEX OF CALF. (UNIVERSITY OF ALBERTA DATA)

	Hereford		Synthetic *	
Age of dam (yr)	Number of calves	Assisted calvings %	Number of calves	Assisted calvings %
2	140	59	190	41
3	73	23	122	6
4+	205	4	348	4
Sex of calf				
Male	215	31	335	18
Female	203	25	325	11
Total	418	27	660	15

* A combination of Charolais, Aberdeen-Angus and Galloway breeds (Berg, 1975).

2.2. Birth weight of calves

In Table 2 the percentage of assisted births from sire progeny groups is regressed on the average birth weight of the groups for six years for Charolais sires and one year each for Limousin and Chianina sires. Most of the regressions are significant and indicate an increase of from 1 to 4% in calving assistance for 1 kg increase in birth weight of calves. The most reliable figure was 1.78% per kg for the pooled results of all Charolais sires.

TABLE 2

REGRESSION OF PERCENTAGE OF ASSISTED BIRTHS ON AVERAGE BIRTH WEIGHT (kg) OF
SIRE PROGENY GROUPS (1)

Year	Number of progeny groups (2)	R^2 x 100	Regression % per kg	SEb
	Charolais sires			
1969	20	20.8	2.02	.92
1970	22	12.8	1.39	.81
1971	21	48.1	3.48	.84
1972	37	25.5	1.80	.53
1973	31	41.5	1.96	.44
1974	23	20.9	1.06	.46
Pooled	154	25.9	1.78	.24
	Limousin sires			
1973	6	13.3	1.01	1.28
	Chianina sires			
1974	6	77.0	3.92	1.08

1) Data obtained from Canadian Charolais Association 'Conception to Consumer' test programme.

2) Each progeny group consisted of a minimum of 20 calves. Dams were a
minimum of 3 years of age.

2.3. Gestation length

Table 3 shows a slight trend in increasing calving assistance with increased length of gestation in the Charolais
Association data.

TABLE 3
INFLUENCE OF GESTATION LENGTH ON CALVING DIFFICULTY (CANADIAN CHAROLAIS
ASSOCIATION DATA)

Gestation length days	Total number of calves	Assisted %
<270	7	0.0
270 - 279	216	5.6
280 - 289	1651	7.1
290 - 299	699	13.7
>300	19	10.3

2.4. Calf weight as a percentage of cow weight

In Table 4 are presented calf weights, cow weights, and calf weights as a percentage of cow weights, by age of dam for the two breed groups from the University of Alberta herds. Calf weights and cow weights were recorded within 2 days of calving. Hereford calves make up a larger percentage of the cow's weight from 2-year-olds and a lower percentage from mature cows than those of the Synthetic group.

TABLE 4

CALF WEIGHT, COW WEIGHT, AND CALF WEIGHT AS A PERCENTAGE OF COW WEIGHT BY BREED AND AGE OF COW. (UNIVERSITY OF ALBERTA DATA).

Breed	Hereford			Synthetic		
Age of cow (yrs)	Calf wt. kg	Cow wt. kg	Calf wt. Cow wt. %	Calf wt. kg	Cow wt. kg	Calf wt. Cow wt. %
2	31.1	328	9.5	33.2	367	9.1
3	32.2	352	9.2	34.4	377	9.1
4+	35.0	428	8.2	37.2	440	8.6
Total	33.2	382	8.7	35.8	410	8.7

In Table 5 the assisted calving percentages are presented for increasing classes of calf weight as a percentage of cow weight. There is a trend to increasing calving difficulty with increasing calf weight percentage but the critical percentage weight is higher for 3-year-old and older cows than for 2-year-olds. The critical percentage seems to be higher in the older cows for the Synthetics (>10%) than for the Herefords (>8%).

2.5. Combinations of variables for predicting calving difficulty

In Table 6 are presented selected relationships to calving difficulty of calf birth weight, cow post-calving weight and calf weight as a percentage of cow weight. Unadjusted birth weight shows no relationship to calving difficulty in these data. Cow weight and calf weight as a percentage of cow weight are about equal in explaining variation in calving difficulty. Twenty to thirty percent of the variation in calving difficulty could be accounted for by the three variables. No improvement

resulted from including only calves greater than 7.0% of cow
weight or from including only calves which required assistance.

TABLE 5

ASSISTED CALVING (%) BY CLASSES OF CALF WEIGHT AS A PERCENTAGE OF COW WEIGHT
BY BREED AND AGE OF COW (UNIVERSITY OF ALBERTA DATA).

Breed	Hereford			Synthetic		
Age	2	3	4+	2	3	4+
Calf wt x 100/cow wt						
< 7.0	0	0	4	22	0	9
7.0 - 7.9	25	0	2	10	0	0
8.0 - 8.9	52	10	8	26	0	1
9.0 - 9.9	60	29	9	39	3	5
10.0 - 10.9	66	27	27	76	16	10
> 11.0	94	67	50	100	14	15

TABLE 6

RELATIONSHIPS OF SELECTED VARIABLES WITH CALVING DIFFICULTIES

	Variables used in regression			R^2 x 100		
Breed	Calf birth wt	Cow wt. post-calving	Calf wt. as % of cow wt.	All calves	Calves ≥ 7.0% cow wt.	Calves requiring assistance
Hereford						
1	*			1	1	0
2		*		19	20	7
3			*	20	22	8
4	*		*	26	27	11
5	*	*	*	29	29	12
Synthetic						
1	*			1	0	5
2		*		10	10	7
3			*	14	17	18
4	*		*	17	19	19
5	*	*	*	21	22	20

3. DISCUSSION

Calving difficulty is largely a problem involving 2-year-old heifers although in some breeds 3-year-olds also have significant trouble. Breed and breed-cross influence the amount and degree of difficulty experienced. Male calves cause significantly more trouble than females. Reduction of calving difficulty requires control of calf and dam influences. Birth weight of the calf is probably the most important calf trait directly influencing calving difficulty. Progeny testing of sires for birth weight of calves would give a useful index of potential calving difficulty, particularly applicable for sires used in crossbreeding. Consideration of breed of dam would be necessary in such a test.

Attempting to reduce calving difficulty by selection in a breed or population requires consideration of calf and maternal traits which are somewhat antagonistic. Progeny testing of bulls as sires and maternal grandsires is possible but probably impractical under most circumstances. Reducing birth weight by sire selection would, in the short run, reduce calving difficulty but in the long run may have negative effects on maternal traits. An index which considers calf weight as a percentage of cow weight, using all parities, adjusted for age of dam and sex of calf, might be useful in reducing calving difficulty. The object would to reduce selectively calf weight as a percentage of cow weight rather than merely reducing calf weight as would be done by direct emphasis on birth weight.

REFERENCES

Berg, R.T. 1975 The University of Alberta Beef Breeding Project, 54th
    Annual Feeders' Day Report. pp. 30-42. The University of Alberta,
    Edmonton, Alberta, Canada.

Laster, D.B. and Gregory, K.E. 1973 Factors influencing peri- and early
    postnatal calf mortality. Journal of Animal Science, 37: 1092-1097.

Long, C.R. and Gregory, K.E. 1974 Heterosis and breed effects in preweaning
    traits of Angus, Hereford and reciprocal cross calves. Journal of
    Animal Science 39: 11-17.

Philipsson, J. 1976 Studies on calving difficulties, stillbirth and assoc-
    iated factors in Swedish cattle breeds III. Genetic parameters. Acta
    Agriculturae Scandinavica 26: 211-220.

Tong, A.K.W., Wilton, J.W. and Schaeffer, L.R. 1976 Evaluation for ease of
    calving for Charolais sires. Canadian Journal of Animal Science 56:
    17-26.

DISCUSSION

J.L. Foulley *(France)*

Dr. Liboriussen, do you think that your estimated birth weight calculated by regression analyses is correct and don't you think that the coefficient used may perhaps be too low? We have the same problem in France in progeny testing, because the ratio is not a result of growth; it is the ratio in the population at the beginning.

T. Liboriussen *(Denmark)*

Yes, that is certainly the weakness in this data, but in fact we don't have the actual birth weight. I am quite sure that the average birth weight is correct, while the variation in birth weight is not. In these studies we calculate a variation in birth weight of 20%, whereas normally only 13% is observed.

F. Pirchner *(West Germany)*

Professor Smidt presented some data from which it is clear that there are much more birth difficulties in Red Danish than in Black-and-White. How does this agree with the differences you have mentioned?

T. Liboriussen

I have to correct that. It did not show that they had more birth difficulties, but rather that they have a higher mortality and that is certainly not the same. Actually the Red Danish are inbred and this is also a reason why they have a high calf mortality. As soon as crossbreeding is performed the calving performance of the Red Danish cow is better than that of the Black-and-White cow.

R.D. Politiek *(The Netherlands)*

How many bulls are used per breed?

T. Liboriussen

Four to five except Blonde d'Aquitaine where there are only three.

J.L. Foulley

Dr. Berg, I want to point out that Taylor does not suggest using the ratio of calf weight to weight of cow but the ratio of calf weight to pelvic opening or an estimate of pelvic opening of the cow which is not exactly the same criterion you suggested.

R.T. Berg (Canada)

The situation concerning pelvic measurements is a bit controversial in North America. But I think it would be worthwhile to actually measure pelvic opening. As you mention there are many problems with pelvic opening because the age and the growth of the cow influence it.

J.L. Foulley

I think we can use either a linear dimension after pelvic opening or body weight to the power 0.4.

R.T. Berg

If you use body weight for the prediction of pelvic opening you might as well stick to body weight.

R.W.J. Plenderleith (UK)

Dr. Hanset gave the cost of operation against the cost of the calf on the market. Has he any figures on the additional cost arising for feeding the cow an extra time to get her back in calf? Has he any figures for the reproductive behaviour of cows after Caesarean section and does he know if the efficiency of the production of the extra weight and double muscling compares favourably?

R. Hanset (Belgium)

It is true that reproduction of the cow after Caesarean is affected as is also indicated by the increased calving interval. However, in respect to longevity of the cow this does not matter too much since in this type of production system the farmer wants to cull the cow when still young. It should have no more than three calvings because thereafter the value of the cow drops

sharply. So in these herds there is a high proportion of heifers
compared with older cows. But it is difficult to make a general
balance including the drawbacks, not only of the Caesarean
sections, but also of impaired fertility thereafter. I am
afraid I have no exact figures on this problem.

R.W.J. Plenderleith

How efficient a converter is this animal (Belgian Blue-
White, double-muscled) for producing meat compared to other
breeds?

R. Hanset

Certainly not worse; it compares favourably in food
conversion with other breeds. There is no inferiority of the
double-muscled on this aspect.

R.W.J. Plenderleith

On what basis do the farmers decide to do or not to do a
Caesarean if 100% of the calves born are double-muscled? You
mentioned that Caesareans are running up to 50 - 70%.

R. Hanset

That is probably very subjective and probably also relates
to the age of the cow.

B.G. Lowman *(UK)*

So the decision is made when the cow has started calving.
Why does not the farmer have surgery done three days before the
cow is due to calve or before it is urgently necessary?

R. Hanset

I also have suggested that Caesareans should be performed
before term but it is the habit to wait until parturition
starts.

R.D. Politiek

How many times can you perform a Caesarean on a cow?

R. Hanset
     About three times.

J. Philipsson *(Sweden)*
     Dr. Langholz, many studies have shown that poor fertility
is observed following calving difficulties and stillbirths.
Do you therefore think it is a correct interpretation that
crossing gave no poorer fertility than straight breeding despite
the higher frequencies of calving difficulty and stillbirths?
Did you consider this when you analysed your data?

H.J. Langholz *(West Germany)*
     No, we did not consider this point. What I wanted to
compare was to straight breed a cow versus crossbreed a cow.
These were the same cows and the question you are raising only
applies to the situation after parturition, asking what happens
to cows after a heavy or easy delivery. I could not find any
data on this and also Dr. Smidt did not present figures on
what is going to happen to the subsequent performance of cows
after difficult deliveries in respect to reproduction and milk
yield. Perhaps we might be able to make use of computer
programmes in the near future, also due to the data we are
getting from organisations; Bavaria is most advanced in this
respect. In general there is no adequate information in our
country and perhaps you can give me some help.

J. Philipsson
     I am not prepared to give any figures but I think there
are several studies showing that poor fertility occurs after
difficult calving. There are figures for Germany and I think
van Dieten published some already in 1963.

A. Osinga *(The Netherlands)*
     Dr. Liboriussen gave some information about the frequency
of malpresentation in crosses. How far is this due to
incompatibility of the size of the foetus and the space in the
birth canal of the dam and how far to the birth weight of the
calf?

T. Liboriussen

The only thing I can say - and I have to guess - is that it is probably a question of space in the uterus during the last days of pregnancy. Actually we have seen this rather frequent mortality associated with malpresentation occurring in cases when the foetus is not especially heavy. Perhaps long legs or something like that is the reason.

A. Osinga

So you think there are two reasons for difficult births - either malpresentation or too heavy calves which would be two completely separate things.

T. Liboriussen

No, I would not say that they are completely separate; I am sure that there are connections.

M.A.N. Taverne *(The Netherlands)*

I think there is a definite correlation between the two. Just imagine what happens when the foetus enters the pelvic opening. The most important parameter - at least at the moment of presentation - is the diameter of the head of the calf lying on the front legs. If it does not fit into the pelvic opening for any reason, the calf goes back again and tries again. With new uterine contractions the situation of malpresentation can occur. In sheep it has been very nicely demonstrated, especially when the Texel breed (meat production) and a less improved breed are compared, that there is also a high correlation between birth weight and malpresentation.

T. Liboriussen

The Chianina also has a high frequency of malpresentation. The material I have is not large enough to allow distinction between the different kinds of malpresentation.

M.A.N. Taverne

I do miss a good control in these studies. It has already been mentioned, especially by Dr. Liboriussen, that there are three different groups of calving, an easy, a difficult, and a very difficult one. Our speaker from Canada on the other hand says there are only two groups, no interference with parturition or some help. I would say that at least in farms in the Netherlands the farmer starts to interfere and manipulate as soon as he sees the calf being born. To me this raises the questions, do we have a real good control for calving performances, especially in respect to differences between breeds? I doubt if there are studies where cows were really tested to see if they could calve themselves without any help. I think for some breeds that would be a very difficult and costly experiment with many calves lost, but it would at least give us a good figure about the relation between breeds and the perinatal losses.

R.D. Politiek

I think there are also different methods of sampling the data.

H.J. Langholz

We have tried to use a much more sophisticated recording system than the one I have shown to you. But the results obtained depend on the day of delivery, on the size of the farm, on the location of the farm in relation to the next farm and the other neighbours. For example on a small farm with many neighbours, up to seven persons may assist or observe a normal calving. I rather think that recording Caesareans or veterinary help would give a better record. Otherwise one always has to consider the special structure of the particular farm.

GENETIC FACTORS AND BREEDING FOR CALVING PERFORMANCE. PART 2

Chairman: R. Hanset

A BREEDING STRATEGY FOR REDUCING PERINATAL CALF MORTALITY IN HEIFER CALVINGS

R. Bar-Anan,

Department of Animal Husbandry, Ministry of Agriculture,
Hakirya, Tel Aviv, Israel.

ABSTRACT

In Israeli dairy herds heritability estimates for perinatal calf mortality (PCM) in calvings of heifer mates (HM) and cow mates (CM) were 0.42 and 0.13; the respective genetic standard deviations were 6.02 and 2.38 and the genetic correlation between HM and CM calvings was 0.58. It appears that CM calvings can contribute little to predicting HM calving performances.

Evidence is presented of sire families with persistently high and low PCM of HM. Selecting PCM proven sires for heifer breeding in Israel reduced the PCM incidence immediately by one-third, and then stabilised. The hypothesis is advanced that loci with large and small effects control foeto-pelvic incompatibility.

The genetic correlation between difficult calving of HM and growth rate of some was 0.35 and with milk yield of heifer daughters it was -0.46. Considering this situation, it is suggested that some of PCM proven sires might be exempted from post-performance test culling.

About 350 HM calvings are considered necessary for reliable PCM tests and only yield-proven sires at return to service are now mated to HM and PCM testing. Sire tests are by within herd/season contemporary comparisons. Since only one-fourth of the inseminations are to heifers, there is no need for culling sires on PCM. By nominating part of the yield-proven sires for heifer breeding, 90% of all heifers could be bred by yield and PCM-proven sires.

CURRENT SITUATION IN PRACTICE

The milk-recording system provides data on calving performance and on the calf, and estimates of the sire effect on the
calvings of uniparous heifer mates are obtained by within herd
and three-month-season contemporary comparison. The traits
observed are difficult calving, stillbirths and perinatal
mortality. In the new computer programme the sire effects on
uniparous calvings of heifer daughters and cow mates in second
and third parities will also be estimated.

INTRODUCTION AND REVIEW OF LITERATURE

The objective of previous research has been to estimate the
genetic parameters necessary for devising AI breeding plans by
which dystocia and calf mortality could be reduced. Perinatal
calf mortality (PCM) seems to be a function of several
aetiologically separate factors:

a) Lethal genetic and congenital factors.

b) Malpresentations.

c) Multiple births.

d) Foeto-pelvic incompatibility (FPI).

Lethal anomalies were reported in less that 0.3% for the
calvings in Israel and Germany (Ben-Zwi and Bar-Anan, 1969;
Dreyer and Leipnitz, 1971). Extreme smallness may impair the
propensity for survival. Gilmore and Fechheimer (1969) found
that the highest proportion of stillbirths was among Jersey
female calves. Donald (1963) reported relatively high proportions of dead calves among the extreme weight groups. Koger
et al. (1967) postulated a quadratic relationship between
birth weight and survival. Woodward and Clark (1959) found
that the standard deviation among moribund calves was twice as
large as among viable calves.

Malpresentations were reported by Donald (1963) in 1.6%
of 2002 calvings. Sloss (1970) postulated that malpresentations

might have been a major cause for PCM in multiparous calvings.

The incidence of twin births was about five times higher in cow calvings than in heifers. Twin births involved a high proportion of PCM in all parities (Hendy and Bowman, 1970; Bar-Ann and Bowman, 1974).

Young (1970) postulated that the difference between heifer and cow calvings in the incidence of dystocia and PCM was due to foeto-pelvic incompatibility in heifer calvings. These findings were supported by the observation by Woodward and Clark (1959) that stillborn calves from heifers were heavier and from cows were lighter than the average for live calves. Friedli (1965) found that, especially in heifer calvings, disproportion between the size of the calf and the width of the pelvic canal of the dam caused trauma and subsequent PCM.

The incidence of PCM in heifer calvings was reported to be 2 - 4 times as high as in multiparae (Lindhé, 1966; Dreyer and Leipnitz 1971), the difference apparently being due to FPI.

MATERIAL AND METHODS

The material consisted of 157 - 255 uniparous, full-term calvings of Israeli Friesian dairy cows in parities one to six in the years 1964 to 1970. Estimates of environmental effects, heritabilities and genetic correlations were obtained (Bar-Anan, Soller and Bowman, 1976). Sire values were estimated by contemporary comparison.

RESULTS

The genetic standard deviation in the calvings of HM amounted to about two-thirds of the mean incidence, thus providing the scope for considerable genetic improvement. The h^2 estimates for calvings of HM and CM were 0.042 and 0.013 respectively, Table 1, so that for sire tests of 0.75 repeatability, 282 HM calvings and 921 CM calvings would be required.

TABLE 1

INCIDENCE AND GENETIC PARAMETERS OF PERINATAL CALF MORTALITY (PCM)

Group	Incidence	SDg	h^2	K
Heifer mates (HM)	9.08	6.02	0.042	94
Cow mates (CM)	4.12	2.38	0.013	307
Heifer daughters (HD)	9.08	4.52	0.018	221
Cow daughters (CD)	4.12	0.91	0.004	999

SDg = Genetic standard deviation

$K = V_e/V_s = 4/h^2 - 1$

The concept h^2 describing the genetic proportion in the phenotype seems not very suitable for all-or-none characters. The use of K, defined as V_e/V_s, is suggested when n and r are the number of offspring and the reapeatability required, n for constant R is a multiple of K. Thus, n for 0.5, 0.75, 0.9 = K, 3K, 9K respectively. The term 'balance K' is proposed, since K balances the effects of sire and error: $K. V_s = V_e$ and when $n = K$, $R = 0.5$

The genetic correlations presented in Table 2 and Figures 1 and 2 apparently corroborate the hypothesis of different factors affecting cow and heifer PCM. The genetic correlation between calvings of CM and CD was high, but moderate between calvings of HM and CM and zero between HM and HD, although the calves of HD are the sire's grandsons.

TABLE 2

GENETIC CORRELATIONS

Character	Group	PCM HM	PCM CM	PCM HD	PCM CD	CC growth sons
PCM	CM	0.58*				
PCM	HD	0.04	0.09			
PCM	CD	−0.16	0.71**	0.24		
Growth	Sons	0.25	0.23	−0.06	−0.27	
Milk	HD	−0.34*	−0.10	−0.26	0.32	−0.04

* $P = < 0.05$, ** $P = < 0.01$

For definition of groups, see Table 1.

Number of sire groups; 87 HM, 138 CM, 94 HD, 104 CD, 140 sons, 142 HD milk. Each sire group for PCM > 100 calvings.

GENETIC MORTALITY CORRELATIONS

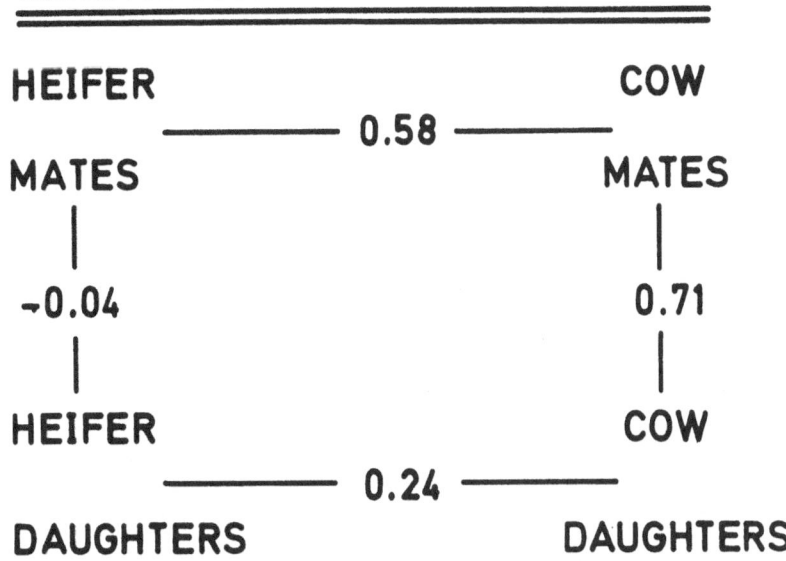

Fig. 1.

GENETIC CORRELATIONS

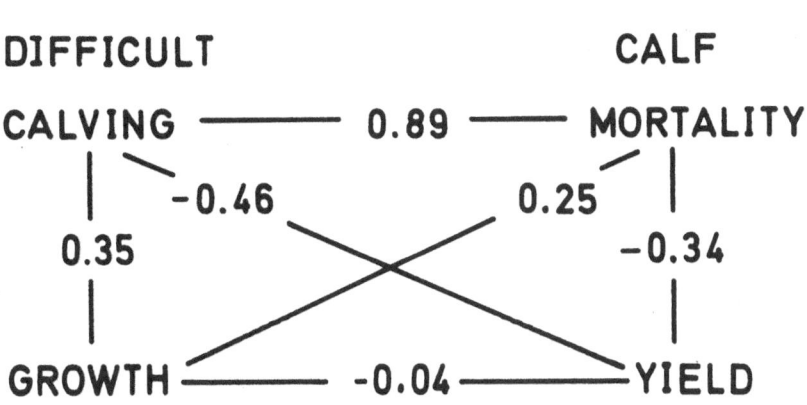

Fig. 2.

There was a negative association of -0.46 ± 0.11 between difficult calving of HM and 305-day yields of HD, which could imply the existence of a streamlined dairy type which is easily born.

On the other hand there was a positive association of 0.35 ± 0.13 between difficult calving of HM and growth rate of sons. Also the association between growth rate and PCM of HM was on the verge of significance.

The finding that the sire variance for PCM was only about 1% of the error variance seems not in agreement with the common belief among cattle breeders that bulls affecting heifer calving performance are easily recognisable, nor with the persistency of this trait in some sire families (Table 3). It is suggested that loci of both small and large effects may be involved in affecting DC and PCM. This hypothesis would explain the observation put forward by Young, Turner and Dolling (1963) that progress in selecting for all-or-none characters may be very fast or very slow.

TABLE 3

CONTEMPORARY COMPARISON (CC) VALUES OF TWO SIRE FAMILIES ON DIFFICULT CALVING (DC) AND PCM OF HM.

Adi 976 & 9 sons	Calv- ings	CC DC	CC PCM	Thor 951 & 8 sons	Calv- ings	CC CD	CC PCM
Adi 976	9203	-3.1	-3.2	Thor 951	125	+8.4	+4.6
Samir 1114	9127	-2.0	-2.3	Haziz 1032	292	+4.8	+1.3
Atid 1184	3057	-1.4	-2.2	Mapal 1064	136	+7.8	+3.5
Saar 1147	1001	-0.9	-1.7	Yuval 1030	128	+1.8	+1.9
Uri 1103	122	-1.4	-3.3	Takoa 1026	122	+5.5	+8.2
Eran 1207	99	-4.8	-3.9	Kochav 1047	117	+2.6	+3.8
Ilal 1160	89	-3.4	-4.5	Ankor 1046	113	+3.8	+0.7
Amiad 1104	81	-1.9	-3.8	Tomer 1024	101	-0.8	-0.1
Eshet 1216	78	-0.8	-3.4	Elem 1060	93	+15.8	+14.7
Abir 1156	51	-6.8	-5.1	8 sons	1102	+5.0	+3.6
9 sons	13 705	-1.8	-2.3	The family	1227	+5.3	+3.7
The family	22 908	-2.3	-2.7				

APPLICATION OF RESEARCH TO PRACTICE

Tables 3 and 4 present the effects of planned heifer matings.
The two sires in Table 3 were widely used as proven bulls together with two sons by each, but only the Adi family was
nominated for heifer matings, ending up with 23 000 calvings of
HM in milk-recorded herds.

TABLE 4

YEARLY LEAST SQUARE MEANS OF DC AND PCM IN HEIFER CALVINGS

Year	Calvings	%DC	%PCM
1964/65	7 474	8.67	11.45
1965/66	7 629	7.74	10.17
1966/67	8 555	5.13	7.08
1967/68	7 986	4.55	6.40
1968/69	9 015	3.79	7.17
1969/70	8 844	2.91	7.08

Table 4 presents the mean values for DC and PCM in heifer
calvings during the period in which planned heifer matings took
effect. In the first two years heifers were randomly mated with
respect to DC and PCM. Starting in 1966 most of the heifer
matings were by PCM proven sires. After an initial reduction
of the PCM incidence by one-third, the incidence rate stabilised.
The lack of further progress may have been due to the heavy use
of proven bulls, which reduced the number of calvings per young
bull tested to less than required for avoiding mistakes. The
values in 1976 were similar to those in the years 1967-1970.

The calving performance of cow mates could hardly contribute
to predicting FPI in heifer mates, so that heifers have to be
allocated for calving tests. If all young bulls were put to
heifers for PCM testing there would be no heifers left for mating
with PCM proven bulls. Therefore, the young bulls are now
mated to cows and the PCM tests are limited to yield-proven sires
mating each at return to service to 400 heifers.

Since only about one-fourth of the inseminations are to heifers, part of the yield-proven sires are selected for heifer breeding. By this plan about 90% of all heifers can be mated to sires, proven for yield and PCM.

It seems advisable that sons of sires causing PCM should not be reared for breeding, and that on the other hand, sons of PCM-proven sires should be exempted from post-performance-test culling. By culling one line on growth and another line on PCM, two within-breed strains could develop, one for breeding heifers and the other for breeding cows. The heifer strain would also be a safeguard against a possible genetic drift away from milk due to the multi-stage system of culling, first on growth and then on milk.

The scope of the sire selection on PCM tests of CM, HD and HM seems small, since reliable tests become available only for a few yield-proven bulls at an advanced age, but such sire values could be important for discerning any segregating major gene of lethal effect.

Economic aspects of difficult calvings

TABLE 5

5 YEAR MEANS OF EFFECTS ON DIFFICULT CALVINGS (1970/71 - 1974/75)

		Increase in:			
Parities	Calvings	Days open	PCM%	Kg milk	Disposal%
Heifers	3511	+7	+47.6	-61	+6.2
Cows	2119	+11	+57.5	-124	+11.0

REQUIREMENTS FOR FUTURE RESEARCH

More accurate recording techniques, such as separating malpresentations from FPI, and PCM due or not due to FPI, could in future decrease the size of test groups.

After the initial progress, due to the use of PCM-proven sires, sire selection does not seem likely to promise further rapid improvement.

Standardising the definitions for the various cases of dystocia and PCM and the general use of within-season contemporary comparisons among heifer mates could provide comparable sire values.

REFERENCES

Bar-Anan, R. and Bowman, J.C. 1974. Twinning in Israeli Friesian dairy herds.
     Animal Production 18, 109-115.

Bar-Anan, R., Soller, M. and Bowman, J.C. 1976. Genetic and environmental
     factors affecting the incidence of difficult calving and perinatal
     calf mortality in Israeli Friesian dairy herds. Animal production 22
     299-310.

Ben-zwi, A. and Bar-Anan, R. 1969. Herd-Book summary. Israel Cattle Breeders'
     Association 32 pp.

Donald, H.P. 1963. Perinatal deaths among calves in a crossbred dairy herd.
     Animal production, 5, 87-95.

Dreyer, D. and Leipnitz, C. 1971. Kälberverluste und Schwergeburten.
     Der Tierzüchter 14, 397-400.

Friedli, V. 1965. Häufigkeit der einzelnen Geburtsstörungen beim Rind
     unter schweizerischen Praxisverhältnissen. Schweizer Archiv für
     Tierheilkunde 107, 497-532.

Gilmore, L.O. and Fechheimer, N.S. 1969. Congenital abnormalities in cattle
     and their general aetiological factors. Journal of Dairy Science
     52, 1831-1838.

Hendy, C.R.C. and Bowman. J.C. 1970. Twinning in cattle. Animal Breeding
     Abstracts, 38, 22-37.

Koger, M., Mitchell, J.S., Kidder, R.W., Burns, W.C., Hemtyes, J.F. and
     Warwick, A.C. 1967. Factors influencing survival in beef calves.
     Journal of Animal science 26, 205.

Lindhé, B. 1966. Dead and difficult births in cattle and measures for their
     prevention. World Review of Animal Production, 2, 53-57.

Sloss, V. 1970. Kausalanalyse der Dystokie des Rindes in Victoria
     (Australien).Dissertation. Justus Liebig Universität, Giessen.

Woodward, R.R. and Clark, R.T. 1959. A study of stillbirths in a herd of
     range cattle. Journal of Animal Science, 18, 85-90.

Young, J.S. 1970. Management of the beef heifer for maximum reproductive
     performance. Veterinary Review, 9, 22. (Sydney University).

Young, S.S.Y., Turner, H.N. and Dolling, C.H.S. 1963. Selection for
     fertility in Australian Merino sheep. Australian Journal of
     Agricultural Research, 14, 460-482.

SELECTION FOR CALVING ABILITY IN FRENCH BEEF BREEDS

J.L. Foulley and F. Ménissier,
Quantitative and Applied Genetics Station
National Institute for Agricultural Research
CNRZ - 78350 Jouy en Josas, France

ABSTRACT

Genetic improvement of calving ability within beef breeds is discussed according to the two main French breeding systems:

1) use of terminal sire breeds,

2) use of beef breeds in purebreeding or in crossing for producing breeding females.

The amount of genetic variability existing for this trait and related factors (birth weight of the calf and size of the dam essentially) is reviewed in these two situations ie on account of paternal (direct effect) and maternal genetic contributions.

With respect to selection of terminal sire lines, it is suggested that more objective selection criteria be applied - at least for lines used on heifers - in the form of selection indexes with restriction on birth weight to avoid a too high indirect increase in this critical trait for calving problems.

In the other situation, specific selection on the maternal component must be carried out particularly in the large-sized and muscled beef breeds. Selection criteria proposed for this purpose must take into account the overall effect of growth and muscularity on beef efficiency in such systems (calving difficulties and mortality of calves, returns and costs of commercial offspring, maintenance costs and culling value of cows). Selection programmes developed in France in this area, such as a progeny test on 20 paternal half-sisters per bull, recorded in station for their breeding qualities, are presented and discussed.

1. INTRODUCTION

At the present time, calving difficulties appear to be an important factor limiting efficiency of beef production in France. It is thus becoming very important to consider this factor in breeding programmes for beef breeds on account of their increasing use on generally small-sized females and because of the indirect consequences on this trait of a very efficient selection for growth and muscle development.

Genetic improvement of calving ability can be undertaken at two different levels:

1) Choice of breeds or strains (especially the male ones);

2) Selection within populations.

As pointed out by Ménissier(1975) for Charolais, Limousin and Maine-Anjou females, the choice of sire line to cross with a given female population leading to production of the heaviest veal or yearling calves without exceeding the critical threshold of calving difficulties, can certainly be planned more objectively.

With regard to selection, which is our topic, we will distinguish, on the basis of the two main French breeding systems, the selection of terminal sire lines from that of beef breeds used in purebreeding or in crossing for producing breeding females. But, before dealing more in detail with selection criteria and methods to be applied for this purpose, it is necessary to review results on the genetic variability of this trait, particularly in French beef breeds.

2. GENETIC VARIABILITY FOR CALVING ABILITY IN FRENCH BEEF BREEDS

As reported by Ménissier (1975), difficult calvings especially result from an anatomical incompatibility between dam and foetus; thus, birth weight and pelvic opening explain around 50 and 10% respectively of the variation of calving score of heifers. These factors depend simultaneously upon the dam and her calf. Therefore, it is fundamental, in the analysis of genetic variability, to distinguish between the following two components:

- a paternal component representing the direct effect of the sire on the birth conditions of his calves through growth genes transmitted to them;

- a maternal component more complex including not only the direct effect of the dam on the size of the calf, but also the uterine influence on the latter as well as the incidence of her own genotype on pelvic opening and the interaction between dam and calf effects.

2.1. Genetic variability of the paternal component (direct effect)

In France, it appears from the results of the field progeny test of AI bulls used in terminal beef crossing that the heritability of the paternal component is very low (1 to 6%) for the calving score as well as for the frequencies of diffcult and very difficult calvings (Table 1). No marked differences exist between the Blonde d'Aquitaine, Charolais and Limousin breeds. As far as the criteria used are concerned, the values tend to be lower when the frequency of very difficult calvings is used which is in agreement with theoretical expectations.

Nevertheless, these heritability estimates are quite similar to published figures, on beef breeds (h^2 = 0.048) in the Charolais breed used in crossing, Belic and Ménissier, 1968; h^2 = 0.069 and 0.126 in the Hereford for all calvings and only for 2-year-old heifers respectively according to Brinks et al., 1973) as well as on dual-purpose or dairy breeds (Philipsson, 1976a). Furthermore, authors mention generally higher values for heifer than for mature cow calvings (Hanset, 1976) as predicted by the theory of heritability of all-or-none traits when the incidence rises (Hill, 1977).

In the same conditions of field progeny testing, birth weight of calves shows a consistently higher heritability (h^2 = 0.19, 0.17 and 0.09 in Blonde d'Aquitaine, Charolais and Limousin respectively), but not as high as the mean values from the literature (around 0.40) reported by Petty and Cartwright

TABLE 1

GENETIC PARAMETERS OF CALVING DIFFICULTIES AND BIRTH WEIGHT IN BLONDE D'AQUITAINE (a) CHAROLAIS (b) AND LIMOUSIN (c) BREEDS (Foulley et al., 1975)

r_p $\diagdown$ h^2 $\diagdown$ r_g		Calving difficulties			Birth weight
		Score	Difficult calvings	Very difficult calvings	
Calving difficulties	Score a	0.043 (0.011; 0.090)			0.885 (0.055)
	b	0.044 (0.020; 0.072)			0.911 (0.024)
	c	0.022 (0.004; 0.040)			0.911 (0.034)
	Difficult calvings (d) a		0.021 (-0.001; 0.061)		
	b		0.060 (0.037; 0.090)		
	c		0.020 (0.004; 0.040)		
	Very difficult calvings (d) a			0.025 (-0.002; 0.67)	
	b			0.021 (0.004; 0.044)	
	c			0.014 (-0.002; 0.036)	
Birth weight	a	0.170			0.191 (0.125; 0.287)
	b	0.316			0.165 (0.127; 0.217)
	c	0.109			0.086 (0.062; 0.117)

- heritability (h^2) estimates are given in the diagonal with their confidence limit intervals within brackets; genetic correlation (r_g) estimates are above the diagonal with their sampling standard deviation and phenotypic correlation (r_p) estimates below.

- data came from field progeny test of AI bulls for veal production including 94 Blonde d'Aquitaine, 256 Charolais and 374 Limousin sires with respectively: 4696, 12 842 and 16 765 pure and crossbred progeny.

(d) - difficult calvings: % of calvings requiring manual assistance, mechanical extraction or Caesarean section; very difficult calvings: % of the two latter types.

(1966) and Preston and Willis (1970) in British breeds. On the other hand, in these French beef breeds there is a very close genetic correlation (r_g = 0.9), although perhaps somewhat over-estimated (Olausson and Rönningen, 1975), between direct effects on birth weight and calving difficulties. Such a high correlation has also been found in dairy cattle by Philipsson (1976b) in Swedish Friesian and Pollak and Freeman (1976) in American Holstein.

The direct effects of morphology of calf and gestation length appear relatively minor when corrected for birth weight according to Ménissier et al.(1974) and Philipsson (1976b) respectively.

2.2. Genetic variability of the maternal component

In beef breeds the heritability of this component of calving difficulty is low but generally higher than that of the direct component. A value of 0.29 was estimated for Charolais cows recorded in pedigree herds by Couteaudier et al. (1971). In the same breed but from 2-year-old paternal half-sisters managed in station, Ménissier (1976) calculated h^2 values of 0.18 and 0.15 for the score and frequency of calving difficulties respectively. In the USA, Willham (1970) and Brinks et al. (1973) obtained figures of 0.09 and 0.13 respectively for Charolais and Hereford females of all ages. These results agree generally well with those obtained in dairy and dual-purpose breeds (Philipsson, 1976a,b). On the other hand, in beef breeds, perinatal mortality is closely linked to calving difficulties. With the exception of other related criteria such as, for instance, vigour score of calf at birth (Ménissier, 1975, unpublished data), this trait does not show a higher heritability than calving difficulty (Dearborn et al., 1973) chiefly as far as the maternal component is concerned (Ménisser, 1975).

With regard to the factors related to calving ability, the results obtained in France indicate the following phenomena (Table 2):

TABLE 2

GENETIC PARAMETERS OF MATERNAL CALVING ABILITY AND RELATED FACTORS IN THE
CHAROLAIS BREED FROM STATION PROGENY TEST DATA (Ménissier, 1976)[1]

| | $h^2 \rightarrow$ | Calving ability[2] | | | | Calf birth weight | |
| | | Score | | Difficult calving[3] (%) | | | |
		0.18		0.15		0.17	
Calf birth weight	0.17	0.67	0.59	0.70	0.51		
Gestation length	0.19	-0.07	0.21	0.08	0.16	-0.11	+0.36
Calving weight	0.47	0.51	-0.08	0.75	-0.03	0.33	0.26
Muscle development score at 18 months	0.48	0.33	-0.01	0.51	-0.01	0.01	-0.05
Weight at 18 months	0.33	0.09	-0.01	0.21	-0.01	0.47	0.05

1) Genetic parameters estimated from 753 heifers sired by 55 bulls and
managed for calving at 2 years.

2) In each column the left hand figure is the genetic correlation, the
right hand figure is the phenotypic correlation.

3) % of calvings requiring manual assistance, mechanical extraction or
Caesarean.

a) Birth weight and gestation length show heritability
estimates similar to those of calving difficulties
(h^2 = 0.17 and 0.19 respectively). But, whereas these are
markedly related genetically as well as phenotypically to
birth weight (r_g = +0.7; r_p = 0.5 to 0.6), their genetic
correlation with gestation length appears to be practically
zero. In addition, gestation length and birth weight as
traits of the dam seem, from a genetic point of view, little
correlated; this is the contrary to what has been observed
about their direct effects in other studies (Philipsson,
1975b).

b) Calving weight, which characterises the dam's size, is
closely linked genetically (but very little phenotypically)

to calving difficulties. It must therefore be concluded that increasing size by genetic means, which is relatively easy to do with such a heritability ($h^2 = 0.33$ in Charolais for 18-month weight), will lead to more calving problems. Monteiro (1969) and Taylor et al. (1975) explain this as follows. The increase in size of the dam is accompanied by increase in birth weight of calves (about the same in relative magnitude as estimated by Monteiro, 1969). Now, calving difficulties depend in fact on the ratio of the dam's pelvic opening (linear dimension) to calf weight. When dam weight increases, the numerator (proportional to weight to the power 0.4) rises relatively much more slowly than the denominator; so the ratio decreases and there are more calving difficulties.

c) However, apart from size, a second important factor has to be considered, ie muscle development which can enhance the unfavourable effect of the size as seen in French beef breeds. Charolais data make clear this specific effect of muscularity since there is a substantial correlation between muscle conformation score at 18 months and frequency of difficult calvings of heifers (+0.51) while calf weight remains unchanged ($r_g = +0.01$).

An exhaustive analysis should take into account the fact that calving performance as a dam trait includes not only genetic maternal (Gm) but also a fraction (50%) of genetic direct influence (Go).

In particular, the study of genetic relationships between these two kinds of effect might lead to a better understanding of the genetic interactions and regulation which may take place in the calf-dam unit. The few available studies on this topic generally reveal negative, but moderate, genetic correlation between direct and maternal effects, for calving score ($r_g = -0.19$ according to Philipsson, 1976b) as well as for birth weight (Philipsson, 1976b; Koch, 1972; Foulley and Ménissier, 1977, unpublished data). These authors obtained estimates of heritability coefficients for direct and maternal effects and genetic correlation between them of 0.295, 0.227 and -0.250 respectively for birth weight in Blonde d'Aquitaine breed. They use paternal half-sisters, maternal grandsire sibs and half aunt-nephew relationships between relatives. Data were obtained from 154 dams (and their progeny) sired by 17 bulls and recorded during 3 years.

3. SELECTION FOR CALVING ABILITY IN FRENCH BEEF BREEDS

In beef breeds, it is difficult to consider selection for calving ability independently of that for other traits. As a matter of fact, as shown previously, calving ability appears, from a genetic point of view, greatly dependent on the general phenomenon of growth. In other respects, growth has many influences on the efficency of beef production: a) directly, return from commercial offspring, culling value of cows and b) indirectly, feeding cost of progeny, maintenance cost of cow herd (Dickerson, 1970). In practice, selection will be different in the case of terminal sire lines (paternal component) from that of strains selected as a source of breeding females (maternal component).

3.1. Selection for terminal sire breeds

In France, the intensive selection practised for about fifteen years on muscle development in beef strains has led to a substantial genetic increase in birth weight and therefore, in calving difficulties. Thus, the mean genetic difference in birth weight between AI bulls selected after field progeny test and all recorded bulls was estimated to be +0.54 and 0.61 kg in the Limousin, Charolais and Blonde d'Aquitaine breeds (Foulley and Gaillard, unpublished). In fact, on account of the previous selection stages, the effective genetic change in birth weight would be higher (about 1.5 to 2% per generation; Foulley, 1976). This indirect effect of selection seems to be quite normal with respect to the genetic correlation values (around 0.7) existing between birth weight and final weight (75-day weight or yearling weight in France) on which selection is based. It then becomes urgent to slow down this genetic evolution for calving difficulties.

Creating a synthetic sire line from breeds with complementary abilities, in particular combining a good growth
potential (Charolais) with a long-bodied,more adapted morphology
(Blonde d'Aquitaine) and a low birth weight (Limousin) constitutes
a first approach.. Such selection programmes, named 'COOPELSO
93' and 'INRA 95' were set up in France a few years ago (Tables
3 and 4).

TABLE 3

DISTRIBUTION OF 'COOPELSO 93' PROGENY TESTED BULLS ACCORDING TO MATING TYPE
(Gaillard, 1977; unpublished data)

Dam Sire	BA	Lm	Ch	BA x Lm (1)	BA x Ch (1)	Total
BA	1	15	2	1		19
Lm	1	1	1	1	2	6
Ch	6	1				7
Total	8	17	3	2	2	32

(1) and reciprocal cross

TABLE 4

BIRTH WEIGHT AND CALVING PERFORMANCE OF COOPELSO 93 COMPARED WITH BLONDE
D'AQUITAINE (BA) LIMOUSIN (Lm), CHAROLAIS (Ch)AND INRA 95 PROGENY TESTED
BULLS (Gaillard, 1977; unpublished data)

COOPELSO 93 Sire- breed (1)	% of: (2)				Difference in mean progeny performance from			
					BA	Lm	Ch	INRA 95
	BA	Lm	Ch	No	EC : BW (3) : (4)	EC : BW	EC : BW	EC : BW
BA	60	28	12	8	1.6 :-0.3	-1.3 : 1.1	2.2 :-2.5	-1.1 :-1.5
Lm	15	65	20	5	6.2 :-1.6	3.2 :-0.3	6.7 :-3.8	3.5 :-2.8
Ch	33	17	50	3	-2.5 : 2.3	-5.4 : 3.7	-1.9 : 0.1	-5.2 : 1.1
All	40.6	37.5	21.9	16	2.3 :-0.2	-0.7 : 1.2	2.8 :-2.4	-0.4 :-1.4

(1) Sirebreed of COOPELSO 93 line

(2) % of Blonde d'Aquitaire, Limousin and Charolais breeds in the COOPELSO
    93 bulls

(3) EC: % of easy calvings

(4) BW: birth weight in kg

A second approach lies in selecting specialised strains particularly devoted to a differential utilisation on females with respect to their age on the one hand and to their size and muscle development on the other. For bulls used on mature cows a very weak (or even no) selection is required, whereas for those used on heifers or females of poor calving performance, the extent to which birth weight could still be increased is seriously limited even for the most favoured French beef breeds such as the Limousin. Therefore a programme is now in progress in this breed for selecting a sire line with restricted birth weight (Table 5).

TABLE 5

SOME RESULTS ON THE CONSTITUTION OF THE LIMOUSIN LINE ('MINIMUM' LINE OR 'ALPHA 16') SELECTED WITH RESTRICTION ON BIRTH WEIGHT (Ferrand, personal communication).

	No. of bulls	Birth weight (kg) (1)	75-day weight (%) (2)
a. AI Limousin bulls available (1974)	95	38.6	103.8
b. Bulls initially selected for the 'minimum' line	15	36.8	104.7
Difference b-a		- 1.8 (-1.4) (3)	+ 0.9

(1) Mean birth weight of progeny

(2) Relative breeding value (% of the mean)

(3) Standardised selection differential.

In these conditions, different selection criteria can be suggested. Selection would be better based on absolute or even relative growth rate (Fitzhugh, 1976) than on final weight since growth rates are automatically less correlated to birth weight than is final weight. It is also possible to select on birth and final weight by independent culling levels in such a way as to limit genetic increase in birth weight. But, for this purpose it is more efficient to use only one selection criterion, ie, a suitable restricted index (Foulley and Ménissier, 1975). Since there is a non-linear relationship between the reduction of genetic change in final weight and the intensity of restriction on birth weight, it is possible to slow down

substantially the genetic increase in birth weight without
decreasing too much that in final weight; for instance, by
selection on 'yearling weight minus 2.4 times birth weight',
we can expect to reduce by half the genetic response of birth
weight with a corresponding loss of only 7.5% in genetic
improvement of yearling weight versus 25% loss when no genetic
change in birth weight is imposed (Foulley, 1976).

On the other hand, in French conditions where field progeny
testing is planned to get about 60 progeny per bull, it seems
advisable to take into account primarily birth weight rather
than frequency of calving difficulties when selecting for
calving ability. It is not of course, the solution chosen
everywhere nor suggested by everyone (Mahon and Cunningham,
1976; Pollak and Freeman, 1976). Such different practices result
in fact from different constraints in recording facilities
(number of progeny, ease of measurement or computing conditions
of traits, percentage of loss and precision of data) according
to country and type of cattle (dairy or beef). Consequently, it
can be noticed that birth weight for instance could be generally
more easy available (or estimated) than gestation length in beef
herds. Nevertheless, the effective relative interest of all
these criteria should be discussed in more detail with respect
to their influence on calving difficulties and also on other
traits such as calf viability.

3.2. Selection in beef breeds used in pure or crossbreeding for producing breeding females

3.2.1. As shown previously by correlation values between
direct and maternal genetic effects, the genetic influence of
a parent on the birth conditions of his progeny and that on
calving performance of his daughters are almost independent.
Thus, choosing bulls only on their direct effects (Go), as
practised in terminal sire breeds, will be inadequate for
improving calving ability in beef breeds producing breeding
females; they should be selected simultaneously for direct and
maternal effects, theoretically for Go/2 + Gm in specialised
dam or parental (if female is crossed) lines and for Go + Gm in

purebreeding. In practice, no difference exists between the two; selection is generally carried out in two stages, on direct effects (Go) then on maternal component (Go/2 + Gm) with different intensities and variable coefficients for Go and Gm according to breed and selection unit.

3.2.2. Selection goals and criteria. In such beef production systems (suckler cow herds) the economic efficiency depends chiefly upon two synthetic parameters; 1) reproductivity of cow herd and 2) productive efficiency,which contribute respectively to the cost of weaned calves and profit due to commercial offspring and culled cows sold (Moav, 1966). Increasing this profit implies improving growth rate to the commercial age and muscle development. However, increased cow size leads to better culling value but also to higher maintenance costs. As far as the influences of size and muscle development on calving ability are concerned, they have been proved to be quite unfavourable. Thus goals seem to be somewhat inconsistent with each other. A selection optimum has to be found. Different approaches can be suggested.

1. We can compute as per Dickerson et al. (1974) an economic function evaluating net efficiency of lean beef production from weaning to slaughter including expected changes in cow-herd costs from increases in birth and mature weight. Knowing the genetic and phenotypic parameters of this function and the main recorded traits, we can then calculate by the conventional selection index theory, a selection criterion which maximises the genetic improvement in this function. Thus, according to Dickerson et al. (1974), by selecting on yearling weight minus 3.2 times birth weight, versus selecting on yearling weight alone, genetic change is reduced by 55% for birth weight and 27% for mature weight; however, 90% of the potential improvement in yearling weight is retained and that in net efficiency is increased by 6%.

Such a procedure supposes that the genetic parameters are known and that we are able to find the right economic function, especially the effects of unit increases in birth weight and

birth size on weaned calf cost. Apart from problems arising
from non-linear effects of some factors such as birth weight
with respect to calving difficulties and calf mortality, it may
be difficult to establish functions with enough accuracy for
complex and heterogeneous production systems. Then, the following rule of thumb way may be more relevant.

2 We might aim to increase, or at least not change, muscle
development and weight at the commercial age and, in addition,
impose some appropriate restriction on genetic change in
traits involved in calving difficulties, calf mortality and
cows maintenance cost. Such is Fitzugh's (1976) procedure
which computes the selection indices maximising genetic
gain in 12-month weight while holding constant mature
weight or mature and birth weight. We might also try to
increase by selection the ratio of pelvic opening (Taylor
et al., 1975) or body weight to the power 0.4 (Monteiro,
1969) at a given age, to birth weight. But as pointed out
by Taylor et al. (1975) selection on this ratio might reduce
size and also muscle development. Moreover, it is essential
not to decrease birth weight relatively to mature weight too
much in order to avoid more calf mortality, since calves
born relatively immature may have poorer viability (Fitzhugh,
1976).

3.2.3. Selection schemes. In France, this selection of beef
strains for producing breeding females has been practised for a
few years through specific and integrated programmes in the
Charolais, Limousin and Blonde d'Aquitaine breeds. The originality
of such schemes is well justified when we realise the following:

- The numerical importance of specialised suckler herds in
purebreeding (about 1.3 million cows);

- The proporotion of beef cows artificially inseminated
(20 to 70% depending on breed);

- The selection pressure applied in French beef breeds on
muscle development at the expense of fitness traits.

The first step consists in a choice of ancestry of young
bulls from planned matings in a nucleus of élite cows. The

TABLE 6

SOME RESULTS ABOUT SELECTION PRACTISED IN THE CHAROLAIS, BLONDE D'AQUITAINE AND LIMOUSIN BREEDS AFTER STATION PROGENY TESTING ON THEIR HEIFER PROGENY (2O PER SIRE).

Mean ($\bar{x}$) and selection differential (d)

Breed	18-month weight (kg) $\bar{x}$: d	Muscle development score (pt) $\bar{x}$: d	Conception rate (1) (%) $\bar{x}$: d	% difficult calvings at 2 years $\bar{x}$: d	Gestation length (days) $\bar{x}$: d	Birth weight of calf (kg) $\bar{x}$: d	120 day weight of calf (kg) $\bar{x}$: d
Charolais (2) 5/16	407.3 : +9.2 : (+1.15)	60.1 : -1.4 : (-0.67)	82.6 : +2.8 : (+0.32)	40.4 : -9.3 : (-0.80)	283.4 : -0.6 : (-0.45)	37.5 : +0.3 : (+0.25)	123.8 : +1.6 : (+0.38)
Blonde d'Aquitaine (2) 5/12	383.5 : +10.9 : (+1.42)	62.1 : +1.0 : (+0.40)	72.3 : +1.2 : (+0.14)	16.8 : -3.0 : (-0.24)	291.1 : -0.5 : (-0.02)	38.6 : +0.1 : (+0.10)	114.5 : -0.4 : (-0.04)
Limousin (2) 8/20	371.8 : +11.6 : (+2.28)	59.5 : +1.0 : (+0.55)	66.3 : +6.0 : (+0.60)	13.3 : 0.0 : (0.0)	286.3 : -0.5 : (-0.34)	34.5 : -0.3 : (-0.24)	107.7 : +2.6 : (+0.63)

(1) Pregnant cows/inseminated cows in natural conditions

(2) Number of selected bulls/number of progeny tested bulls for the last two recorded groups

(d) Standardised selection differential within brackets

choice of these bull dams is made in particular on account of
their breeding values from field performance recorded data.
Calving ability of females is characterised by the number of
unassisted calvings and a genetic estimation of the maternal
component of birth weight.

A second selection is practised on young bulls after individual station performance testing. A choice on their pelvic
opening relative to size has been planned at this stage. A
first progeny testing in field conditions follows and leads to
culling the worst bulls with too high birth weight (direct
effect) for their progeny.

After this, bulls are subjected to a station progeny test
on the breeding qualities of their heifer progeny. Ten Charolais
and 6 Blonde d'Aquitaine bulls every year and 10 Limousins
every two years are compared within their breed. About 20
daughters per sire are recorded in the station from weaning to
30 months of age and observed through a first calving at two
years. Apart from its interest in selection for earlier
reproduction, this calving at two years contributes to reducing
the generation interval and permits a better expression of
genetic variability especially for maternal calving ability.
Besides, calving performance receives an equally important
place in selection as muscle growth potential, fertility and
suckling performance (Table 6). However, as indicated previously,
selection efficiency would certainly be improved if index
criteria better adapted to the different objectives (as those
involved in growth) could be applied.

4. CONCLUSION

In French beef breeds, calving difficulties finally appear
as the price paid for concentrating all selection opportunity
on a few traits which do not include fitness traits.

It is essential, in spite of new solutions which may be
proposed by other disciplines, to be aware of the need for

developing genetic methods for the future if we wish to avoid
the underlying causes of the calving problems remaining or
even increasing.

In addition to planning more objectively the choice of
sire breeds, it would be necessary to reconsider the overall
selection goals and criteria for terminal sire breeds and for
dam breeds or purebreeds.

In France, some selection programmes are now in progress
in this way as:

- Creation of the synthetic lines 'COOPELSO 93' and
'INRA 95' and selection of a Limousin line with restricted
birth weight;

- Setting up of a specific scheme of beef breeds for improving breeding qualities in Charolais, Limousin and
Blonde d'Aquitaine.

To be efficient, such methods need simultaneous efforts
from selection structures and research in order to improve
selection criteria and spread of the genetic change to the
overall population.

Finally, it is important to emphasise that all European
countries, including those with a majority of dairy or dual-
purpose cattle, are concerned by this new direction of selection
since the necessary reconversion of milk into beef production
will unavoidably lead to the use of this specialised beef
livestock.

REFERENCES

Belic, M. and Ménissier, F. 1968. Etude de quelques facteurs influençant
les difficultés de vêlage en croisement industriel. Annales de
Zootechnie 17, 107-142.

Brinks, J.S., Olson, J.E. and Carroll, J.E. 1973. Calving difficulty and
its association with subsequent productivity in Herefords. Journal of
Animal Science 36, 11-17.

Couteaudier, J.F., Regis, R. and Ménissier, F. 1971. Possibilités de sélection
de l'aptitude au vêlage en race Charolaise. X Congrès international
de Zootechnie, Fédération européenne de Zootechnie, Paris-Versailles,
juillet 1971, 10pp.

Dearborn, D.D., Koch, R.M., Cundiff, L.V., Gregory, K.E. and Dickerson, G.E.
1973. An analysis of reproductive traits in beef cattle. Journal of
Animal Science 36, 1032-1040.

Dickerson, G.E. 1970. Efficiency of animal production. Moulding the biological
components. Journal of Animal Science 30, 849-859.

Dickerson, G.E., Künzy, N., Cundiff, L.V., Koch, R.M., Arthaud, V.H. and
Gregory, K.E. 1974. Selection criteria for efficient beef production.
Journal of Animal Science 39, 659-673.

Fitzhugh, J.A. (Jr). 1976. Analysis of growth curves and strategies for
altering their shape. Journal of Animal Science, 42, 1036-1051.

Foulley, J.L. 1976. Some considerations on selection criteria and optimisation
for terminal sire breeds. Annales de Génétique et de Sélection animale
8, 89-101.

Foulley, J.L. and Ménissier, F. 1975. Note sur certaines possibilités de
sélection des taureaux de croisement terminal avec limitation du
poids à la naissance: cas du contrôle de descendance en ferme sur la
production de veaux de boucherie en race Charolaise. Institut national
de la Recherche agronomique, 21 pp. ronéoté.

Hanset, R. 1976. Cattle selection in Belgium. In: EEC seminar on 'Optimisation
of cattle breeding schemes'. Dublin, 26-28 November 1975 CEC, EUR 5490, pp
263-272.

Hill, W.G. 1977. Comments on statistical efficiency in bull progeny testing
for calving difficulty. Livestock Production Science 4, 203-207.

Koch, R.E. 1972. The role of maternal effects in animal breeding. VI.
Maternal effects in beef cattle. Journal of Animal Science 14, 979-986.

Mahon, G.A.T. and Cunningham, E.P. 1976. Statistical efficiency in bull
progeny testing for calving difficulty. Livestock Production Science
3,147-153.

Ménissier, F. 1975. Calving ability in French beef breeds: an analysis of
components and breeding improvement. Bulletin technique du Département
de Génétique animale Institut national de la Recherche agronomique,
France, No. 21, 57-102.

Ménissier, F. 1976. Comments on optimisation of cattle breeding schemes:
beef breeds for suckling herds, a review. Annales de Génétique et
de Sélection animale 8, 71-87.

Ménissier, F., Bibé, B. and Perreau, B. 1974. Possibilités d'amélioration
des conditions de vêlage de trois races à viande francaise. Annales
de Génétique et de Sélection animale 6, 69-90.

Monteiro, L.S. 1969. The relative size of calf and dam and the frequency of
calving difficulties. Animal Production 11, 293-306.

Moav, R. 1966. Specialised sire and dam lines. I. Economic evaluation of
crossbreds. Animal Production 8, 193-202.

Olausson, A. and Rönningen, K. 1975. Estimation of genetic parameters for
threshold characters. Acta Agriculturae Scandinavica 25, 201-208.

Petty, R.R. (Jr) and Cartwright, T.C. 1966. A summary of genetic and
environmental statistics for growth and conformation traits of young
beef cattle. Department Technical Report No. 105, Texas A and M
University.

Philipsson, J. 1976a. Studies on calving difficulty, stillbirth and associated
factors in Swedish cattle breeds. I. General introduction and breed
averages. Acta Agriculturae Scandinavica 26, 151-164.

Philipsson, J. 1976b. Studies on calving difficulty, stillbirth and associated
factors in Swedish cattle breeds. III. Genetic parameters. Acta
Agriculturae Scandinavica 26, 211-220.

Pollak, E.J. and Freeman, A.E. 1976. Parameter estimation and sire evaluation
for dystocia and calf size in Holsteins. Journal of Dairy Science
59,1817-1824.

Preston, T.R. and Willis, M.B. 1970. Intensive beef production. Pergamon
Press, Oxford, 567 pp.

Taylor, St. C.S., Monteiro, L.S. and Perreau, B. 1975. Possibility of reducing
calving difficulties by selection. III. A note on pelvic size in
relation to body weight of cattle. Annales de Génétique et de
Sélection animale 7, 49-57.

Willham, R.L. 1970. Variation in calving scores given Charolais cows. Journal
of Animal Science 31, 171.

RELATIONSHIP BETWEEN PERFORMANCE TEST DATA AND CALVING PERFORMANCE OF TEST BULLS

K. Osterkorn, H. Kräusslich, G. Averdunk and A. Göttschalk

Institut für Tierzucht, Vererbungs- und Konstitutionsforschung
der Universität München, Veterinärstrasse 13, D-8000 München
and
Bayerische Landesanstalt für Tierzucht, Prof.-Zorn-Strasse 6,
D-8011 Grub, Post Poing, BRD

ABSTRACT

Correlations between growth-rate, weight for age, body measurements and rate of veterinary aid as well a rate of stillbirths are positive and significant. Correlations between performance test results and gestation period are not significant.

Correlations differ significantly between sexes. They are higher for male calves than for female calves.

Selection on gain parameters has negative side effects on calving performance. Therefore, selection after performance test should be based on an index in which birth weight or starting weight is kept constant.

INTRODUCTION

The pre-selection of bulls for beef potential using a performance test and the subsequent progeny testing of bulls for milk potential is a widely used selection procedure in AI breeding programmes. The primary aim of the performance test is to increase growth. But one of the consequences of selecting for growth rate is an increase in birth weight and an associated increase in calving difficulties. Pollak and Freeman (1976) found genetic correlations between dystocia and calf size of 0.97 ± 0.07 (Midwest) and 0.89 ± 0.17 (Select Sires). It is therefore important to increase growth with minimum increase in birth weight, so that calving problems are minimised. According to the "General Recommendations on Procedures for Performance And Progeny Testing for Beef Characteristics" (edited by H. Kräusslich, 1974) selection should be based on a growth or net growth index during the test period with birth weight or starting weight constant. Simpler approaches include the selection of bulls according to growth rate during part of the test or according to final test weight.

The purpose of this study is to estimate the correlations between growth characters, body measurements and skeletal measurements of test bulls and the calving performance of calvings after test inseminations.

MATERIAL

Performance test data
From May 1971 to July 1976, 1477 Fleckvieh bulls[*] were performance tested at Bavarian test stations up to an age of 420 days. Out of these, 623 bulls were selected as AI test bulls. Performance test data of all tested bulls and of the selected bulls are listed in Tables 1a and 1b. The selected bulls have higher values in all measurements. Selection differentials amount to approximately 0.5 standard deviations for growth data, 0.4 standard deviations for body measurements and
[*] German Simmental.

0.1 - 0.4 standard deviations for skeletal measurements (circumference of cannon bone - 0.4 standard deviations).

TABLE 1a

GROWTH AND BODY MEASUREMENTS OF PERFORMANCE TESTED BULLS

	Total			Selected test bulls		
	n	x̄	SD	n	x̄	Deviation[1)
Weight at 420th day	1477	558.9	46.7	623	581.6	23.7
Daily gain since birth	1477	1235	112	623	1290	56
Daily gain during test period (112-420 days)	1477	1312	137	623	1371	64
Circumference of chest	1477	191.5	6.5	623	193.8	3.0
Spiral measurement	1477	242.2	8.5	623	246.2	4.2
Height at withers	1477	126.6	3.0	623	127.8	1.3
Height at rump	1477	132.0	3.3	623	133.2	1.4
Chest depth	1477	67.1	2.5	623	67.9	0.8
Thigh measurement	1451	199.3	7.6	623	201.9	2.9

1) Deviation from a rolling average of contemporaries.

TABLE 1b

SKELETON MEASUREMENTS OF PERFORMANCE TESTED BULLS

	Total			Selected test bulls		
	n	x̄	SD	n	x̄	Difference
Width of forehead	895	22.0	1.0	472	22.0	0.2
Width of chest	897	48.2	2.7	474	49.0	0.8
Width of hips	897	47.0	2.3	474	47.2	0.2
Width of pelvis	897	50.9	2.5	474	51.7	0.8
Circumference of Cannon bone	896	21.4	1.0	473	21.8	0.4

Calving performance

A total of 1 888 960 calvings were recorded. After eliminating multiple births and births where sex of the calf was not recorded, the total number of analysed calvings amounts to

1 829 960. The average values of the stillbirth percentage,
the veterinary aid percentage, length of gestation period and
the influence of age of dam and sex of the calf are shown in
Table 2. Stillbirths are defined as calves born dead. The
parameters on veterinary aid and gestation period are self-
explanatory. The influence of sex of calf on all three calving parameters and of age of cow on stillbirths and veterinary
aid are highly significant.

TABLE 2

CALVING PERFORMANCE (TOTAL MATERIAL) INFLUENCE OF AGE OF THE DAM AND
SEX OF THE CALF

	n	Stillbirths %	Vet aid. %	Gestation period days
Male	938 747	2.15	3.37	288.1
Female	890 620	0.93	1.45	286.7
1st lact.	446 822	2.81	4.22	286.5
2nd lact.	364 281	1.21	1.92	287.4
≥3rd lact.	1 018 264	1.13	1.84	287.8

ADJUSTMENT OF CALVING PERFORMANCE DATA

Apart from the significant influences of sex of calf and
age of dam, calving performance data are biased by the factors
year-season and region. Therefore, calving performance data
have to be adjusted for the abovementioned systematic factors.
Adjustment factors are calculated according to the following
model:

$$Y_{ijklm} = \mu + a_i + b_j + (ab)_{ij} + c_k + d_e + e_{ijklm}$$

a_i = lactation number (first, second, third and higher
       lactations)

b_j = sex

c_k = region (four regions)

d_e = year and season (five years x four seasons)

The following analyses are carried out with adjusted
calving performance data.

RESULTS AND DISCUSSION

Correlations between calving performance data

Correlations between calving performance data of test
bulls with more than 200 recorded calvings were calculated.
Table 3 shows a significant positive correlation between veterinary aid percentage and length of gestation period. Correlations between stillbirth percentage and veterinary aid percentage are zero. Thus, there is no difference between partial
correlations and direct correlations.

TABLE 3

PHENOTYPIC CORRELATIONS BETWEEN CALVING PERFORMANCE PARAMETERS

		Direct correlations			Partial correlations		
	n	r_{12}	r_{13}	r_{23}	$r_{12.3}$	$r_{13.2}$	$r_{23.1}$
Male and female calves (adjusted)	353	0.064	0.112	0.312++	0.031	0.097	0.308**
Male calves only	350	-0.012	0.036	0.259++	-0.022	0.040	0.260**

1=stillbirths, 2=veterinary aid, 3=gestation period; ** = $p < 0.01$

Correlations between performance test data and calving
performance data are listed in Table 4a.

Correlations are calculated for the absolute values recorded during performance test (weight, gain, body measurements)
and for the deviations from a rolling average of contemporaries. The two correlation coefficients are rather similar, yet
there are some exceptions. Most correlations are positive,
but some are near zero. Gain parameters and body measurements
are more highly correlated with the veterinary aid percentage
than with the stillbirth percentage. This is to be expected
since positive selection on performance test results increases

the size of the calves of the selected bulls and there exists
a higher correlation between veterinary aid percentage and calf
size than between stillbirth percentage and calf size. Skeletal measurements are, however, more closely correlated with
stillbirth percentage than with veterinary aid percentage.
This is especially significant for the parameter width of pelvis.

TABLE 4A

PHENOTYPIC CORRELATIONS BETWEEN ADJUSTED CALVING PERFORMANCE DATA AND
PERFORMANCE TEST DATA

		n	Stillbirths	Veterinary aid
Weight at 420th day	abs[1]	353	0.090	0.120
	dev	353	0.102	0.126
Daily gain since birth	abs	353	0.089	0.122
	dev	353	0.103	0.127
Daily gain during test period	abs	353	0.082	0.085
	dev	353	0.054	0.091
Circumference of chest	abs	353	0.087	0.066
	dev	353	0.030	0.064
Spiral measurement	abs	353	0.076	0.102
	dev	353	0.090	0.086
Height at withers	abs	353	0.094	0.107
	dev	353	0.093	0.121
Height at rump	abs	353	0.110	0.107
	dev	353	0.094	0.121
Chest depth	abs	353	0.028	0.095
	dev	353	0.050	0.062
Thigh measurement	abs	349	0.068	0.079
	dev	349	0.043	0.094
Width of forehead		244	0.029	-0.024
Width of chest		245	0.117	0.099
Width of hips		245	0.158	0.118
Width of pelvis		245	0.131	0.052
Circumference of cannon bone		244	0.058	0.079

1) Deviation from the rolling average of contemporaries.

TABLE 4b

REGRESSION OF ADJUSTED CALVING PERFORMANCE DATA ON PERFORMANCE TEST DATA

| | | Stillbirths | | Veterinary aid | |
		Total	Males only	Total	Males only
Weight at 420th day	abs	0.0021	0.0045*	0.0039*	0.0074**
	dev	0.0032	0.0048	0.0054*	0.0089**
Daily gain since birth	abs	0.0009	0.0019*	0.0017*	0.0032**
	dev	0.0013	0.0020	0.0023*	0.0037**
Daily gain during test	abs	0.0006	0.0016*	0.0009	0.0021*
period	dev	0.0006	0.0013	0.0013	0.0022
Circumference of	abs	0.0142	0.0326*	0.0147	0.0358*
chest	dev	0.0062	0.0148	0.0182	0.0228
Spiral measurement	abs	0.0095	0.0174	0.0173	0.0351*
	dev	0.0141	0.0202	0.0186	0.0333
Height at withers	abs	0.0309	0.0410	0.0494*	0.0956*
	dev	0.0349	0.0409	0.0618*	0.1007*
Height at rump	abs	0.0319*	0.0469	0.0436*	0.0924**
	dev	0.0325	0.0373	0.0573*	0.0967*
Chest depth	abs	0.0116	0.0254	0.0575	0.0837
	dev	0.0278	0.0340	0.0471	0.0637
Thigh measurement	abs	0.0108	0.0129	0.0172	0.0408*
	dev	0.0078	0.0179	0.0234	0.0497*
Width of forehead		0.0324	0.0162	-0.0374	0.0120
Width of chest		0.0494	0.0803	0.0566	0.0654
Width of hips		0.0648*	0.1015*	0.0658	0.0851
Width of pelvis		0.0567*	0.0781	0.0303	0.0778
Circumference of cannon bone		0.0626	0.1803	0.1153	0.2607

* = p 0.05
** = p 0.01

Table 4b shows the coefficients of the regressions of adjusted calving performance data on performance test data. All regression coefficients are higher for the male sex than in the adjusted data for both sexes, and higher for veterinary aid percentage than for stillbirths. From the regression of veterinary aid percentage on weight at 420 days it follows that at a selection differential of one standard deviation, the expected increase of veterinary aid amounts to about 0.2% for both sexes and to about 0.35% for male calves. Except for one, all regression coefficients are positive.

In the Bavarian Fleckvieh AI breeding programme, all test bulls are intensively selected on gain parameters (after performance tests on stations or according to field data).

If there really exists a regression of calving performance on gain of test bulls, a gradual increase of the veterinary aid percentage and of the stillbirth percentage is to be expected in the total population. This increase should be more pronounced for male calves and for veterinary aid percentage. Figures 1 and 1a show the time trend for veterinary aid percentage and for the stillbirth percentage for the total Fleckvieh population from 1971 to 1976. This time trend is in good agreement with the abovementioned expectations.

In addition, the Figures show a rather regular seasonal pattern with the highest values of veterinary aid percentage and stillbirth percentage during the period October to December, and the lowest values during the period July to September. Schwark and Oehler (1973) found that calves born during winter were 0.7 kg heavier and gestation period was 1.11 days longer compared to calves born during summer. Wilson et al. (1976), however, found that calves born from October to December 1973 were on average 0.81 kg lighter than calves born from April to June 1974. Gestation period, however was longer for calves born from October to December 1973 (0.45 days). In our material, pregnancies ending in the period October to December are about 0.4 days longer than those ending in the period July to September. A non-biological explanation for higher veterinary aid percentage during the winter season may be that farmers have more time available for taking care of their cattle during winter than during summer and will therefore call veterinarians more frequently during the winter. Another explanation may be seasonal differences in feeding and management.

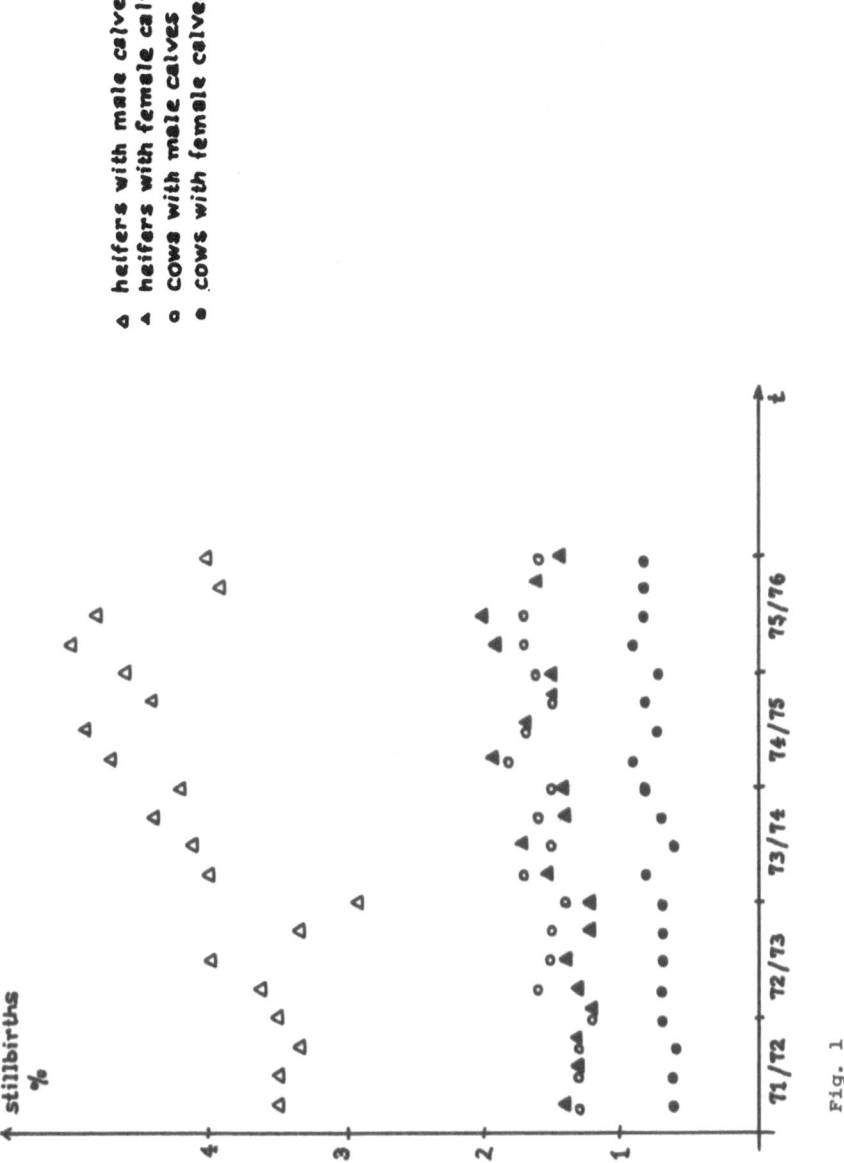

Fig. 1

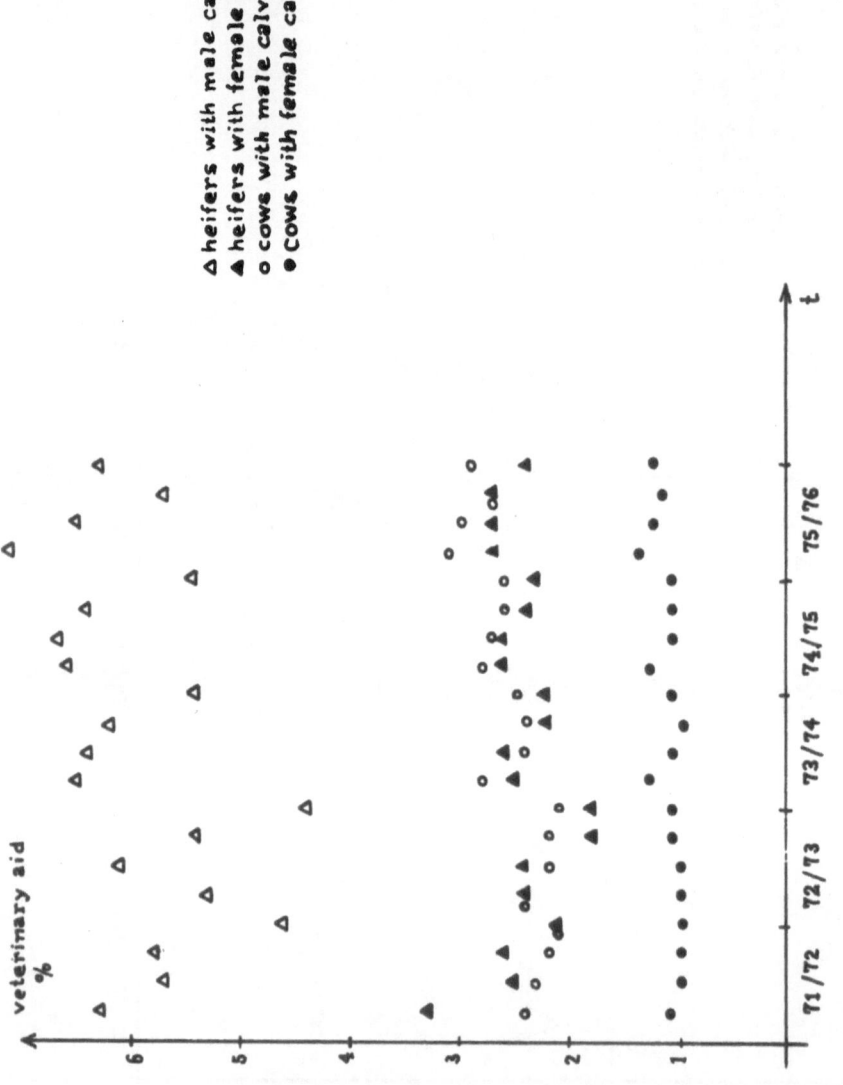

Fig. 1a

In the Bavarian AI breeding programme, bulls with a low stillbirth and veterinary aid percentage (first results are recorded from calves out of test inseminations) are recommended for insemination of heifers. As the figures show it was not possible to compensate completely the negative side effect of the selection on gain by these recommendations.

CONCLUSIONS

Negative side-effects of the pre-selection of AI test bulls for beef potential using a performance test are an increase of stillbirths and veterinary aid percentages.

From all recorded gain parameters, gain during test period has the lowest correlation to calving performance. Selection after performance test should be based on this parameter or better on an index in which birth weight or starting weight is kept constant.

Correlation and regression coefficients between the recorded body and skeletal measurements and calving performance are in the same direction and of the same magnitude as correlations and regressions between gain parameters and calving performance. Thus, it is doubtful whether the introduction of a second selection index based on body and skeletal measurements is worthwhile.

REFERENCES

Kräusslich, H. (Editor) 1974. General recommendations on procedures for
    performance and progeny testing for beef characteristics. Livestock
    Production Science 1: 33-45.

Pollaĸ, E.J. and Freeman, A.E. 1976. Parameter estimation and sire
    evaluation for dystocia and calf size in Holsteins. Journal of
    Dairy Science 59: 1817-1824.

Schwark, H.J. and Oehler, H. 1973. Die Geburtsmasse des Kalbes als Ergebnis
    des intrauterinen Wachstums and die ursachen ihrer Variabilität.
    Archiv fur Tierzucht 16: 71-82.

Wilson, A.,Willis, M.B. and Davison, C. 1976. Factors affecting calving
    difficulty and gestation length in cows mated to Chianina bulls and
    factors affecting the birth weight of their calves. Animal Product-
    ion   22: 27-34.

BREEDING FOR CALVING PERFORMANCE

J. Philipsson

Department of Animal Breeding, Agricultural College,
S-75007 Uppsala 7, Sweden.

ABSTRACT

Investigations on Swedish dual-purpose cattle have revealed considerable breed differences in calving performance, stillbirth rate and related characters.

In the present investigation, the relative importance of genetic factors, divided into characters relating to the calf and those concerning the dam, was analysed. A heritability of 10% was found for calving performance of both the direct and the maternal character at first calving. For stillbirth rate, values were on average 3%. A close genetic relationship was found between calving performance and birth weight as direct characters, but for the maternal characters it was considerably weaker. Correlations between stillbirth rate and birth weight were generally lower, as the relationship was non-linear. Estimations of the genetic correlations between direct and maternal effects gave values close to zero, or negative, for the characters investigated, indicating an antagonistic relationship between the genetic make-up of the cow and the calf.

Long-term intrabreed selection should be based primarily on daughter-group results, which best express ease of calving in relation to inherited birth weight of the calf. Genetic variation in calving performance at second and subsequent calvings is generally slight and only partly correlated to the results at first calving. Progeny testing should therefore be based on heifers. A well planned progeny testing and breeding programme is necessary in order to achieve a substantial improvement in calving performance through selection and differential use of bulls with heifers and cows.

INTRODUCTION

Although the problems of calving difficulty and stillbirth have a long history in veterinary medicine, their influence in the area of breeding was not acknowledged in many countries until various beef breeds were tested for the purpose of cross-breeding. However, the genetic problems are far from limited to crossbreeding. Van Dieten (1963) early showed a great variation within a breed between bulls as sires of both calves and dams.

The seriousness of calving difficulty is not limited to the great losses of calves but also concerns generally lowered fertility of the dams and increased culling of cows. One Swedish investigation (Philipsson, 1976d) shows a 25% impaired fertility and a 70% higher culling rate during the first lactation of heifers giving birth to a dead calf compared with those giving birth to a live calf. Of the overall economic consequence of calving difficulty and stillbirth, half could be attributed to calf losses and the other half to costs or losses concerning the dam.

Structural changes in agriculture leading to relatively less supervision of individual animals generally argues for robust animals and consequently calving performance should be seriously considered in future breeding programmes.

In the following, various genetic parameters and aspects of breeding schemes will be discussed, mainly based on experience gained from investigations on Swedish dairy and dual-purpose breeds.

GENETIC PARAMETERS

The various causes of calving difficulty and stillbirth represent a wide range of mechanisms. Knowledge of the relative importance of these factors, their heritability and inter-connections and the relationships between calf and dam

characteristics is important in order to design efficient
breeding programmes in respect of calving performance.

Breeds

The importance of genetic factors may firstly be demonstrated by breed differences. Stillbirth rate at first
calving has been reported to range from 3 to 16% for various
dairy and dual-purpose breeds (Lindhé, 1967; Remmen, 1975).

The present Swedish investigation also shows considerable
differences between breeds as regards calving difficulty,
stillbirth rate, malpresentation of foetus and labour of dam
at parturition (Table 1). Calf mortality also varied between
breeds, independently of calving performance.

TABLE 1

CALVING PERFORMANCE AND RELATED CHARACTERS IN SWEDISH DAIRY BREEDS
(PHILIPSSON, 1976a)

Breed	Calving difficulty, %	Still-birth rate, %	Birth weight, kg	Gesta-tion length days,	Malpre-sentat-ion of foetus %	Weak labour of dam, %
Swedish Friesian						
cows	4.8	2.5	43.0	278.5	4.7	8.4
heifers	15.7	6.5	39.9	277.5	5.7	4.8
Swedish Red-and-white						
heifers	7.8	3.2	32.2	280.6	3.7	4.5
Swedish Polled						
heifers	11.8	9.2	24.0	284.5	6.5	9.5

Heritability

It has also been firmly established by several investigations
that the heritability of calving performance, both as a trait
of the calf and of the dam, ranges from about 3 to 15% at first
calving. However, it is considerably lower at subsequent parities

(Tables 2 and 3). For stillbirth rate in heifers the heritability is usually less than 5% and, in the case of cows, often below 1%. Although calving performance is scored subjectively, it usually shows a greater heritability than the more accurately defined stillbirths. The Swedish data also underline the importance of a good method of recording calving performance. Scoring calves in three classes instead of two generally gave higher heritabilities.

TABLE 2

ESTIMATES OF HERITABILITY FOR CALVING PERFORMANCE AND STILLBIRTH RATE WITH HEIFERS AS DAMS

Author	Calving performance as a trait of:		Stillbirth rate as a trait of:	
	calf	dam	calf	dam
Bar-Anan (1972) direct estimation	0.042	0.018	0.036	0.012
indir. estimation	0.045	0.028	0.045	0.023
Brinks et al. (1973)	0.126	neg.		
Cloppenburg (1966)	0.043		0.00	
Freeman (1975)	0.17			
Hansen (1972)		0.10-0.32		
Lindhé (1974)			0.16	0.020
Philipsson (1976b)	0.08-0.10	0.04-0.11	0.05	0.01
Schlote et al. (1975)	0.03-0.07	0.02-0.08		
Vogt-Rohlf and Lederer (1975)	0.045	0.035	0.016	0.024

Although heritability values are comparatively low, it should be stated that the genetic variance is relatively high as was shown by Bar-Anan et al. (1976). One of the great advantages of AI breeding programmes is that they also allow progeny testing for characters of low heritability with reasonable accuracy. For a character with a heritability of 3% a progeny test based on 200 individuals will have a repeatability of 60% ($r_{TI} = 0.78$). A heritability of 10% gives an accuracy of 84% (r_{TI} = 0.91).

TABLE 3

ESTIMATES OF HERITABILITY FOR CALVING PERFORMANCE AND STILLBIRTH RATE WITH COWS AS DAMS

Author	Calving performance as a trait of:		Stillbirth rate as a trait of:	
	calf	dam	calf	dam
Bar-Anan (1972) direct estimation	0.007	0.003	0.013	0.002
indirect estimation	0.009	0.005	0.014	0.002
Cloppenburg (1966)	0.037		0.007	
Freeman (1975)	0.05-0.08			
Lindhé (1974)			0.005	0.008
Philipsson (1976b)	0.009		neg.	
Schlote et al. (1975)	0.01-0.04	0.00-0.04		
Vogt-Rohlf and Lederer (1975)	0.008	0.006	0.005	0.002

Relationship between heifer and cow effects

Considered from the breeding planning point of view, the great difference in heritability between estimates in heifers and cows is important. In the present Swedish investigation no significant differences whatsoever were detected between bulls when evaluating calving difficulty and stillbirth in cows. Although calving difficulty and stillbirth constitute a serious problem as regards heifers, it must be noted that a substantial number of dead calves are also born at subsequent calvings. Not only do the estimates of heritability indicate a slight genetic variation in calving performance and still-birth rate at later parities, but breed statistics too show that interbreed differences are found mainly in heifers and only to a minor extent in cows.

There are several circumstances which could explain these differences between heifers and cows, apart from the fact that cow results are affected by selection due to the calving performance as heifer. A variety of factors influence calving performance and stillbirth rate but the relative importance of these factors is not the same at the first as at subsequent calvings. For instance, the parturition process is more prolonged in first calvers than at subsequent calvings. The Swedish data also showed that weak labour at parturition and malpresentation of the foetus were more closely connected with calving difficulty and stillbirth at later parturitions than at the first. Furthermore, many more of the stillborn calves of heifers were born at difficult calvings than were those of older dams of the same breed.

The involvement of inheritance in the various factors may also vary greatly. This means that calving performance and stillbirth rate, from a genetic or physiological point of view, may not be considered as identical characters in heifers and in cows. This may be an important reason for the low genetic variation found in calving performance and stillbirth rate when measured in cows, although birth weight and gestation length show the same heritabilities in both heifers and cows.

Changes in stillbirth rate during recent years in Swedish cattle breeds have been estimated at 1% per annum in heifers and 0.2% in cows. Although this decrease is small and the causes are not analysed, they suggest a correlation of about 0.3 between heifer and cow results when the effects of different variances for heifers and cows are considered. Roughly the same value of the relationship between heifers and cows may be calculated from the increased stillbirth rates as reported by Remmen (1975) for the Dutch MRY breed. These results are also supported by the figures given by Pijnenburg (1974) and by Bar-Anan et al. (1976). The latter authors estimated the genetic correlation between effects in heifers and cows to about 0.5.

These findings underline the importance of progeny testing bulls of heifers in order to obtain repeatable results when choosing certain bulls for heifers or for long-term genetic selection.

Genetic correlations between calf and dam effects

In the present investigation particular attention was
paid to the genetic correlations between calf and dam effects.
These were estimated from data where daughter group results
were given together with results for the bulls as sires of
calves. In such data it is possible to eliminate the part of
dam variance that is caused by calf effects and covariance
between calf and maternal influence. These affect the
estimates of both the heritability of maternal effects and
the genetic correlation between calf and maternal effects.

Table 4 shows values in the region of -0.5 for both birth
weight and gestation length and -0.2 and 0.1 for calving
performance and stillbirth rate respectively. These
correlations emphasise that quite different characteristics of
the calf and the dam cause problems at parturition. When one
also considers that positive genetic relationships usually
exist between dam and calf size, the low or even negative
genetic relationships between corresponding calf and dam
characteristics seem plausible. A big calf generally causes
greater calving difficulty but is also genetically correlated
to a bigger dam with its correspondingly larger pelvis. This
will contribute to an easier calving following random mating
with a bull of the same population.

TABLE 4

GENETIC CORRELATIONS BETWEEN CALF AND MATERNAL EFFECTS IN SWEDISH
FRIESIAN HEIFERS (PHILIPSSON, 1976b)

Character	Correlation
Calving performance	-0.19
Stillbirth rate	0.07
Birth weight	-0.53
Gestation length	-0.56

The antagonistic relationship between the genotype of the
calf and the genetic maternal influence indicates that selection

for these characters will not be as effective in the long run
as the heritabilities alone suggest. Selection based on the
direct characters will be accompanied by a negative response
on the maternal effects. These effects seem to be of the
order of 35% for calving performance. Kräusslich and Gottschalk
(1975), Vogt-Rohlf and Lederer (1975) and Bar-Anan et al. (1976)
also concluded from their investigations that the correlations
between the results of the same bull as sire and maternal
grandsire of the calves are weak. Progeny testing will
therefore be necessary for both the calf and dam effects if
improvements are desired in both these characters.

Genetic correlations between different characters

In Table 5, genetic correlations are given for both direct
and maternal traits. The direct genetic correlation between
calving performance and stillbirth rate is about 0.8, while
the maternal correlation is 0.6.

TABLE 5

ESTIMATES OF GENETIC CORRELATIONS BETWEEN DIFFERENT CHARACTERS. CORRELATION
BETWEEN CALF EFFECTS ABOVE THE DIAGONAL AND BETWEEN MATERNAL EFFECTS
BELOW THE DIAGONAL. SWEDISH FRIESIAN HEIFERS (PHILIPSSON, 1976b)

Character	Calving performance	Stillbirth rate	Birth weight	Gestation length
Calving performance		0.82	0.98	0.34
Stillbirth rate	0.64		0.41	0.26
Birth weight	0.60	0.30		0.45
Gestation length	0.08	-0.01	0.43	

There is a strong direct genetic correlation between
calving performance and birth weight of 0.98. The maternal
correlation is weaker, meaning that the dam compensates to some
extent for a higher birth weight of the calf by having a larger
body and pelvic opening. The correlations between stillbirth
rate and birth weight are considerably weaker than those
between calving performance and birth weight. This is explained

by the fact that there are higher stillbirth rates at low as
well as at high birth weights, while calving difficulty is
more linearly correlated with birth weight. The correlations
with gestation length range between O and 0.4. In this case,
fewest problems usually occur after average gestation periods
or slightly shorter, and most after long gestations. Remmen
(1975) found slightly higher values for the phenotypic .
correlations between progeny group means.

One of the most common causes of calving difficulty is
incompatibility in size between calf and pelvic opening of the
heifer. As correlations in the region of 0.4-0.7 (eg Nielsen,
1965; Bellows et al., 1971, Krahmer and Jahn, 1971) have been
shown between various external body measurements and pelvic
opening in heifers it may be of interest to correlate calving
performance to external body dimensions of the heifer.

TABLE 6

GENETIC CORRELATION BETWEEN BODY DIMENSIONS AND FREQUENCY OF CALVING
DIFFICULTY IN SWEDISH FRIESIAN HEIFERS (PHILIPSSON, 1976c)

Body measurement	Correlation
Withers height	0.10
Chest girth	-0.18
Hip width	-0.19
Thurl width	0.17
Pin bone width	-0.13
Rump length	-0.24
Vertical distance:	
hip-thurl	-0.18
Sacrum-thurl	-0.41

The genetic correlations given in Table 6 show that no
close relationships are found. This is not surprising, however,
as there are many different factors that influence calving
performance. Nevertheless substantial genetic relationships
with calving performance of the order of 0.2-0.4 are found for

198

several rump measurements. They indicate that low - placed
thurls in relation to hip bones and sacrum ie a roof - shaped
rump, is favourable for the calving procedure, as also is a
long rump with widely spaced hip bones in relation to width
between between thurls.

In practice these relationships mean that the traditional
way of judging conformation of cows at shows or at registration
for herdbooks, where flat and square rumps are desired, prefers
cows that have a slight but significant disposition to calving
difficulty.

The genetic relationships between birth weight, pelvic
size and calving performance reflect the incompatibility in
size between calf and dam. It may then be of interest to
compare different breeds as regards stillbirth rate in relation
to birth weight. In Figure 1 the stillbirth rates at different
birth weights as given by Remmen (1975) for the Dutch MRY
heifers and by Philipsson (1976b) for Swedish Friesian heifers,
are illustrated. The mean birth weights for these breeds are
38 and 40 kg and the stillbirth rates 14.5 and 6.5%. The
curves show that the great difference in stillbirth rate is
apparently not a question of generally poorer viability in the
MRY breed, as the mortality rate is about the same for both
breeds when the calves weigh around 28 - 35 kg. The stillbirth
rates then diverge rapidly as the birth weights increase. The
problem of calving difficulty and stillbirth is therefore rather
a question of deficiency in maternal ability in the MRY breed
vis-à-vis the Swedish Friesian.

As the standard deviation of the birth weight of most
breeds amounts to about 5 kg it is important that the maternal
propensity for easy calving and for giving live calves is good
in a fairly wide range of birth weights around the breed
average. It is quite possible to reduce the mean birth weight
by either continuous selection or by use of specific bulls
with heifers in order to shift the average birth weight away
from the steep part of the curve. However, selection for reduced

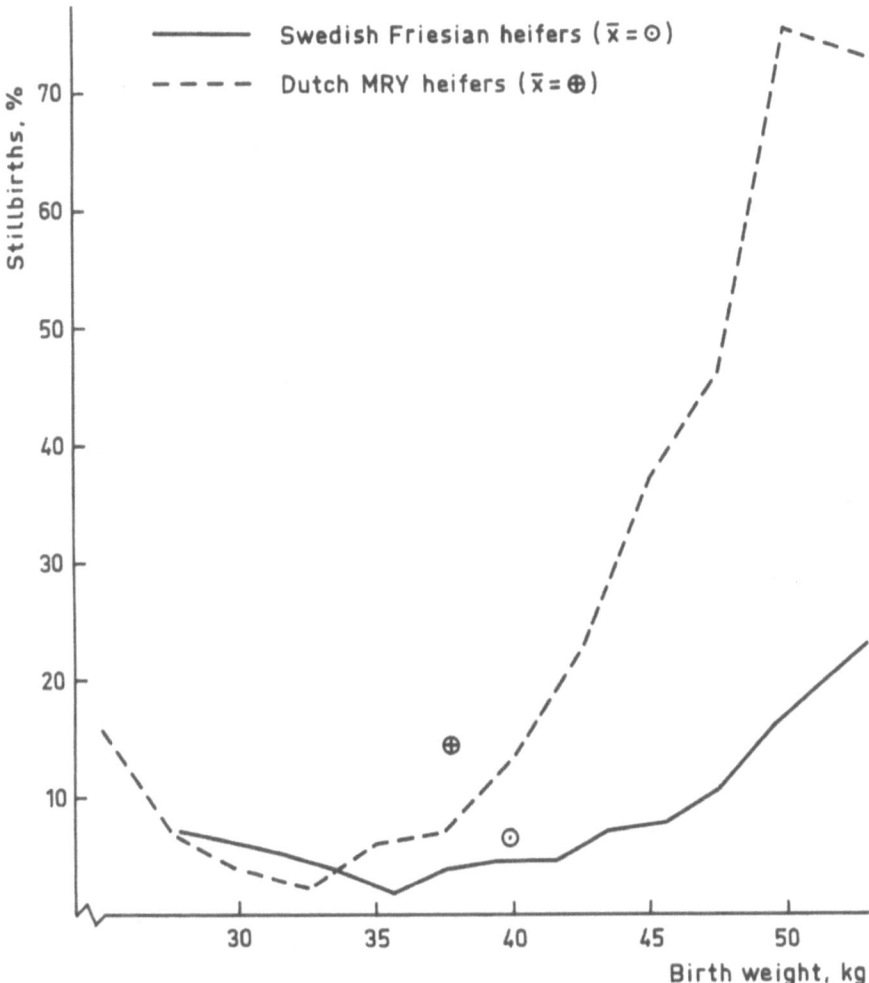

Fig. 1. The relationship between stillbirth rate and birth weight in Swedish Friesian heifers (Philipsson, 1976b) and in Dutch MRY heifers (Remmen, 1975).

birth weight will lose some of its effectiveness depending on the negative relationship between calf and dam effects. Also, because of the non-linear relationship between birth weight and mortality and the poorer growth capacity of small calves,

it is not very attractive to select for lower birth weight
alone. Selection for better maternal ability is an alternative
that allows for greater variation in birth weight without
increasing the risk of calving difficulty. Selection based on
daughter group results seems most efficient in improving
maternal ability in relation to the inherited birth weight and
may to some extent be combined with the use of bulls siring
small calves to heifers.

BREEDING SCHEMES

    In order to get some idea of the efficiency of different
testing schemes - selection and strategy of use of bulls -
four different breeding programmes were compared. In three
alternatives, selection was performed among bulls, in one case
based on calf characters and in two cases on daughter results.
In all four alternatives, progeny testing for calf characters
is performed but different strategies are used for the progeny
testing and the choice of bulls for heifers. The various
strategies do not involve any genetic selection but only a
differential use of bulls with heifers and cows.

    The assumptions for the calculations are chosen to fit a
population of 0.5 million cows of which 30% are heifers; 300
bulls are tested per year and 70% of all inseminations are done
with semen of progeny tested bulls. These circumstances
approximate to the present situation in Swedish cattle breeding.

    The genetic parameters assumed are given in Table 7.

    It should be noted that a genetic correlation of 0.4 between
heifer and cow results is assumed as well as no correlation at
all between calf and dam effects. The four alternatives are
specified as follows:

    Alternative 1: Progeny testing for calf effects in
    heifers. For 70% of the heifers the best 35% of the bulls
    are chosen. After progeny testing for dam effects in

heifers the best 75% of the bulls are selected. Progeny group sizes : 150 calves and 150 daughters.

Alternative 2: Progeny testing for calf effects in cows. Only progeny tested bulls (the best 45%) are used with heifers. Selection for dam effects as in Alternative 1. Progeny group sizes: 500 calves and 150 daughters.

Alternative 3: Progeny testing and use of bulls as in Alternative 2 but selection for calf effects instead of dam effects.

Alternative 4: The best 7% of bulls from preliminary progeny testing for milk are progeny tested for calf effects in heifers and the best 40% of the bulls are then chosen for heifers. No genetic selection. Progeny group size: 300 calves. Because of the lag prior to obtaining milk production results before testing for calving performance, a 0.5 year longer generation interval for progeny tested sires is assumed. This effect on the genetic progress for milk production corresponds to the decreased selection for milk in Alternatives 1 - 3. Thus, Alternatives 1 - 4 all give the same genetic progress in milk production.

TABLE 7

PARAMETERS ASSUMED FOR COMPARISON OF DIFFERENT BREEDING PROGRAMMES FOR CALVING PERFORMANCE

Parameter	Calving difficulty heifers	cows	Stillbirth rate heifers	cows
Population frequency	0.15	0.05	0.075	0.025
Phenotypic variance	0.1275	0.0475	0.0694	0.0244
Heritability				
Alternative a	0.10	0.01	0.05	0.005
Alternative b	0.05	0.01	0.02	0.005
Genetic correlation				
heifer-cow effects	0.4		0.4	
calf-dam effects	0		0	

The results are summarised in Table 8 and presented as
relative frequencies of calving difficulty and stillbirth after
one generation of selection in relation to the original values
of the whole population.

TABLE 8

RELATIVE FREQUENCIES OF CALVING DIFFICULTY AND STILLBIRTH AFTER ONE
GENERATION FOR DIFFERENT BREEDING SCHEMES

Breeding scheme	Calving difficulty differential use of bulls	total effects	Stillbirth rate differential use of bulls	total effects
1 a	.88	69	88	71
1 b	93	80	94	84
2 a	100	81	100	82
2 b	101	88	101	91
3 a	100	88	100	90
3 b	101	91	101	93
4 a	83	83	84	84
4 b	89	89	92	92

Although the results are based on a number of assumptions
that may be varied, some important features may be noted.
Furthermore, the assumption of a normal distribution of the
breeding values of the bulls may be questioned, although
Swedish data indicate that possible discrepancies are of minor
importance.

It is of great value that a differential use of bulls with
heifers v cows without any real selection will give immediate
effects and may reduce the frequencies by 10 - 15% in the best
of the given examples. It should be noted, however, that if
the best bulls are used with heifers, the worst will be used
with cows, if no selection is applied, and that the number of
calvings out of cows is more than twice as great as those out
of heifers. The negative effect of this is especially important
in the case where progeny tests are based on cows, which reduces

the positive effects in heifers. The total number of difficult births or stillborn calves will not be changed in these cases.

The selection response is expected to reduce frequencies by 10 - 20% in one generation and underlines the importance of long-term selection in combination with differential use of bulls for both heifers and cows.

REFERENCES

Bar-Anan, R. 1972. Heritabilitätsschätzungen einiger Abkalbemerkmale.
Züchtungskunde 44: 360-367.

Bar-Anan, R., Soller, M. and Bowman, J.C. 1976. Genetic and environmental
factors affecting the incidence of difficult calving and perinatal
calf mortality in Israeli-Friesian dairy herds. Animal Production
22: 299-310.

Bellows, R.A., Gibson, R.B., Anderson, D.C. and Short, R.E. 1971. Pre-
calving body size and pelvic area relationships in Hereford heifers.
Journal of Animal Science 33: 455-457.

Brinks, J.S., Olson, J.E. and Carroll, E.J. 1973. Calving difficulty and
its association with subsequent productivity in Herefords. Journal
of Animal Science 36: 11-17.

Cloppenburg, R. 1966. Geburtsverlauf bei Nachkommen von schwartzbunten
Bullen einer westfälischen Besamungsstation. Dissertation, Inst. für
Tierzucht und Haustiergenetik der Georg August-Universität, Göttingen.

Freeman, A.E. 1975. Management traits in dairy cattle. Dystocia, udder
characteristics related to production, and a review of other traits.
26th Annual Meeting of the European Association for Animal Production,
Warsaw, 23rd - 27th June, 1975.

Hansen, M. 1972. Kaelvningsforløb samt relationer mellem dette og nogle
kødproduktionsegenskaber hos RDM og SDM. Licentiatafhandling i
kvaegets avl. Den Konglige Veterinaer- og Landbohøjskole, København.

Krahmer, R. and Jahn, W. 1971. Ein Beitrag zur Bedeutung der Beckenmessungen
beim weiblichen Deutschen Schwarzbunten Rind während der Wachstumsperiode
unter besonderer Beachtung der Abhängigkeit der Beckeninnenmasse von
den Beckenaussenmassen. Archiv für Tierzucht 14: 41-54.

Kraüsslich, H. and Gottschalk, A. 1975. Die Erfassung der Kälbersterblichkeit
und des Geburtsverlaufs in der Besamungszucht. Der Tierzüchter 27: 8-10.

Lindhé, B. 1967. Studier över frekvensen dödfödda och missbildade kalvar
inom svenska nötkreatursraser. SHS meddelande 13.

Lindhé, B. 1974. Improvement in beef-breeding by selection. 1st World
Congress on Genetics applied to Livestock Production, Madrid, 1: 655-
669.

Nielsen, J. 1965. En undersøgelse over sammenhaengen mellem krydsets og
baekkenindgangens dimensioner hos køer. Forsøgslaboratoriets årbog
1965, pp 240-261.

Philipsson, J. 1976a. Studies on calving difficulty, stillbirth and associated factors in Swedish cattle breeds. I. General introduction and breed averages. Acta Agriculturae Scandinavica 26: 151-164.

Philipsson, J. 1976b. Studies on calving difficulty, stillbirth and associated factors in Swedish cattle breeds. III. Genetic parameters. Acta Agriculturae Scandinavica 26: 211-220.

Philipsson, J. 1976c. Studies on calving difficulty, stillbirth and associated factors in Swedish cattle breeds. IV. Relationships between calving performance, precalving body measurements and size of pelvic opening in Friesian heifers. Acta Agriculturae Scandinavica 26: 221-229.

Philipsson, J. 1976d. Studies on calving difficulty, stillbirth and associated factors in Swedish cattle breeds. V. Effects of calving performance and stillbirth in Swedish Friesian heifers on productivity in the subsequent lactation. Acta Agriculturae Scandinavica 26: 230-234.

Pijnenburg, W. 1974. Doodgeboorte bij MRY - kalveren. Mimeograph 76 pp.

Remmen, J.W.A. 1975. Calving performance and calf mortality. 26th Annual Meeting of the European Association for Animal Production, Warsaw, 23rd - 27th June, 1975.

Schlote, W., Buchsteiner, R. and Wortmann, H. 1975. Calving performance in cattle (heritability estimates). 26th Annual meeting of the European Association for Animal Production, Warsaw, 23rd - 27th June, 1975.

Van Dieten, S.W.J. 1963. Mortaliteit van kalveren bij de partus à terme van M.R. IJ. - Runderen. Dissertation, Faculteit des Diergeneeskunde, Rijksuniversitet, Utrecht.

Vogt-Rohlf, O. and Lederer, J. 1975. Möglichkeiten einer Nachkommenschaftsprüfung auf Kalbeverluste und Schwergeburten an Hand von Feldmaterial. 26. Jahrestagung der Europäischen Vereinigung für Tierzucht, Warschau, 23. bis. 27. Juni 1975.

SIRE EVALUATION FOR DYSTOCIA IN DUTCH CATTLE BREEDS

R.D. Politiek

Department of Animal Breeding, Agricultural University,
Postbox 338, Wageningen, The Netherlands.

ABSTRACT

Breeding for calving performance is of economic importance in many countries. A careful and well defined sampling method of calving data, a correct analysis of these data (mixed models for correction of environment influences) and construction of an index for selection on ease of calving are very important. The suggestion is made to ask for an international working group to develop on the basis of the present knowledge and experience an effective and easily comparable system.

Dystocia is defined as prolonged and/or difficult parturition. Records on calf size, dystocia score and stillbirth are collected systematically on AI bulls which are progeny tested by mating to first calved heifers. Dystocia data obtained for second calvings are used to predict calf losses for first calvings.

The frequency of stillbirths in Dutch Friesian cattle in 1975 was 5.5% (heifers 10%, second calvers 3%, older cows 2%) and in Dutch Red-and-White (MRY) 6.5% (heifers 15%, second calvers 4%, older cows 2%).

Dystocia data obtained during 1969 - 73 for MRY AI centres in the Province of North Brabant for 30 232 heifer and 109 306 cow calvings have been analysed by Pijnenburg (student thesis, University of Wageningen, 1974). The results of the analysis are presented in Table 1.

Age at first calving and season of calving are both important factors with respect to the occurrence of stillbirths. In winter months (December, January) about 18%, and in summer months (May, June) about 10% stillbirths are found. The production level of the herd also had an influence on stillbirth.

Calf size and sex of calf are the most important factors in dystocia (Table 2). In second calvings the frequency of stillbirths is about 4%, but is higher after a birth weight of about 45 kg. In second calvings the score for dystocia combined with birth weight gives a rather good estimation of the stillbirth percentage in heifers. A correlation of +0.6 between daughter groups from AI sires tested in first and second calving was found.

The direct influence of the sire on stillbirth of calves is important (variation 5 - 30%). A preliminary evaluation of the direct influence of sires can be based on a test sample of 500 - 750 calf births in second calvers. In a group of proven bulls a certain percentage can then be recommended for mating with heifers; in order to reduce stillbirth incidence subsequently the calving performance of daughter groups (about 150 heifers from the above testing scheme) can then give a base for final selection for transmission of ease of calving.

The indirect maternal characteristic measured by calving performance of daughter groups is also quite pronounced (variation 5 - 30%) and is virtually independent of the direct influence of the sire.

1. INTRODUCTION

Dystocia is defined as prolonged and/or difficult parturition. Records on calf size, dystocia score and still-birth (which here includes death within 24 hours after birth) collected on AI bulls from the Dutch Red-and-White (MRY) breed (Van Dieten, 1956, 1963) and the Dutch Friesian breed (Politiek, 1963), demonstrated an important direct effect of the sire on the stillbirth frequency in first calving heifers. By re-commendation of the best bulls to be used on heifers Van Dieten (1963) succeeded in decreasing the incidence of still-births in heifers from 16% to 8% during a seven year period; later on the frequency of stillbirths increased again to a level of about 15%.

The increased level of stillbirths is one reason to look more closely at genetic and environmental effects on dystocia. The second reason is that young AI bulls are nowadays routinely tested on about 750 heifers in milk. This means that only the calving performance of second calvers are available to predict the calving difficulty in heifers.

Pijnenburg (1974) analysed dystocia data obtained between 1969 and 1973 for MRY AI centres in the Province of North Brabant. We intend to present some results from this study. Recent publications, eg Ménissier (1975), Pollak and Freeman (1976), Philipsson (1976), Remmen (1976), demonstrate that breeding for calving performance is of economic importance in many countries. A careful and well defined sampling method for calving data, a correct analysis of these data and construction of an index for selection on ease of calving are very important. This seminar can stimulate the discussion on how to reach an effective system for sire evaluation and selection.

2. DYSTOCIA IN THE DUTCH MRY BREED

2.1. Material

Pijnenburg (1974) analysed material of 30 232 heifer and
109 306 cow calvings collected between 1969 and 1973 for MRY
AI centres in the Province of North Brabant. Information on
dam (heifer, cow), sire, insemination date, calving date
(gestation length) were collected. In addition the farmer gave
on the birth card information on ease of calving with a
subjective score for dystocia, namely easy, normal, heavy, or
veterinarian help with Caesarean delivery or dismembered calf.
The calf (male or female, single or twin) was scored for
viability (alive, stillborn, dead within 24 hours). Weight of
calf was estimated in classes, namely light, normal, heavy; but
also an estimated weight in kilograms was given. Any physical
abnormalities of the calf were described, or named in the case
of a known lethal. Only calves born as singles and after a
normal gestation length (260 - 290 days) were used in the
calculations.

2.2. Age at first calving and season of calving

Age at first calving and season of calving are both important factors with respect to the occurrence of stillbirths.
In Table 1 data on the incidence of stillbirths and dystocia at
first calvings within age classes are given for three different
months.

The deviations from the overall mean of 15.1% in stillbirths
and 16.5% in dystocia for the various months were: Jan: +4.1,
+5.8; Febr: +2.6, +3.2; March: +0.7, +0.4: April: -1.4, -1.6;
May: -2.1, -2.4; June: -3.0, -3.2; July: -2.4, -3.4; Aug: -2.6,
-3.9; Sept: -1.4, -2.9; Oct: -0.1, -1.1; Nov: +1.8, +3.1; Dec:
+3.7, +5.7.

Both parameters were significantly influenced by the month
of calving. In the younger age classes a higher incidence of
stillbirths was found in heifers calving under two years of
age and in heifers calving during the winter months. Calving

at an age of about two years was more frequent during winter but
in autumn calving was more frequent at an age of about 2½ years;
and therefore an interaction effect occurred.

TABLE 1

INFLUENCE OF AGE AT FIRST CALVING AND SEASON OF CALVING ON STILLBIRTHS AND
DYSTOCIA IN THE MRY BREED (Pijnenburg, 1974)

Age class at first calving (days)	<750	750-800	800-850	850-900	900-950	950-1000	>1000
January:							
Number of calvings	520	748	624	278	249	274	533
Stillbirth %	21	20	17	16	16	11	13
Dystocia %	25	24	22	16	21	15	14
April:							
Number of calvings	218	419	344	288	146	54	217
Stillbirth %	13	12	12	11	7	18	12
Dystocia %	17	14	14	8	8	13	10
September:							
Number of calvings	150	171	430	778	846	380	410
Stillbirth %	15	12	10	11	10	12	11
Dystocia %	15	11	9	11	9	9	8

2.3. Production level of herds and stillbirths

Material from 1971 and 1972 was divided into three production classes, namely low, with less than 537 g fat per cow
per day; medium; and high with more than 613 g fat per cow per
day. Herds with a higher production level showed an increased
incidence of stillbirth and dystocia. Each production class was
subdivided into two subpopulations: herds of younger cows
(average age at calving about 4 years; subpopulation 1) and herds
of older cows (average age about 5 years; subpopulation 2).

Average birth weight did not differ between production
levels and average age at calving of herds. There was a
tendency for higher incidence of stillbirths and dystocia in
herds with younger cows. Stillbirth percentage and dystocia

percentage increased to some extent (NS) with production level; this may indicate an influence of the condition of the cow. In older cows dystocia percentage increased with production level, namely 2.9; 4.7; 5.3; (P<0.001) in subpopulation 1 and 3.3; 3.6; 4.5 (P<0.1) in subpopulation 2.

TABLE 2

INFLUENCE OF PRODUCTION LEVEL OF HERDS ON STILLBIRTH AND DYSTOCIA

Production level	Low		Medium		High	
Subpopulation	1	2	1	2	1	2
Heifer calvings n	460	459	799	819	666	884
Stillbirths %	14.8	12.6	15.5	13.3	18.0	15.1
Dystocia %	14.8	14.2	15.1	15.9	17.3	16.1
Birth weight kg	36.4	36.6	36.4	36.7	36.3	36.7

Herds in free stalls showed a 2.2% higher percentage of stillbirths (n = 768; 16.5%) than herds in stanchion stalls (n = € 959; 14.3%). In older cows stillbirth percentage was 5.0 in free stalls compared with 4.1% in stanchion stalls (P<0.05) and dystocia percentage 5.0 compared with 4.0 (P<0.01). Level of production is about 10% higher in freestall herds (603 g/fat/cow/day) than in stanchion stall herds. Generally the level of nutrition for the former is higher than for the latter. The nutritional status or condition of the dam is one of the factors that may influence the ease of calving and the incidence of stillbirths (Philipsson, 1976).

2.4. Calf size and sex of calf

Birth weight has a predominant influence on stillbirth of the calf. Male calves are about 3 kg heavier than female calves. Compared with females (100), in males stillbirth and dystocia occurred for heifers at levels of 190 and 230, for second calvers at levels of 170 and 270 and for older cows at levels of 130 and 270. Also, at the same estimated birth weight, male calves are more frequently stillborn. In the estimated weight class of 36 - 38 kg for example, stillbirth percentage in heifers was 5 for females and 10 for males and in weight class 39 - 41 kg it was 12 for females and 20 for males.

In Table 3 the material is subdivided in birth weight classes and parity of the dam (1, 2, 3 and more calvings).

TABLE 3

CALF SIZE AND STILLBIRTH WITHIN AGE CLASSES (Pijnenburg, 1974)

Birth weight class (kg)	<29	30-32	33-35	36-38	39-41	42-43	44-47	48-50	>50
Heifers n	496	1950	3928	2310	2930	510	540	140	-
Stillbirth % (total)	17	9	8.5	8	16	33	45	70	-
Subdivided into:									
death at birth	12	6	5	5	8	12	18	21	-
death within 24 hours	3	2	1.5	1	2	3	5	7	-
Caesareans	2	1	2	2	5	16	18	33	-
Dismembered calves	-	-	-	-	1	2	4	9	-
Second calvers n	214	854	2814	2275	5670	1303	1660	636	186
Stillbirth % (total)	13.9	5.4	3.4	2.2	3.4	4.9	8.0	21.5	37.7
Subdivided into:									
death at birth	7.9	4.4	2.5	1.5	2.3	3.0	4.8	11.9	12.9
death with 24 hours	6.0	1.0	0.7	0.7	0.8	1.0	0.8	2.5	3.8
Caesareans	-	-	0.2	-	0.3	0.8	2.3	5.7	16.6
Dismembered calves	-	-			-	0.1	0.1	1.4	4.3
Older cows n	328	1238	4512	4065	12712	3595	3144	1969	610
Stillbirth %	19.5	5.5	3.6	1.2	2.3	1.5	3.9	7.5	21.5
Subdivided into:									
death at birth	14.0	4.4	2.6	0.8	1.6	1.0	2.7	4.2	9.7
death within 24 hours	5.5	1.5	0.9	0.4	0.6	0.4	0.7	1.5	3.4
Caesareans	-	-	0.1	-	0.1	0.1	0.5	1.5	6.9
Dismembered calves	-	-	-	-	-	-	-	0.3	1.5

In heifers, stillbirths increased considerably after weight class 39 - 41 kg and Caesarean deliveries were used frequently in the weight class 44 - 47 kg and over.

In second calvers the frequency of stillbirths and dystocia are 1.5 times and 1.9 times higher than in older cows. This demonstrates an age-class effect also after first calving. Small calves (below 30 kg) were in all age classes frequently still-born. They can better be excluded from the calculation of stillbirth percentage in relation to dystocia.

In Figure 1 stillbirth in relation to birth weight, sex of calf and age of the dam is shown.

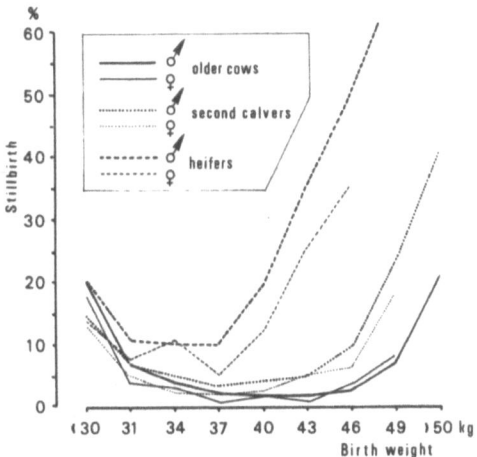

W.H. 56 - 1977 - 336

Fig. 1. Still births in relation to birth weight, sex of calf and age of dam.

3. SIRE EVALUATION FOR DYSTOCIA

Dystocia can be considered as a trait of either the calf or the dam. The former expresses the direct influence of the sire of the calf on calf size, on gestation length and via the placenta also on the calving performance of the dam (Osinga, 1976). The latter expresses the calving results of the daughters of the bull = pelvic opening of the dam, preparation at calving and the maternal effect on calf size. Parity of the dam and

214

environmental effects also play a role (Ménissier, 1975).

On the basis of a scheme of Ménissier (1975) we give: in Figure 2 a diagram of the effects on dystocia and the subsequent results on stillbirths, fertility of the dam etc.

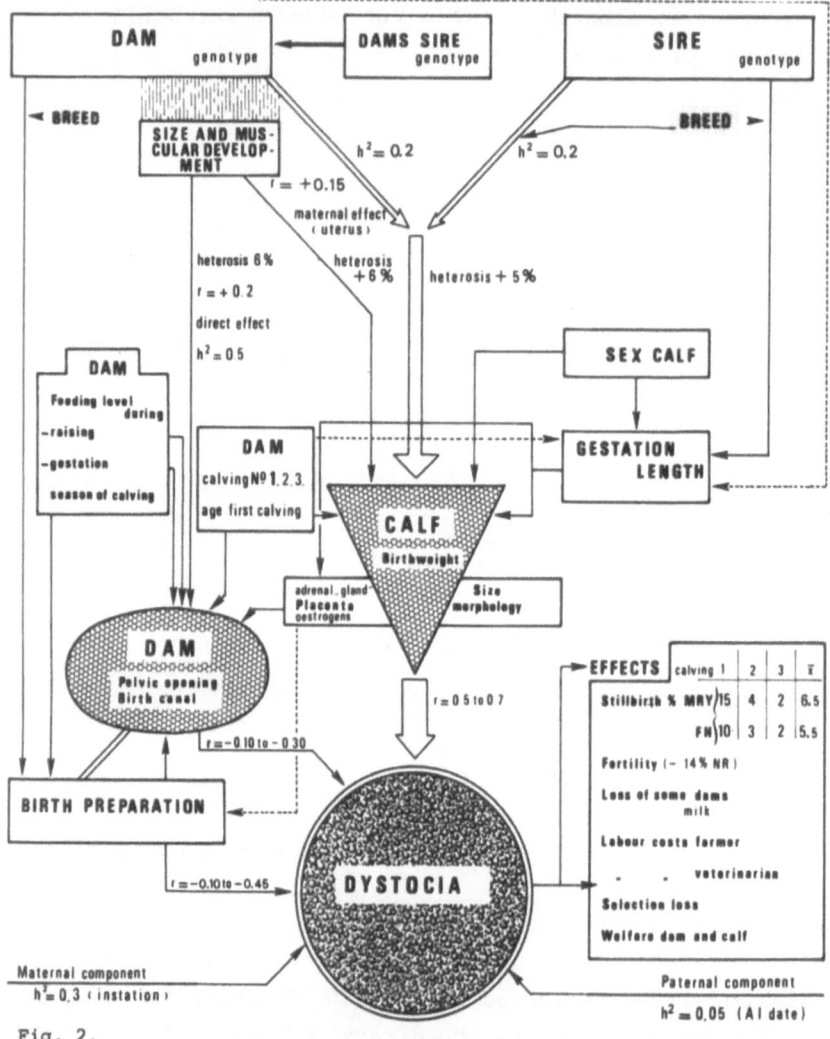

Fig. 2.

3.1. Direct effect of the sire on dystocia and stillbirth

In our country van Dieten (1956, 1963), introduced a complete birth registration of calves in order to make recommendations concerning the use of certain AI bulls for

inseminations of heifers. Initially these measures did prevent calf losses in first calving from 16 to 8% (van Dieten,
1963), but later the losses increased again to a level of 15%.
The loss of calves is economically very important, eg Remmen
(1976), Philipsson (1976).

The heritability estimate for dystocia is low, eg according
to Ménissier (1975) 5%, Philipsson (1976) 1 - 5%, Pollak and
Freeman (1976) 18%, 8% and 5% in 1st, 2nd and 3rd calvers
respectively.

In our material (Pijnenburg, 1974) we could test 50 AI
bulls with at least 100 offspring from heifers and also at
least 100 from second calvers. We found an average stillbirth
percentage of 15 in heifer calvings with a variation between
sires of 4 - 26% and a dystocia percentage of 16 (variation
5 - 30%). In second calvers the frequency of stillbirths was
much lower, but still 4% (variation 1 - 11%) and dystocia 6%
(variation 1 - 13%). These sire effects were both significant
(P<0.01). An important question was to what extent the ranking
of the AI bulls on the basis of second calvers agreed with the
ranking based on first calvers. The incidence of dystocia and
stillbirths in second calvers is lower than in heifers and we
suggest excluding small calves (<30 kg) from the calculation.
Also the correction for environmental effects is very important
as suggested, for example, by Pollak and Freeman (1976) as is
also, of course, the number of calves.

In our data we used uncorrected averages of bulls with
restricted number of calves. The rank correlations (Spearman)
between stillbirth percentage in heifers and different dystocia
aspects in second calvers are presented in Table 4.

Dystocia score and birth weight of calves from second
calvers gave a rather good indication (r = 0.47, 0.45) for
stillbirth percentage in heifers. Stillbirths in second
calvers gave a poor indication (r = 0.25), but a correction for
small calves could perhaps increase this value. In this material

there was a small negative correlation with gestation length in second calvers. Some bulls gave an average gestation length over 281 days with heavy calves(eg,41.3, 40.7 kg) and a high percentage of dystocia and stillbirths but a bull with an average gestation length of only 274 days also gave heavy calves and dystocia. Remmen (1976) found a higher frequency of still-births in calves with an average gestation length over 280 days. A weighted score (namely 2 x dystocia rate + 2 x birth weight + 1 x stillbirth rate) gave a rank correlation of +0.6 with stillbirth percentage in heifers. A small group of four recommended bulls per year on the basis of the birth records from heifers stayed also within the first four places on the list of 16 bulls in four different years, with only two exceptions. Five bulls per year scored as dangerous for heifers ranked also high on the list in four different years (always above average).

TABLE 4

RANK CORRELATIONS OF STILLBIRTH PERCENTAGE IN HEIFERS WITH DYSTOCIA
PARAMETERS IN SECOND CALVERS

Second calvers:	Dystocia	Still-birth	Birth weight	Gestation length	Weighted score [1]
Heifers stillbirth %	0.47	0.25	0.45	-0.25	0.6

[1] See text.

In the Netherlands young AI bulls are nowadays systematic-ally tested on heifers in milk (p = 0.2). This category of young cows is easy to detect in the dairy herd. About 750 first inseminations yield about 600 registered births of calves in second calvers and later a progeny group of 50 - 100 daughters with completed lactations in milk-recorded herds. Recommendation of AI bulls to be used on heifers can then only be based on second calvers. Therefore improvement of sampling methods and a correct evaluation of these data is very important. Further research on the components of dystocia and perhaps the use of new parameters (Osinga, 1976) is needed.

3.2. Effect of the sire on dystocia and stillbirth in daughter groups

The indirect maternal characteristic measured by the calving performance of daughter groups is also pronounced with a variation of 5 - 30%. Till now this aspect has not been systematically tested in our country. A daughter group of about 150 heifers from the above mentioned testing scheme can give a reasonable basis for a final selection for transmission on ease of calving. The data of Pijnenburg (1974) showed a small positive effect, namely the daughters of 20 AI bulls recommended for insemination of heifers gave about 2% less stillbirths in daughter groups than the 24 bulls detected as 'dangerous'. Pollak and Freeman (1976) found $r = +0.19$ (NS), Philipsson (1976) $r = -0.2$. The direct and indirect influences of the bull on calving performance are virtually independent of each other and both breeding aspects are necessary to evaluate on AI bull for dystocia. More research is needed on this aspect.

4. DISCUSSION

In the Netherlands the frequency of stillbirths increased during recent years. Remmen (1977) recorded in 1976 calf losses in 230 MRY herds and 129 Dutch Friesian herds in the Province of North Brabant (about 50 calves per herd). He found in the MRY herds 6.2% stillbirths, 0.8% deaths within 24 hours, 2.5% deaths within a month and 9.5% total loss. In the Dutch Friesian herds 4.4% stillbirths, 1% deaths within 24 hours, 3.7% within a month and 9.1% total loss occurred. To what extent is this high percentage of stillbirths of about 15% in heifers of the MRY breed and 10% in the Dutch Friesian breed a failure of the birth registration system? Some AI bulls are recommended for use on heifers, and others can be used on older cows only. This procedure can be improved by a more extended and accurate sampling method, and a better evaluation of the data, eg mixed models for correction of environment influences as proposed by Pollak and Freeman (1976). These recommended matings are not expected to change gene frequencies for calving difficulties. Daughter groups should then be included in the registration programme.

A combination of factors may have caused an increase in stillbirths.

- Selection for growth rate and size in bulls has given about a 1 kg increase in birth weight of calves from 1967 to 1973.

- The production level of milk recorded herds has increased from 4 400 kg in 1968 to over 5 000 kg milk in 1976 (Annual Report, 1976). The nutritional status of the dam has probably increased (freestall housing and more concentrate feeding).

- Increased herd size with less individual care and assistance at birth may have caused some extra losses.

- Age at first calving has decreased somewhat. Average age at calving is 4.6 years; age at first calving is 2.1 years in Dutch Friesian, 2.3 in MRY (Annual Report, 1976).

The suggestion is made to ask an international group of experts to develop on the basis of the present knowledge and experience an effective and easily comparable system of sire evaluation.

REFERENCES

Annual milk recording report 1976. C.M.D. Arnhem.

Ménissier, F. 1975. Calving ability in French beef breeds: analysis of
components and breeding improvement. Bulletin technique du département
de Génétique Animale, INRA No. 21: 57-102.

Osinga, A. 1976. Differences in the level of urinary oestrogens in late
pregnancy between cattle breeds and breed crosses. Proc. VIIIth
International Congress Reproduction and AI, Cracow.

Philipsson, J. 1976. Studies on calving difficulty, stillbirth and
associated factors in Swedish cattle breeds. I, II, III, IV, V.
Acta Agriculturae Scandinavica 26, 151-174, 211-234.

Pijnenburg, W. 1974. Doodgeboorte bij MRY kalveren. Student thesis,
Agricultural University, Wageningen.

Politiek, R.D. 1963. De volledige geboorteregistratie van kalveren bij een
KI-vereniging. Veeteelt en Zuivelberichten 6, 511.

Pollak, E.J. and Freeman, A.E. 1976. Parameter estimation and sire
evaluation for dystocia and calf size in Holsteins. Journal of Dairy
Science 59, 1817-1824.

Remmen, J.W.A. 1974. Een onderzoek naar mogelijkheden om perinatale sterfte
bij het rund te beperken. Dissertation, Utrecht.

Remmen, J.W.A. 1977. Personal communication.

Van Dieten, S.W.J. 1956. Calf birth registration in connection with animal
reproduction. Proc. 3rd International Congress of Animal Reproduction
Cambridge, Series 3, pp 53-56.

Van Dieten, S.W.J. 1963. Mortaliteit van kalveren bij de partus à terme van
MRY runderen. Dissertation,Utrecht.

INVESTIGATIONS ON THE RELATIONSHIPS OF BODY MEASUREMENTS AND WEIGHT OF HEIFER AND CALF TO CALVING DIFFICULTIES IN GERMAN SIMMENTAL (FLECKVIEH) CATTLE.
PRELIMINARY RESULTS OF THE EEC PROJECT NO. 320 OF THE BEEF PRODUCTION PROGRAMME.

W. Schlote and H. Hässig

Institut für Tierhaltung und Tierzüchtung, Universität
Hohenheim, Postfach 106 (06300), D-7000 Stuttgart 70, BRD.

ABSTRACT

In order to study the reasons for calving difficulties, to find auxiliary traits related to calving difficulties and to estimate phenotypic and genetic parameters for them a field investigation of Fleckvieh heifers is currently being conducted in the south-west of Germany. The calves have been sired by young bulls on progeny test for milk production. All sires are tested for individual performance of growth in central stations. Between 10 and 30 calves per sire and their heifer dams are measured. Body measurements and weight of calf are taken between the second and the eighth day post partum by members of the Institute. Preliminary evaluations include 180 pairs of heifer and calf. Incidence of easy, normal, difficult and surgical calvings was 16, 36, 45 and 3%, respectively. Most of the essential correlations were low or moderate, correlations among body measurements generally being high. Estimates of heritability for calving performance are slightly higher than reported elsewhere.

1. INTRODUCTION

Due to selection for muscling and growth, calving performance has become a major problem in many dual-purpose and
beef breeds of cattle. Since heritability for this trait is
low, progeny tests are costly in terms of calf losses, and
selection for calving performance is counterbalanced by
selection for growth and muscling, a field experiment was conducted in order to study relationships among weight, body
measurements, calving performance and several other traits and
to estimate population parameters for these traits.

2. MATERIAL AND METHODS

Fleckvieh(Simmental) calves and their heifer dams are
measured by two members of the Institute on farms in southern
Germany between 1 and 10 days after birth. The calves were
sired within the insemination scheme for the progeny test of
young bulls for milk production. All sires were tested for
growth on central stations. The calves are weighed by a
portable balance that can be suspended in the barn. Muscling
scores were from 1 to 5 points with half-point intervals. The
bulls were weighed and measured at 330 days of age on the
performance testing station.

Regression programmes of the SPSS package were used to
calculate simple product-moment correlations and multiple
correlations by the forward selection procedure. Least
squares analyses were performed by the mixed model programme
of Harvey (1972) for the estimation of heritabilities. The
models included sex of calf as fixed, and sires as random
effects. To study the nature of the sex effect a regression
for calf weight was also considered.

3. RESULTS

The calculations are based on the data of 185 calf-heifer
pairs. Results for calving performance are summarised in Table 1.

It is obvious that the two types of classification lead to different results (for instance 22% difficult calvings vs 45%).

TABLE 1

CALVING PERFORMANCE

Official classification	Incidence % (n)		Experimental classification	Incidence % (n)	
1. Normal	75	(139)	1. Without assistance	16	(29)
2. Difficult, heavy calf	17	(31)	2. Slight assist. (not more than 2 persons)	36	(67)
3. Difficult, other reasons	5	(10)	3. Heavy assist. (more than 2 persons or calf/puller)	45	(84)
4. Surgical intervention, heavy calf	3	(3)	4. Surgical intervention	3	(5)
5. Surgical intervention, other reason		(2)			

However, the experimental classification seems to provide more information as the criteria are more objective and can be better identified.

In 165 (97%) of 170 observed calvings the calves were in the normal forward position; 5 calves (3%) were born in the backward position. In 13 cases (7%) correction of head or legs or turning of the calf was necessary. Veterinarians assisted at 17 calvings (9%).

Calf weights were different for calving performance (scores 1 to 3) and sex (Table 2). Differences between calving scores 1 and 2 were much smaller than between scores 2 and 3. Weight may not be the most important factor for calving problems of female calves, as average weight for calving score 2 was lower than for 1. However, speculations are premature, as the number of animals is still quite small.

TABLE 2

LEAST SQUARES MEANS FOR CALF WEIGHT (kg)

Calving score	Sex of calf ♂	♀	Total
1	40.4	38.6	39.5
2	42.7	37.1	39.8
3	46.1	42.2	44.2
Total	43.0	39.3	41.2

The traits recorded, their means ($\bar{x}$) and standard deviations (SD) are listed separately for heifers, calves and bulls (Tables 3, 4 and 5). Correlations with date of measurements are given for heifer and calf traits (Tables 3 and 4). As the number of animals recorded so far is small, the numerical results should not be taken too seriously. However, there is the problem of correcting the data for date of measurement, since the animals were measured between 1 and 10 days after calving. Correlations for heifer traits indicate a general decrease of measurements after calving, for calf traits an increase. The considerable negative correlation for circumference of cannon bone of the calf may be due to swollen forelegs because of extraction of the calves.

Correlations of heifer traits with calving performance (Table 3) are low to moderate. The highest values were estimated for heart girth (0.22), body length (0.20) and width of chest (0.19), whereas the exterior pelvic measurements had disappointingly low correlations. However, it was surprising that width of pins, a measurement that is quite difficult to take, had the highest correlation among them. It was also expected that gestation length would be more closely correlated with calving performance than was calculated from these data.

For calf traits the correlations with calving performance (Table 4) were in general higher than for heifer and bull traits. Most closely correlated with calving performance were calf weight,

(0.38) and muscling of shoulder (0.31): heart girth, chest depth, width of hips and muscling of hindquarters followed (all 0.26). The high correlations of the muscling scores were not expected since they are highly subjective characteristics.

TABLE 3

HEIFER TRAITS: MEANS, STANDARD DEVIATIONS, CORRELATIONS WITH DATE OF MEASUREMENT AND CALVING PERFORMANCE

Trait	Unit	$\bar{x}$	SD	Correlation with date of measurement	Correlation with calving performance
Age at calving	months	30.1	3.1	-	-0.03
Gestation length	days	286.5	4.6	-	0.05
Width of head	cm	20.9	0.8	-0.06	0.04
Length of head	cm	49.5	2.1	-0.14	-0.02
Heart girth	cm	194.1	7.8	-0.07	0.22
Withers height	cm	125.5	3.9	-0.13	0.11
Body length	cm	146.7	6.4	0.03	0.20
Width of chest	cm	46.2	3.6	-0.05	0.19
Width of shoulder	cm	51.3	3.7	-0.03	0.17
Chest depth	cm	69.5	2.6	-0.03	0.10
Width of hips	cm	50.0	2.3	-0.13	0.06
Width at thurls	cm	47.9	2.3	-0.08	0.07
Width of pins	cm	15.9	1.4	-0.04	0.13
Rump length	cm	49.2	2.1	-0.06	0.02
Height at hips	cm	132.6	4.0	-0.20	0.11
Height at pins	cm	117.6	4.1	-0.19	0.06
Difference between heights at hips and at pins	cm	15.0	3.7	0	0.05
Circumference of cannon bone	cm	19.5	1.0	-0.11	0.17
Muscling of hindquarters	points	3.2	0.6	0.03	0.14

TABLE 4

CALF TRAITS: MEANS, STANDARD DEVIATIONS, CORRELATIONS WITH DATE OF MEASURE-
MENT AND CALVING PERFORMANCE

Trait	Unit	$\bar{x}$	SD	Correlation with date of measurement	Correlation with calving performance
Body weight	kg	42.1	5.6	0.12	0.38
Width of head	cm	12.1	0.7	-0.02	0.25
Length of head	cm	24.1	1.0	0.04	0.19
Heart girth	cm	76.7	3.4	0.12	0.26
Withers height	cm	70.8	2.8	-0.08	0.16
Body length	cm	62.0	3.3	0.04	0.22
Width of chest	cm	15.1	1.1	0.01	0.08
Width of shoulder	cm	17.9	1.5	0.0	0.15
Chest depth	cm	26.9	1.4	0.12	0.26
Width of hips	cm	15.3	0.9	0.02	0.26
Width at thurls	cm	20.3	1.2	-0.04	0.24
Circumference of cannon bone	cm	11.5	0.7	-0.16	0.25
Muscling of hind-quarters	points	3.2	0.6	0.10	0.26
Muscling of shoulder	points	3.0	0.6	0.13	0.31

The correlations of bull traits with calving performance
(Table 5) were low. Highest correlations were found for heart
girth (0.15) and body weight (0.12). The lower correlations
than those for heifer traits could be explained by maternal
effects and by the youth of the bulls which have not grown
as close to their mature size and weight when they are measured.
It should be borne in mind that there are as yet only 11 bulls.

Multiple correlations of heifer, calf and bull measurements
with calving performance (Table 6) show that combinations of
traits explain considerably more of the variation of calving
performance than single traits alone. The relatively steady
increase when additional traits were included was not expected.

TABLE 5

BULL TRAITS: MEANS, STANDARD DEVIATIONS, CORRELATIONS WITH CALVING PERFORMANCE

Trait	Unit	$\bar{x}$	SD	Correlation with calving performance
Body weight	kg	453.7	23.1	0.12
Heart girth	cm	180.0	3.4	0.15
Withers height	cm	121.7	2.2	0.06
Body length	cm	142.0	3.7	0.02
Width of chest	cm	43.5	2.1	0.03
Chest depth	cm	62.1	1.5	0.08
Width at thurls	cm	47.9	2.9	0.04

TABLE 6

MULTIPLE CORRELATIONS WITH CALVING PERFORMANCE FOR HEIFER, CALF AND BULL TRAITS*

Heifer traits	R	Calf traits	R	Bull traits	R
1. Width of chest	0.24	1. Body weight	0.37	1. Heart girth	0.15
2. Body length	0.26	2. Width of shoulder	0.41	2. Width of chest	0.21
3. Width of hips	0.31	3. Muscling of shoulder	0.43	3. Width at thurls	0.24
4. Rump length	0.32	4. Width of chest	0.45	4. Withers height	0.25
5. Muscling of hindquarters	0.33	5. Heart girth	0.45	5. Body length	0.26
6. Length of head	0.34	6. Width of hips	0.46	6. Body weight	0.30
7. Withers height	0.35	7. Chest depth	0.46	7. Chest depth	0.31
8. Width of head	0.36	8. Withers height	0.47		
9. Circumference of cannon bone	0.36	9. Width at thurls	0.47		
16. Height at hips	0.37	10. Muscling of hindquarters	0.47		
11. Width at thurls	0.38	11. Length of head	0.47		
12. Width of pins	0.38	12. Body length	0.47		
13. Heart girth	0.38	13. Circumference of cannon bone	0.47		
14. Chest depth	0.39	14. Width of head	0.47		
15. Difference of heights at hips and pins	0.39				
16. Width of shoulder	0.39				

* Traits are listed in consecutive order of inclusion by the forward procedure.

Combinations of all traits of heifer, calf and bull yielded high correlations with calving performance (Table 7). The independent variables accounted for 41% of the total variation of calving performance at the maximum. Inconsistencies in the increase of the correlation coefficient indicate that the forward procedure may not be the best method and rather a backward or stepwise elimination should be used. Again the steady increase of the correlation coefficient is to be noted.

TABLE 7

MULTIPLE CORRELATION WITH CALVING PERFORMANCE, ALL TRAITS CONSIDERED *

Trait	R	Trait	R
1. Calf body weight	0.37	20. Calf heart girth	0.60
2. " width of shoulder	0.41	21. Heifer heart girth	0.60
3. Heifer width of chest	0.44	22. " chest depth	0.61
4. " rump length	0.46	23. Calf width of head	0.61
5. Calf width of chest	0.47	24. " muscling of hind-quarters	0.62
6. " muscling of shoulder	0.49		
7. Heifer body length	0.49	25. Heifer width of head	0.62
8. Bull heart girth	0.50	26. Calf width at thurls	0.63
9. " width at thurls	0.52	27. Heifer width of hips	0.63
10. " width of chest	0.53	28. Bull chest depth	0.63
11. " withers height	0.53	29. Calf width of hips	0.63
12. Calf body length	0.54	30. Heifer circumference of cannon bone	0.64
13. Heifer height at pins	0.55	31. " length of head	0.64
14. Bull body length	0.55	32. Calf length of head	0.64
15. " body weight	0.56	33. " chest depth	0.64
16. Heifer withers height	0.57	34. " circumference of cannon bone	0.64
17. " difference of height at hips and pins	0.57		
18. Heifer width at thurls	0.58	35. Heifer width of shoulder	0.64
19. Heifer width of pins	0.59		

* Traits are listed in consecutive order of inclusion by the forward procedure.

Calf traits cannot be used for prediction of calving performance. Therefore, multiple correlations for combinations of heifer and bull traits are listed separately (Table 8). The

maximum correlation is of the order of the maximum correlation
of calf traits alone.

TABLE 8

MULTIPLE CORRELATION WITH CALVING PERFORMANCE, HEIFER AND BULL TRAITS*

Trait	R	Trait	R
1. Heifer width of chest	0.24	13. Heifer height at hips	0.44
2. Bull heart girth	0.28	14. " width of head	0.45
3. " width at thurls	0.31	15. Bull chest depth	0.46
4. " body length	0.32	16. Heifer width of pins	0.46
5. Heifer circumference of cannon bone	0.33	17. " width at thurls	0.47
		18. " chest depth	0.47
6. " width of hips	0.35	19. " heart girth	0.48
7. " body length	0.37	20. " muscling of hindquarters	0.48
8. " length of head	0.38		
9. " withers height	0.39	21. " rump length	0.48
10. Bull body weight	0.40	22. " height at pins	0.48
11. " width of chest	0.41	23. " width of shoulder	0.48
12. " withers height	0.43		

* Traits are listed in consecutive order of inclusion by the forward procedure.

Preliminary genetic analyses were performed to get some
indications on the level of heritability (Table 9). The
estimate for calving performance is higher than the usual
estimates which could be expected because of the high incidence
of difficult calvings and more careful data collection. Other
estimates are lower than expected. Again, the small numbers
should be borne in mind.

The data for calving performance were analysed using two
statistical models, one with sex as a fixed effect, the other
including sex as a fixed effect and a regression on calf weight.
The resulting F-values for the sex effect of 3.995(*) vs. 0.002
(NS) show clearly that the sex effect of the calf is due to
calf weight.

TABLE 9

HERITABILITY ESTIMATES FOR CALVING PERFORMANCE AND CALF TRAITS[1]

Trait	$h^2 \pm SE_h 2$	Trait	$h^2 \pm SE_h 2$
Calving performance [2]	0.13 ± 0.16	Chest depth [3]	0.22 ± 0.21
Body weight [2]	0.18 ± 0.18	Width of hips [3]	0.13 ± 0.18
Width of head [3]	-0.04 ± 0.11	Width of thurls [3]	0.03 ± 0.14
Length of head [3]	0.14 ± 0.18	Circumference of cannon bone [3]	0.08 ± 0.16
Heart girth [3]	0.08 ± 0.16		
Withers height [3]	0.09 ± 0.16	Muscling of hind-quarters [3]	0.02 ± 0.14
Body length [3]	0.28 ± 0.23	Muscling shoulder [3]	-0.10 ± 0.09
Width of chest [3]	0.08 ± 0.08		
Width of shoulder	0.13 ± 0.13		

(1) Half-sib analysis, 11 sires

(2) Stillbirths included, 180 calves

(3) Without stillbirths, 164 calves.

The preliminary analysis has shown low or, at best, moderate correlations of single traits with calving performance. However, combinations of traits yielded correlations of considerable order. Points for discussion on this study are the necessity of correction of data for date of measurement; analysis of meaningful conformation description traits or proportion traits; classification of calving performance; inclusion of additional traits for data collection and additional influences for data analysis.

Complementary small-scale studies on internal pelvic measurements were conducted at the experiment station of the University. However, the data are not yet analysed.

REFERENCE

Harvey, W.R. 1972. Outline for use of LSMLMM. Mimeo.

COMPARISON OF THE MAIN EUROPEAN CATTLE BREEDS USED IN INDUSTRIAL CROSSING ON FRENCH FRIESIAN DAIRY COWS: PRELIMINARY RESULTS ON CALVING DIFFICULTIES (+)

F. Ménissier, J. Sapa, J. Gogue, B. Bonaiti and J. Frebling.
Station de Génétique quantitative et appliquée,
Centre National de Recherches Zootechniques - INRA -
78350 Jouy en Josas - France.

ABSTRACT

Industrial crossing on dairy cows with large-sized and large-muscled breeds (more or less specialised beef breeds and dual-purpose breeds) represents an efficient method for increasing meat production. Therefore, according to the Common Research Programme of the EEC, a series of comparisons have been made between various paternal genotypes crossed with French Friesian cows. Among the 18 genotypes studied (Table 1), 8 were more specialised for 'terminal crossing' and 9 were more fitted for production of breeding females ('breeding type'); the Friesian breed was used as a reference. The results referred to in this report concern only calving difficulties and they have been obtained from 533 calvings of multiparous cows during the first of the two replicates planned. Estimates for each paternal breed concerning calving difficulties, frequency of assisted calvings, frequency of dystocic calvings and birth weight, are given in Table 2 and Figure 1. It can be noticed that: 1) All breeds used increase the weight of calves more or less markedly, but calving difficulties only become frequent with a birth weight of over 42 - 45 kg. 2) British breeds (small sized) do not reach this birth weight, except perhaps the English Hereford. 3) Italian breeds cause calving difficulties either due to their size (Chianina) or to their morphology (Piemontese). 4) French beef breeds show large differences between breeds, but also between types. 5) The synthetic French lines give results similar to those of the paternal breeds of which they are composed, with variations probably due to their morphology. These first results emphasise the necessity of using an adequate sample of sires from each breed, notably in the case of breeds with multiple selection of goals.

(+) EEC Contract number 301 - Beef Production and INRA number 652 011, realised in collaboration with two Selection Units: MIDATEST and Auvergne-Limousin-Charentes.

1. OBJECTIVE OF THE EXPERIMENT

Industrial crossing with cows of dairy breeds represents one of the most efficient means for increasing meat production, while at the same time maintaining or improving their milk production (Anderson and Lindhé, 1973; Cunningham, 1974; Cunningham and McClintock, 1974; Elsen and Mocquot, 1976). In terms of meat production, the advantage of crossbreeding depends especially on the higher slaughter value of the paternal breed used as compared with the maternal dairy breed (Cunningham, 1974; Elsen and Mocquot, 1976). Thus when using dairy cows of various sizes but with a moderate muscle development, the breeds most fitted for this industrial crossing are most certainly the large-size and large-muscled beef breeds. This explains the success of European beef breeds in countries where such crossbreeding is practised (Charolais, Blonde d'Aquitaine and Limousin in France, Hereford and Angus in Great Britain, Piemontese in Italy).

One of the factors limiting this industrial crossing is the occurrence of calving difficulties related to the increase in the size of the calves at birth. These difficulties are even more harmful as they have an important repercussion on the milk production of high-yielding cows. Therefore, for industrial crossing in a given herd of dairy cows, the problem is to find a particular breed with a large muscle development but in which the weight of calves at birth is compatible with a very low frequency of dystocic calvings.

As a matter of fact, the choice of the paternal breed for industrial crossing with dairy cows depends, not only on the weight of the calves and the complementarity with the maternal dairy breed, but especially on the utilisation of female F_1 progeny (intensive or on-pasture fattening, once-bred heifers as suckler cows in intensive or extensive systems, etc (Ménissier et al., 1975). Thus, industrial crossing can be made with more or less specialised beef breeds, but also with large-sized dual-purpose breeds (especially Simmental).

Consequently, according to the Common Research Programme of the EEC on beef production (Tayler, 1976), we have carried out a comparison between various breeds or genotypes used in industrial crossing with dairy cows of the French Friesian type. Eighteen paternal genotypes were used in this investigation (Table 1); eight were selected for terminal crossing, nine were selected for another goal (designated the 'breeding type'). Among the breeds of the terminal crossing type, we chose three French beef breeds (Charolais, Blonde d'Aquitaine, Limousin) specifically selected on their growth and muscle development, and the three French synthetic lines selected on the same criteria (Coopelso 93 and INRA 95 1st and 2nd lines). (Bibé et al., 1977). To this terminal crossing group, we also added the Piemontese breed (Italy) and the dual-purpose Blanc-Bleu (Belgian Blue-White), in which the double-muscle trait is very common. In the breeding group, we chose the three French beef breeds (Charolais, Blonde d'Aquitaine, and Limousin) selected on the breeding qualities of their daughters (Ménisser, 1976), British beef breeds (South Devon, English and American Hereford), an Italian beef breed (Chianina) and the more current large-sized dual-purpose breeds (Maine-Anjou, Tachetée de l'Est or French Simmental). In addition to these 17 paternal breeds, we also used the Friesian breed (type NRS) as purebred reference genotype. For each of these paternal breeds or types, six sires were sampled at comparable stages of selection - among the sires retained after performance testing in the case of the terminal crossing type, and among the best or the most used sires in the case of the breeding type (Table 1).

The experiment was started at the end of 1975 (choice of sires). We only report results on calving difficulties recorded in about three-quarters of the calvings in the first of the two replicates planned for this comparative study.

2. ANIMALS AND METHODS OF ANALYSIS

For this first replicate (3 sires/breed), 1 790 artifical inseminations were made on dairy cows of the French Friesian type in the south of France (6 departments) where industrial

TABLE 1
SIRE BREEDS AND CHARACTERISTICS OF THE SIRES USED IN INDUSTRIAL CROSSING ON FRENCH FRIESIAN DAIRY COWS.
(Preliminary results, EEC experiment number 301-beef)

Sire breed or genotype	Abbreviation	Performances of sires chosen (index)				Inseminations		Choice of sires (a)	
						Number AI	%non-returns 60 - 90d	Goal	Particularities
Terminal crossing type		On performance test:				($\bar{m}$ = 67.6%)			
		Final weight	Conformation:						
Charolais	Ch$_T$	103	111			102	71.6	3 bulls per breed, among the performance tested bulls.	Crossbred:Ch, BA, Lm
Blonde d'Aquitaine	BA$_T$	103	109			103	68.9		
Limousin	Lm$_T$	106	114			102	65.7		
Coopelso 93	C93	109	101			101	69.3		
INRA 95 (BA, Ch)	195 (1)	107	110			104	60.6		Crossbred and double-muscled (Ch, BA)
INRA 95 (MA, BA, Lm)	195 (2)	103	105			103	76.7		(MA, BA, Lm)
Blanc Bleu Belge	BB					101	60.4		.Partly double-muscled
Piemontese	Pm	(no results)				77	67.5		.Partly double-muscled
Breeding type		On progeny test:				($\bar{m}$ = 68.9%)			
		Weight and conformation	Fertil-ity	Calving ability	Suckling ability				
Charolais	Ch$_E$	102	98	116	105	109	65.1	3 bulls per breed, among the best tested bulls and used in the breed.	selected on the performance of their daughters.
Blonde d'Aquitaine	BA$_E$	109	110	106	103	105	72.4		
Limousin	Lm$_E$	118	105	92		95	71.6		
Hereford (English)	HeGB	102	(not tested)			96	68.3		.Selected for crossing
Hereford (American)	HeUS	(not tested)				100	73.0		.Imported to France
South Devon	SD	(not tested)				102	61.8		-
Chianina	Cn	(not tested)				98	74.2		-
Maine-Anjou	MA	101	(not tested)			98	67.3		-
Simmental (French)	Sm	(indexed on milk production)				91	67.0		-
Reference Friesian (NRS type)	Fr	(indexed on milk production)				103	56.4		NRS type

(a) For the 2nd replicate, 3 other bulls per breed have been chosen according to the same criteria.

crossing is already well developed. These inseminations were
performed from 15 March to 15 June 1976 by two selection units
(Midatest and Union Auvergne-Limousin-Charentes). The conception
rate was high (67.8% of non-returns at 60 - 90 days, Table 1)
and purchase of 760 to 780 calves could be considered. The
analysis was made on the first calves obtained after these
inseminations, ie 570 calves purchased 1 to 2 weeks after birth:

Born before 16 January:	56 ♂ + 61 ♀	= 117 calves
Born from 16 to 31 January:	106 ♂ + 104 ♀	= 210 calves
Born from 1 to 15 February:	101 ♂ + 91 ♀	= 192 calves
Born from 16 to 28 February:	25 ♂ + 26 ♀	= 51 calves
Total	: 228 ♂ + 282 ♀	= 570 calves

The mortality of the calves was relatively low and not
analysed; we have only examined calving difficulties. Among
these 570 calves, twins (28) or premature calves (2) as well
as those from heifers (7) were eliminated from the analysis.
As regards the remaining 533 calvings, the calving difficulties
were analysed from the information supplied by the farmer on
the calf birth declaration paper (national model). The
difficulties were scored according to the following notation:

1 = calving without assistance;

2 = calving with little assistance;

3 = calving with much assistance or forced extraction;

4 = calving with surgical intervention (Caesarean operation
    or embryotomy).

These calving difficulties have been estimated according to
3 criteria:

Mean score for calving difficulties

Mean frequency of assisted calvings (2 + 3 + 4/1 + 2 + 3 +4)

Mean frequency of dystocic calvings (3 + 4/1 + 2 + 3 + 4).

For 478 calves (90% of those purchased) the weight was measured by the farmer immediately after birth.

On the basis of this information, we estimated the mean effects of each paternal breed taking into account the sex effect of the calves as well as the unequal number per sex and per sire breed. Effects of season and region were not considered in our analysis.

3. RESULTS AND DISCUSSION

Table 2 shows the estimates of the weight and calving difficulties of the calves at parturition for each paternal breed. Figure 1 indicates the relationship between the weight of the calf and frequency of difficult or dystocic calvings according to sire breed. Taken as a whole, the situation of the different breeds is rather similar to that which could be expected from the performance of each breed in purebreeding or from the experimental results obtained in crossbreeding with dairy cows (Andersen et al., 1976) and with British beef breeds (Smith et al., 1976). Several observations can be made.

1. All breeds used increase the weight of the calf and the calving difficulties, but for each of these two criteria, the variations between breeds are large. In particular, as long as the mean weight of the calf does not reach 42 kg, the frequency of assisted calvings (Figure 1) remains low (less than 15%). Beyond this threshold, the frequency rapidly increases and reaches about 50% for a weight of 46 kg. The frequency of dystocic calvings only begins to rise between 42 and 45 kg; the threshold effect is then less marked. This threshold value of 45 kg for dystocic calvings is the same as that found for adult cows in the female crossbreeding stock (Belic and Ménissier, 1968), or in Friesian females only (Ménissier and Foulley, 1975), used for industrial crossing, as well as in Swedish Friesian cows (Philipsson, 1977).

236

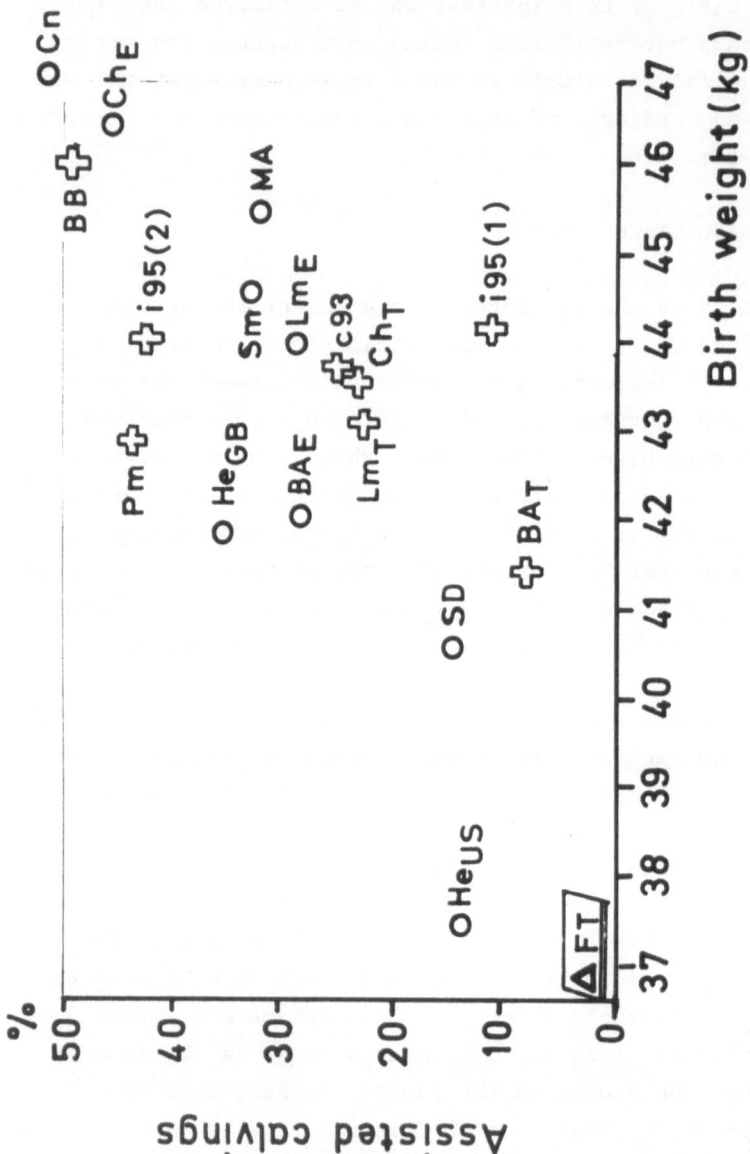

Fig. 1. (a)

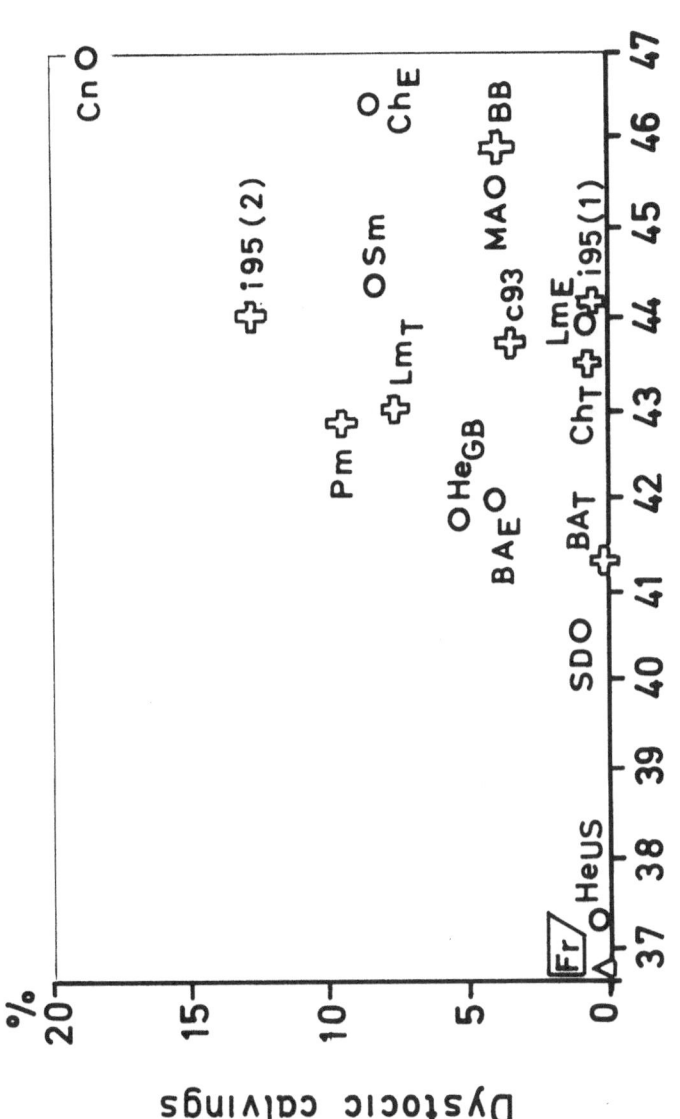

Fig. 1. (b)

Fig. 1. Relationship between sire breed effect (a) on frequency of calving difficulties and birth weight of calves, in beef crossing on French Friesian dairy cows (preliminary results, EEC experiment number 301-beef).
(a) Fr = Friesian; Ch = Charolais, BA = Blonde d'Aquitaine, Lm = Limousin, MA = Maine-Anjou, Sm = French Simmental, Cn = Chianina, Pm = Piemontese, SD = South Devon, He = Hereford English (GB) or American (US) Hereford, BB = Blanc Bleu Belge, C93 = Coopelso 93, 195 = INRA 95, from Ch, BA (1) or MA, BA, Lm (2) Terminal crossing type (T) or Breeding type (E).

TABLE 2

COMPARISON OF DIFFERENT SIRE BREEDS IN INDUSTRIAL CROSSING ON FRENCH FRIESIAN DAIRY COWS [a]. (Preliminary results EEC experiment number 301-beef).

Sire breed or genotype	Birth weight		Calving difficulties: [b]			
	Number	Mean (kg)	Number	Average	score	Assisted calvings (%)	Dystocic calvings (%)
Terminal Type							
Charolais	31	43.6 + 6.6	32	1.24	+0.22	23.5 +20.4	0.4 + 1.2
Blonde d'Aquitaine	34	41.4 + 4.4	36	1.08	+0.06	8.0 + 4.9	0.0 + 0.7
Limousin	25	43.1 + 6.1	26	1.31	+0.29	23.1 +20.0	7.7 + 8.5
Coopelso 93	31	43.7 + 6.7	32	1.30	+0.28	26.2 +23.1	3.4 + 4.3
INRA 95 (BA, Ch)	27	44.2 + 7.2	28	1.12	+0.10	11.6 + 8.5	0.2 + 1.1
INRA 95 (MA, Lm, BA)	39	44.0 + 7.0	45	1.57	+0.55	42.6 +39.5	12.9 +13.1
Blanc bleu belge	20	46.0 + 9.0	24	1.58	+0.56	49.5 +46.5	4.0 + 4.9
Piemontese	20	42.8 + 5.8	20	1.53	+0.51	43.7 +40.5	9.7 +10.5
Breeding Type							
Charolais	23	46.5 + 9.5	24	1.54	+0.52	45.8 +42.7	8.3 + 9.2
Blonde d'Aquitaine	39	42.1 + 5.1	49	1.32	+0.30	28.2 +25.1	4.0 + 4.8
Limousin	25	44.0 + 7.0	29	1.29	+0.27	28.7 +25.6	0.3 + 1.1
Chianina	24	47.0 +10.0	27	1.78	+0.76	52.1 +49.0	18.6 +19.4
Maine Anjou	21	45.5 + 8.5	23	1.36	+0.34	32.3 +29.2	3.7 + 4.5
Simmental (French)	30	44.4 + 7.4	36	1.41	+0.39	32.6 +29.2	8.2 + 9.0
South Devon	16	40.6 + 3.6	18	1.16	+0.14	14.7 +11.6	0.9 + 1.7
Hereford (English)	37	41.8 + 4.8	40	1.40	+0.38	35.3 +32.2	5.1 + 5.9
Hereford (American)	22	37.5 + 0.5	28	1.14	+0.12	13.8 +10.7	0.0 + 0.7
Reference Friesian (NRS type)	14	37.0 0	16	1.02	0	3.1 0	0.0(c) 0
Average	478	43.1 kg -	533	1.34 points -		28.6% -	4.8% -

(a): Least squares estimates adjusted for the sex of calves (50% males-50% females); + 2.5 kg birth weight, + 12.8% assisted calvings, + 3.3% dystocic calvings, between male and female calves; followed by difference from Friesian.

(b): Score: 1 = without assistance, 2 = little assistance, 3 = large assistance, 4 = surgical intervention; % assisted calvings = (2 + 3 + 4/total) 100; % dystocic calvings = (3 + 4/total) x 100.

(c): The mean square value is -0.8%.

2. The small sized breeds, represented here by the British beef breeds (Hereford and South Devon) increase the birth weight of calves without substantially increasing the frequency of difficult calvings. The threshold of calving difficulties is not reached, except perhaps in the case of Hereford of the English type (Figure 1). The English Hereford bulls chosen were selected by the Milk Marketing Board on their growth potential (selection on crossbred Friesian progeny in station). On the other hand, the Hereford bulls of the American type have not been tested and their semen was imported to France several years ago. Thus with Herefords of the English type, the calves reach the limit of mean weight producing an increase in the frequency of assisted calvings (42 kg) and are close to the weight limit leading to dystocic calvings. While these data on English Herefords are similar to those observed by Andersen et al. (1976) and Philipsson (1977), our estimates are generally higher than those found in Great Britain (Allen, 1977; various testing reports of the MMB). Likewise, these crossbred English Hereford calves are heavier than the crossbred South Devon calves, contrary to observations made elsewhere (Smith et al., 1976; Kilkenny and Stollard, 1976).

3. The large-sized Continental breeds generally produce heavier calves than the previous breeds and consequently tend to induce more calving difficulties.

Dual-purpose breeds (Maine-Anjou, Simmental and Blanc-Bleu Belge) belong to those producing the heaviest calves at birth (44.5 to about 46 kg), leading to 30 - 50% assisted calvings and some dystocic calvings (3 to 8%). Among these three breeds, the Blanc-Bleu Belge gives the heaviest calves at birth and the greatest number of assisted calvings. Crossbred Maine-Anjou calves, although heavier than Simmentals do not cause a higher percentage of dystocic calvings; their poorer conformation may account for this fact.

Italian beef breeds (Chianina and Piemontese) belong to those leading to the highest percentage of assisted calvings (52 and 44% respectively) and dystocic calvings

(19 and 11% respectively). Chianina bulls gave the heaviest
calves of the whole experiment (+10.0 kg as compared with
Friesian) and thus confirm the first results of a comparison
with the other large beef breeds (Wilson et al., 1976;
Andersen et al., 1976; Anonymous, 1976). Despite their longer
body, their calves have exceeded the threshold beyond which the
frequency of dystocic calvings increases. Crossbred Piemontese
calves with a mean birth weight of 42.8 kg have caused many
assisted and dystocic calvings; in contrast to the Chianina
breed this might be due to the fact that they are more compact
(greater muscle development).

French beef breeds (Charolais, Blonde d'Aquitaine and
Limousin) show large variations between breeds with respect
to weight of calf and calving difficulties. The within-breed
variations between types are as important as those between
breeds. Bulls of the breeding type generally produce heavier
calves (especially Charolais) than those of the terminal
crossing type, resulting in a greater number of assisted
calvings; this is opposite to what is known about the effects
of selection on muscle development and growth potential (Vissac
et al., 1972; Ménissier, 1975; Foulley and Ménissier, 1976).
Let us however mention that not only are the numbers of calvings
and of bulls studied in this analysis small, but the bulls of
the breeding type have already been selected on their progeny
(with a high selection pressure) whereas those of the terminal
crossing type were only chosen on performance testing (and with
a low selection intensity, especially in the case of Charolais -
Table 1). These differences in the selection efficiency might
be corrected later by comparison with the National Reference
sire group (Foulley and Gaillard, 1975). With the same type of
bull, Charolais bulls generally produce the heaviest calves and
therefore cause more calving difficulties than the other two
breeds. The weight of crossbred Charolais calves of the
breeding type is similar to that of crossbred Blanc-Bleu Belge
or Chianina calves and their number of assisted calvings is the
same whereas the frequency of dystocic calvings is not as high
as in Chianina bulls. Crossbred Limousin calves are heavier than

crossbred Blonde d'Aquitaine calves contrary to observations
made previously (Foulley et al., 1975; Ménissier, 1975;
Andersen et al., 1976). Here again, the Limousin bulls were
certainly chosen with a higher selection intensity and pressure
than in the case of Blonde d'Aquitaine.

The specialised French lines for crossbreeding (Coopelso
93 and INRA 95) give almost the same results as the French
beef breeds constituting these synthetic lines. The bulls of
the two INRA 95 lines, although producing calves with the same
birth weights, have a different incidence of calving difficulties.
Those of the line containing the most Blonde d'Aquitaine (about
50%) did not cause more calving difficulties than South Devon
bulls, whereas those of the other line (about 25% Blonde
d'Aquitaine) brought about as many calving difficulties as in
the case of Piemontese bulls.

4. CONCLUSION

For the moment, these first results obtained with few
samples (30 calvings/breed) and a restricted number of bulls
(3 bulls/breed) only show the general trends between breeds.
In particular the effects of paternal breeds on calving
difficulties are primarily related to the size (birth weight)
of the calves produced rather than to their conformation. In
crossbreeding with Friesian cows, these paternal breed effects
on the size of the calves have only a repercussion on the
frequency of difficult calvings at mean weight of 42 - 45 kg.

The magnitude of the differences observed between type of
bulls within the breeds confirms the fact that, in comparative
studies and especially in the case of breeds with multiple
selection goals, it is very important to use a suitable sample
of bulls representative of each breed.

REFERENCES

Allen, D. 1977. Ease of calving in beef herds. Winter conference, 10 - 13th
January, Cambridge, British Cattle Breeders Club. Digest number 32,
62-67.

Andersen, B. Bech, Liboriussen, T., Thysen, I., Kousgaard, K., and Buchter,
L. 1976. Crossbreeding experiment with beef and dual-purpose sire
breeds on Danish dairy cows. Livestock Production Science, 3, 227-228.

Anderson, J. and Lindhé, B. 1973. Optimum use of beef semen in dual-purpose
or dairy breeds. Acta Agriculturae Scandinavica, 23, 102-108.

Anonymous, 1976. Germ plasm evaluation programme. US Meat Animal Research
Center, USDA, Clay Center, Nebraska 68933. Progress report number
4, ARS-NC-48, June 1976.

Belic, M. and Ménissier, F. 1968. Etude de quelques facteurs influençant
les difficultés de vêlage en croisement industriel. Annales de
Zootèchnie, 17, 107-142.

Bibé, B., Frebling, J., Ménissier, F. and Vissac, B. 1977 Double muscled
sires for terminal crossing: French experiments. In: 2nd Seminar on
Genetics in the EEC programme of coordination of research on beef
production 'Crossbreeding experiments and strategy of breed utilisation
to increase beef production'. 9-11th February 1976, Verden. CEC
EUR 5492e, pp 72-96.

Cunningham, E.P. 1974. The economic consequences of beef crossing in dual-
purpose or dairy cattle populations. Livestock Production Science,
1, 133-139.

Cunningham, E.P. and McClintock, A.E. 1974. Selection in dual-purpose cattle
populations: Effect of beef crossing and cow replacement. Annales
de Génétique et Sélection Animale, 6, 227-239.

Elsen, J.M. and Mocquot, J.C. 1976. Optimisation du renouvellement des
femelles dans les troupeaux laitiers soumis au croisment terminal.
Annales de Génétique et Sélection animale, 8, 343-356.

Foulley, J.L. and Gaillard, J. 1975. Utilisation du lot témoin dans les
schémas de sélection des taureaux de races à viande pour le croisement
industriel: principes et résultats. Bulletin technique du Département
de Génétique animale INRA No. 19 60-68.

Foulley, J.L. and Ménissier, F. 1976. A few reflections on the genetic
improvement of the calving ability of Limousin breed. Technical
meeting of the International Limousin Council, Oklahoma City-USA,
25 September 1976, 13pp.

Foulley, J.L., Ménissier, F., Gaillard, J. and Nebreda, A.M. 1975.
Aptitudes maternelles des races laitières, mixtes, rustiques et à
viande pour la production de veaux de boucherie par croisement
industriel. Livestock Production Science, 2, 39-49.

Kilkenny, J.B. and Stollard, R.J. 1976. Calf birth weights in beef breeding
herds and in the relationship between birth weight, calf mortality and
calving difficulties. BSAP Winter Meeting 1976. 11pp (Mimeograph).

Ménissier, F. 1975. L'aptitude au vêlage des races à viande françaises:
l'origine des difficultés de vêlage et leur amélioration génétique.
VI Journées du 'Grenier de Theix', l'exploitation du troupeau de vaches
allaitantes, novembre 1973, supplément du bulletin technique, CRVZ
de Theix (octobre 1974), 139-170.
Also in English, in: 'Optimum breeding plans for beef cattle'. Bulletin
Technique du Département de Génétique Animale, INRA, Number 21: 57-102.

Ménissier, F. 1976. Comments on optimisation of cattle breeding schemes:
beef breeds for suckling herds. In 'Optimisation of cattle breeding
schemes' - EEC Seminar on 'Genetics', 26-28th November 1975, Dublin.
Annales de Génétique et Sélection Animale 8, 71-87.

Ménissier, F. and Foulley, J.L. 1975. Amélioration génétique des difficultés
de vêlage pour les taureaux de races à viande sélectionnés en vue du
croisement terminal: détermination des objectifs de sélection et
utilisation planifiée des taureaux (non publié).

Ménissier, F., Vissac, B. and Frebling, J. 1975. Optimum production plans
for milk and meat: specialised beef herds and beef crossing in dairy
herds: 25ème Réunion Annuelle, Fédération Europeenne de Zootechnie,
Copenhague, août 1974. In: 'Optimum breeding Plans for beef cattle',
Bulletin technique du Departement de Génétique Animale, INRA, No.21, 2-56.

Philipsson, J. 1977. Studies on calving difficulty, stillbirth and
associated factors in Swedish cattle breeds. VI Effects of crossbreeding
Acta Agriculturae Scandinavica 27, 58-64.

Smith, G.M., Laster, D.B. and Gregory, K.E. 1976. Characterisation of
biological types of cattle. 1. Dystocia and preweaning growth.
Journal of Animal Science 43, 27-36.

Tayler, J.C. 1976. Beef production in the EEC and the coordination of
research by the Commission of the European Communities, Livestock
Production Science 3, 305-318.

Vissac, B., Molinuevo, H.A. and Ménissier, F. 1972. Note sur l'évolution de la race charolaise sous l'effet de la sélection. 22ème Réunion annuelle, Fédération Européenne de Zootechnie, juillet 1971, Versailles, Annales de Génétique et Sélection Animale, 4, 128-129 (Abstract).

Wilson, A., Willis, M.B. and Davison, C. 1976. Factors affecting calving difficulty and gestation length in cows mated to Chianina bulls and factors affecting the birth weight of their calves. Animal Production 22, 27-34.

DISCUSSION

B. Hoffmann *(West Germany)*

This afternoon practically all speakers used the words 'dystocia' and 'stillbirth', I wonder whether they were all talking about the same thing. I also would like to hear the opinion of our clinicians here; what do they consider to be dystocia, and what is stillbirth? Stillbirth in my opinion has many causes; were all these causes included in the figures we have seen or were we talking about a selected stillbirth only due to calving difficulties? Dr. Politiek clearly related dystocia to sires and even classified bulls as dangerous and recommendable. Couldn't you further specify and simply classify dangerous bulls as bulls having large calves and recommendable bulls as bulls having smaller calves?

R.D. Politiek *(The Netherlands)*

Your question, Dr. Hoffmann, asking for a definition of dystocia and stillbirth is indeed important. Stillbirth in our case means death at birth or within the first 24 hours after parturition due to a heavy calf. In calving difficulties we included also Caesareans and dismembered calves. With this definition of stillbirths, a rather high percentage originates from parturitions in heifers with heavy male calves (48 to 50 kg). The farmer discovers that the calf cannot be born normally and he then calls a veterinarian to make a Caesarean in the hope the calf is still alive. When we talk about bulls as being dangerous or not dangerous in respect to prevention of stillbirth, environmental factors should be eliminated as far as possible. Also light calves with a birth weight of 30 kg or less are better excluded. These calves are quite often stillborn or so weak that they die within 24 hours. There are many approaches in different countries. That is why I suggested more uniformity and clearer classification of our observations in respect to calving difficulties. The combination of parameter, eg birth weight, stillbirth, extended gestation length can give the best estimation for calving difficulties.

B.G. Lowman *(UK)*

Can I make a comment on that last point about excluding stillbirths in light calves from the statistics? I think in a calf mortality study everything that dies should be included.

R.D. Politiek

Yes of course I agree, but when a ranking of bulls on dystocia or calving difficulties is to be made, this special category should be excluded. Of course there are two different aspects, whether we talk about easy calving or giving birth to light calves.

H. Kräusslich *(West Germany)*

I think we are faced with two situations. The first one is when we use experimental data. Here we have to define stillbirth and dystocia. The second situation is when we have to use field data and there we depend on reliable information from the farmer. We have experienced with our farmers that the only useful information is whether the calf was born alive or dead, or whether the veterinarian was called or not. So I think we have to restrict our field data to reliable information. Maybe the farmers are differently educated in the various countries, but in our country it must be very, very simple.

R.D. Politiek

I agree to a large extent with Dr. Kräusslich, but I think our farmers are in general more clever than we think they are. For instance the value of the calf which is expressed in kg live weight delivered, is quite well known by the farmers and it has been shown that the estimated weights of 250 or 300 calves born during progeny testing agrees well with the average weight. When differentiation is only made between two classes of animals it is very difficult to make the right statistical evaluations. It is better to have more classes, perhaps with some mistakes, but a group of 300 to 400 calves gives a good possibility for correct estimations.

H.J. Langholz *(West Germany)*

In our country recording is done through milk recorders. They come to the farm once a month and, of course, the farmer does not remember exactly what happened three or four weeks ago when a cow has calved. Therefore I would say that under these conditions we have to stick to a very simple system, like the one Dr. Kräusslich suggested. If we are going to make it more sophisticated and more intensive we have to change the whole system and maybe the AI services will have to follow up the calvings. Otherwise I think we would not improve the system.

R. Bar-Anan *(Israel)*

It comes out here very clearly, that there at least two different syndromes which may cause the death of the calf. One cause is difficult calving; that means the parturition itself as an effect of pelvic incompatibility. The other cause is the genetic one; the calf is not fit to survive, being too small or genetically unfit. When you combine difficult calving and early mortality in one index you get an estimate of pelvic incompatibility. The genetic correlation between difficult calvings of heifer and cow mates was only 0.6%, that means a difference in the frequency of the factors involved. Therefore we now do routine evaluations of all calvings, separately for heifer and cow mates and daughters. I think all the effects we have discussed concerning the calf, length of pregnancy, etc are contributing factors and not the only ones. Sometimes you get a bull which has big calves and no problems and you don't know why and sometimes you have a bull which has small calves and they die and again you don't know why. So as with many things; when you don't know the etiology, you have to select for the results.

G. Averdunk *(West Germany)*

I have a question to Dr. Bar-Anan. You mentioned the suspicion that there are some genetic effects maybe due to only a few genes. If that is true, shouldn't you find in your large material with older cows some sire by grandsire interactions

supporting this theory, and should you then not only avoid
difficult calving by progeny testing bulls on heifers but also
use it in your selection programme?

R. Bar-Anan

Of course the scheme I have presented has limitations
compared with the scheme of Dr. Philipsson. I said we only
nominate services, but we don't improve genetically. As a
matter of fact we do, since sires which have a very bad effect
are not used as paternal grandsires. Sometimes a very good
bull causes a lot of calving trouble and we don't rear sons from
him. On the other hand, we take bulls which are very good and
very easy on calving and we use a lot of their sons and so I
think the system is also achieving genetic improvement.

B. Hoffmann

We have talked a lot about the mechanical problems of
parturition; that is, how does the size of the calf fit the size,
of the pelvis in the cow. I think we should exclude this from
further discussion. It may be very important but to me it
looks controllable within a population. I would like to mention
another topic, that is crossbreeding and how is gestation length
related to calving problems. Is the calf born too early due
to reduced gestation length or is the calf born too late due to
prolonged gestation? And in this respect the problem of the
early viability of the calf could be further discussed.

H.O. Gravert (West Germany)

I would like to put a question, possibly to Dr. Politiek
and perhaps also to Dr. Philipsson or somebody else. What we
really are interested in is the frequency of stillbirth and
this figure can be obtained fairly easily from field data. It
is much more difficult to get any evaluation of the calving
itself. What is the economic gain in having the more sophisticated registration on ease of calving when compared to the
registration of the frequency of dead calves?

H.J. Langholz

There is a difference; what we are really faced with after calving difficulties - and not so much after the stillborn calf - is, as at least Dr. Philipsson has pointed out, the subsequent (most likely) impaired fertility in the cow. Our information is also very poor about the influence on the dam's milk production and the fattening performance of the calf itself. I think we should at least mention here that we have to differentiate between the kind of live birth and then we can come back to the question of Dr. Gravert.

J. Philipsson *(Sweden)*

I really don't know if I can answer Dr. Gravert's question. In general it very much depends on how your recording system is working; if it is good it may work with just stillbirth rate recorded. The calculation I have done shows that the plan I have outlined is realistic under Swedish conditions. That means that one could decrease the total number of stillborn calves in one generation by about 30%. That is the best figure I can give. As for the economic consequences the calculations I have done show that about half of the costs may be referred to calf losses and another half of costs are due to effects on the dam. We have a high culling rate, poor fertility and so on, which very much affects the economics.

R. Bar-Anan

I didn't give the figures, but I have records from several thousand calvings showing the effect of difficult calving on the performance of the dam afterwards. About 100 kg less milk and about 10% more culling have been observed.

R.D. Politiek

Also in our conditions we observe a reduced non-return rate by about 14% and also more heifers are not presented for first insemination again. So we indeed get important losses after difficult calving. Also the labour costs arising from dystocia have to be calculated; especially when herd size increases this aspect is important. You ask how we can prevent

dystocia. I think the best way would be that farmers having
the possibility to select from a number of proven bulls, chose
bulls for their heifers which are especially selected for easy
calving. The farmer has the trouble of catching the heifers for
AI. In bigger herds till now natural service is more usual
for heifers. If the gain in this category - cost at calving,
better fertility afterwards and a higher breeding value of the
calves - is calculated it is worthwhile to catch the heifers.
Another point is to have available bulls proven on the calving
performance of daughter groups. I think the farmer wants to
use bulls causing less calving problems in first calving heifers.
We should evaluate this total complex and also make some kind
of proposal for an index, meeting the point Dr. Gravert raised
about how to proceed and differentiate. A practical and
efficient system should be chosen. For example Dr. Langholz
mentioned that when the milk recorder comes 2 - 3 weeks after
parturition the farmer no longer recalls the details. So we are
thinking of another, computerised system under the responsibility
of the AI centres, where a computer card is sent out to the
farmer about a fortnight before the expected birth of a calf.
The farmer fills in the birth card; some points are very simple,
like single or twin, sex, born dead or alive, etc. Of course
some points are more subjective, like a score for ease of
calving, estimated weight of calf, but classification and
recording done in such a way in my opinion should yield
sufficient information.

W. Oxender (USA)

What percentage of stillbirth is non-preventable and at
what point do we become alarmed about stillbirth? We have seen
all kinds of figures but I think there is a figure that we
probably cannot fall short of and I wonder if anybody has
information on this. What is the base line when working on
stillbirth?

R. Bar-Anan

As an average figure and base line for perinatal mortality
we take about 7% in heifers and 4% in cows. But it is also

possible to have a bull or sire family which gets down to 4% in heifers. A calf mortality rate of 15% or, after more than 100 calvings, of 12%, following inseminations with one bull is an alarm signal. Our farmers are very keen to get for their heifers the bull which is very good on calving. He doesn't mind so much whether the bull is a little bit better or not on yield; he doesn't mind young bulls, but he wants to be sure to get for his heifer a bull which is proven for easy calving. And if you can come up with a workable breeding scheme through this conference, I think every farmer in this community will thank you.

B. Hoffmann

One question to Dr. Philipsson. You mentioned the maternal ability on stillbirth; could you specify this a little bit more? One question to Dr. Bar-Anan. You mentioned that farmers in Israel specially select bulls for heifers; what is the difference between the calves born from heifers and cows? For example are they of the same size?

R. Bar-Anan

I don't have birth weight figures, because we don't weigh the calves. But when we did weigh progeny groups the differences were only few kilograms at one year of age.

J. Philipsson

The definition of the material ability I have used is simply the capacity to give birth to a living calf. Of course there are lots of reasons for not giving birth to a living calf from the dam's point of view. One of the reasons is difficult calving, but also a poor uterine environment may affect the viability of the calf. The causes can be divided into a maternal and a direct (calf) component.

H.J. Langholz

I want to bring the discussion back to the question of what shall we do, and I see there two perspectives. One is to breed for the trait and the other is to organise the use of the

bulls. The model calculations of Dr. Philipsson, seem to show
that the second approach is more effective. I get the impression
from today's discussion, that there is an overall negative
correlation between calving and fattening performance. And the
question is, especially when we discuss crossbreeding strategies,
to what extent are we able to find sires which are very good
in this trait and also giving us good prospects in fattening
performance and carcass quality? I want to draw attention to
the work done in industrial farming in the German Democratic
Republic (GDR). There they have very large industrial dairy
farms and they are testing Charolais bulls extensively in
crossbred matings. Teams of 10 - 20 performance tested beef
bulls are tested on about 200 calvings for calving performance
and subsequently some of the male offspring are progeny tested
for beef production. The rate of dystocia is not as high as we
see it here, which may be an effect of the lower feeding level
in the GDR. But big differences between bulls from 0.9 up to
30% of dystocia are observed. And the bull we need has about
1.0% of dystocia, almost no change in birth weight and the
highest ranking in performance and progeny testing, the latter
being done under industrial farming conditions with about 25
offspring. I am afraid that with the structure we have in
western countries it is very difficult to introduce such a
testing scheme. Maybe that, if we put our facilities together
in the EEC countries, we could come up with a similar approach
(see Table)?

J.L. Foulley (France)

Since there is not a genetic correlation of unity it is
possible to select for reduced birth weight and simultaneously
for increase in yearling weight. They are not incompatible.
It is theoretically possible to reduce the genetic improvement
on birth weight by half and maintain about 75 or 80% of progress
in yearling weight. I think it is the same thing Dr. Langholz
presented in this specific example.

H.O. Gravert

I think there is another possibility and it certainly will

TABLE

RECENT RESULTS OF TESTING CHAROLAIS BULLS FOR COMMERCIAL CROSSBREEDING ON FRIESIANS UNDER INDUSTRIAL FARMING
CONDITIONS OF NEUBRANDENBURG, GDR (Meisel et al., 1977)

Sire	Dystocia (%)	Birth weight (kg)	Daily gain: Performance test * (g)	Progeny test ** (g)	Carcass weight (kg)	Valuable cuts (%)	Kidney and pelvic fat (kg)	Grade
Charmell	0.9	39.4	1 269	970	251	50.2	5.7	
Chauz	1.1	37.4	1 273	1 106	288	54.8	4.9	I
Challer	1.3	39.4	1 322	1 154	301	57.3	5.9	E
Charta	2.0	39.2	1 130	1 045	279	58.1	5.7	II
Chassi	3.0	38.8	1 224	1 057	296	56.8	4.5	
Chamann	4.5	42.3	1 205	1 055	279	54.8	4.9	II
Chamo	4.6	38.0	1 211	978	260	50.9	5.5	
Chaplin	6.6	41.4	1 223	1 073	279	58.4	6.2	E
Charibo	12.9	43.3	1 100	1 008	260	50.3	7.1	

*   180 - 365 days

** 50 - 450 days.

be discussed tomorrow. Like the Belgian farmers who
deliberately use and accept more Caesareans, some farmers might
also make more use of induced parturition. If this is done we
will find that the regression of weight on gestation length is
higher than from our field data; it is about 500 g per day, ie
the weight gained at the end of gestation. This means that if
parturition is induced about 10 days before the normal end of
gestation, birth weight is about 5 kg less. We can solve a lot
of problems with this system. The only trouble is that there
is no way to estimate birth weight before the calf is born.
If we could estimate the calf weight within the cow, we might
be able to avoid having too light and weak calves.

M. Bosc *(France)*

I think that there was some work done in France on this
point several years ago. It was shown that there is a
correlation of about 0.8 between the total oestrogen level
in the maternal blood at Day 220 of pregnancy and the weight of
the calf at the time of calving. This was done by Terqui and
Ménissier but I don't know of any attempts to use this method
for estimation of the weight of calf at parturition.

H.J. Langholz

I want to add some information from our crossbreeding trial
relating to the suggestion of Dr. Gravert. Calves classified
by the farmer as heavy at birth were carried 14 days longer
than the normal gestation length of the dam breed and, similarly,
calves for which the farmers called for a veterinarian for
assistance were carried 17 days longer, while calves born after
a Caesarean were carried 25 days longer on average. So it might
perhaps be advisable to terminate pregnancy at the average
gestation length of the breed to solve at least part of the
problem.

I.L. Mason *(Italy)*

In England the Aberdeen-Angus is used on Friesian heifers
to avoid difficult calvings and this is one of the reasons
which keeps the Aberdeen-Angus breed alive. This coincides with

what Dr. Bar-Anan said - farmers are concerned with easy
calvings. It strikes me that this also is the way to get easy
calving and a lot of beef. This gives more support for
Cunningham's system of selecting a dual-purpose breed entirely
for milk yield and crossing it with a beef breed for beef
production. Now we must modify that and say selection for
milk yield and easy calving - and these two things seem to be
correlated. That at least would solve half of the problem but
it still doesn't solve the problem of how we are going to deal
with the calving difficulties in beef heifers.

R. Bar-Anan

Mr. Mason, you said that there are two ways to proceed.
One is to take the heifer, which is the latest generation of
genetic improvement, and put it to an Angus bull. This gives
a small calf which according to the Milk Marketing Board
figures is of little value. The second is to take a bull proven
for easy calving which gives about the same percentage of calf
mortality. In England the figure of 7% dead calves from crossing with Angus bulls given by the Milk Marketing Board is
exactly the same as our 7% from heifers bred with proven bulls.
Putting proven bulls on young heifers means to make the best of
the latest generation, and thus to achieve a high rate of genetic
improvement. On the other hand, the old cow not needed for
replacement, is put to a Charolais bull. I would like to stress,
that this is also a procedure in beef breeding. We have a small
beef breeding system in Israel and of course we have the problem
of the dam breed and of the top-cross sire breed. We produced
a crossed breed of Simmental-Hereford origin. It also has some
other blood in it, but we call it Simford and in size it is not
so big, between the Simmental and the Hereford. We use Sinford
bulls from the first two calvings and the big old cow is used
for top crosses. That means the young cow for replacement and
the old cows for crossing as Mr. Mason suggested.

R.D. Politiek

I think Mr. Mason has come up with most interesting
point. We should have a sire line for beef crossing suitable

for heifers and perhaps a sire line for beef crossing for
older cows as we proposed earlier. Following the proposal to
use Aberdeen-Angus bulls for Friesian heifers at least in Holland
the farmers have no trouble with calving difficulties but
have trouble with the value of the calf. That is also a
practical point. In my opinion the point also stressed by Dr.
Langholz, that was to look for a real good sire line for beef
crossing for young cows and a sire line for older cows, is
the kind of project we should try to tackle together. In the
EEC countries a lot of effort is put into breed comparisons.
We should come up with a joint programme to select the very
best bulls within breeding lines. It should be remembered that,
having found some bulls which are really good in performance
for beef production and at the same time give very few calving
difficulties, about 25 000 semen doses can be obtained from
one bull per year. Beef crossing on dairy breeds in Europe
could be made more effective. I think we should also use the
data from the beef breeds in France and Italy, because we gain
quite a lot of important information when data are collected
in the right way. We should try to combine our efforts in a
combined project, covering the different environments from the
Scandinavian countries to France and Italy.

G. Averdunk

Is information already available, maybe in France, on the
ranking of bulls for stillbirth or dystocia in different breeds?
Are there sire by breed-of-dam interactions?

F. Ménissier *(France)*

The difference between bulls concerning calving difficulties
is large. From the genetic point of view it seems very difficult
to reduce calving difficulties and simultaneously increase the
beef value of the cow. There is also the question, is it
necessary to reduce the birth weight of the calf? For example
more calving difficulties are observed when Jersey bulls are
used on Friesian dams than when the Friesian bull is used on
the Jersey dam. Likewise there are more calving difficulties
when Limousin bulls are used on Maine-Anjou dams than when

Maine-Anjou bulls are used on Limousin dams; the birth weight
of the Limousin is 47 kg and for the Maine-Anjou it is about
50 kg.

M.J. Drennan (Ireland)

In recent years, there has been increased usage of
Continental type bulls such as the Charolais particularly on
beef cows both in Ireland and the UK. As expected this has
resulted in an increased incidence of calving difficulties.
While farmers wish to avail themselves of the increased
growth potential of the progeny of Continental type sires
relative to Herefords the increase in calving problems is not
acceptable. It is well documented that there is a tremendous
variation between individual bulls within any one breed in the
incidence of calving problems and therefore the tendency is to
select bulls known to give rise to a low incidence of dystocia.
The question is, by culling for calving problems are you also
culling the bulls with greatest growth potential? What is the
relationship between these two factors?

H. Kräusslich

All the figures have shown that there is a very strong
genetic correlation between the weight of a calf and the
grc h rate but this correlation, as already mentioned by Dr.
Fou ey, is not unity. So it is possible to find bulls with a
higⅉ growth rate and which produce small calves. I think this
is the only solution. As Dr. Bar-Anan suggested, there may
be major genes involved and if this is true we should only look
for exceptional bulls and not worry about heritabilities and
genetic correlations. If it holds generally as Dr. Bar-Anan
found, that there is a bimodal distribution, then all our
genetic correlations and heritability estimates are wrong from
the statistical point of view, because we assume a normal
distribution.

R. Bar-Anan

Very true. We speak of heritability, but these are not
heritabilities in the sense we are used to; because as Dr. Hanset

and I have shown, we changed the population and hence the heritabilities quite significantly from one year to the other. The concept of heritability describing the genetic proportion in the phenotype seems not very suitable for all-or-none characters. The genetic correlation changes, depending on the particular bull or group of animals. Two bulls were imported into Israel from the USA 15 years ago. One was the son of Wisleader Ideal and the other of Winterthur Zeus, two famous American bulls at that time. They were two individuals within the American Holstein population, having in common the black and white colour, but as to ease of heifer calving, these two bulls belonged to two different breeds. Therefore with respect to calving characteristics estimates seem valid for one year only. In a short time the population can be quite different. Dr. Politiek and I both suggested developing within a breed one line for growth and another line for easy calving, one for breeding cows and the other for breeding heifers.

R.D. Politiek

Is it worthwhile to get a small working group together, just to call on ideas and to get a kind of guide line? Is there any support for this idea?

R. Hanset *(Belgium)*

I agree with Dr. Politiek. There is a need, perhaps an emergency, to elaborate a system of recording, which could be universal. It is perhaps a dream but at least in Western Europe the systems of collecting data should be understandable from one country to the other.

B. Hoffmann

Perhaps we can postpone this point till Friday when it will be a central topic in the panel discussion with the session chairmen.

I.L. Mason

Why do male calves give more calving difficulties than females even of the same weight?

G. Averdunk

Perhaps it is due to the fact that more males than
females survive to parturition; in many populations you will
not find a 50 to 50 sex ratio at birth.

PHYSIOLOGICAL ASPECTS OF PARTURITION

Chairman: M. Bosc

HORMONAL MECHANISMS INVOLVED IN CONTROL OF PARTURITION IN THE COW

B. Hoffmann, J. Schmidt and E. Schallenberger.

Institut für Physiologie der Südd. Versuchs- und
Forschungsanstalt für Milchwirtschaft, Technische Universität
München-Weihenstephan, D805 Freising, BRD.

ABSTRACT

*In the cow, proper maintenance of pregnancy throughout the whole
length of gestation is bound up with corpus luteum (CL) function. The
occurrence of parturition is first indicated by a decline of progesterone
values in peripheral plasma, starting 30 - 40 hours before expulsion of
the foetus. Within 20 hours levels decline from 5 - 10 ng/ml to
concentrations below 1 ng/ml.*

*Similarly artificial elimination of the CL in the last two to three
months of gestation will lead to parturition. It has been further demonstrated that in the cow the placenta seems to be an effective barrier
for hormones between the foetal and maternal compartments; thus oestrone is
the main maternal oestrogen while in the foetus it is oestradiol-17a.
Conjugated oestrogens may range between 50 and 100 ng/ml plasma, exceeding
free oestrogens more than 100 times. Progesterone is constantly low in
the foetus and not correlated with maternal levels. Nevertheless the
signal for onset of parturition — and hence luteolysis — seems to originate
in the foetus. In this respect much attention has been given to the
functional status of the foetal pituitary-adrenal axis. Yet it is unlikely
that the direct trigger of luteolysis is the final increased output of
foetal corticoids. Firstly it is not reflected in the maternal circulation
probably due to the restricted placental permeability, and secondly under
in vitro conditions progesterone synthesis in CL from late pregnant animals
is not affected by addition of corticoids. Whether prostaglandins act as
the mediator between foetal corticoid release and luteolysis is still open;
we could observe no effect of PgF_2a on progesterone synthesis in vitro and
in vivo, prostaglandin release seems to coincide with rather than to
precede luteolysis. Further we could show that, due to the depletion of
progesterone depots especially in fat (up to 500 ng/g in the late pregnant*

animal), peripheral plasma values of progesterone probably do not allow definite conclusions on the actual stage of cessation of CL function. In the cow it is also unlikely that the final oestrogen increase observed before parturition is a result of an enhanced conversion of progesterone produced by the CL into oestrogens, since this increase is not altered after ovariectomy. From all observations it may be concluded that in cattle the maternal compartment is basically ready for parturition during the last three months of gestation. So far the exact nature of the signal coming from the foetus and triggering parturition, as well as possible receptor sites in the maternal compartment, are not yet known.

1. INTRODUCTION

Control of parturition in man and domestic animals has been chosen as the main or one of the main topics of several recent international meetings (Ciba Foundation Symposium No. 47, May 1976, "The Foetus and Birth"; VIIth Int. Congr. Animal Reprod. Art. Insem. Krakow, July 1976; 9th Annual Meeting Society for the Study of Reproduction, Philadelphia, August 1976) and in respect to cattle the data available up to this point have also been carefully reviewed (Bosc, 1977; Hoffmann et al., 1977; Thorburn et al., 1977). Thus the purpose of this paper is not to add another review but to survey the available information and to discuss some of the observations in the light of the latest results.

2. THE TWO COMPARTMENTS

One fact that has clearly emerged from experiments performed in recent years is that in the bovine species the dam and the foetus represent two distinct and separate compartments. It has also been demonstrated that within each compartment, depending on the sampling site, differences in hormone level can be measured.

Dam

In the dam quite a uniform picture has been obtained for the hormonal profiles of progesterone, corticoids, oestrogens, prolactin, LH and FSH in the peri-partal phase (for review of literature see Hoffmann et al., 1977). Concerning the free unconjugated hormones the information is summarised in Figure 1. As is shown the decline of progesterone beginning 30 - 40 hours prior to expulsion of the foetus coincides with an increase in corticoid levels. The quotient of cortisol to corticosterone was rather constant around 7.3, indicating that no shift in the corticoid production had occurred. The amount of conjugated cortisol was generally less than 5% of free cortisol while conjugated corticosterone represented 10 - 20% of the amount

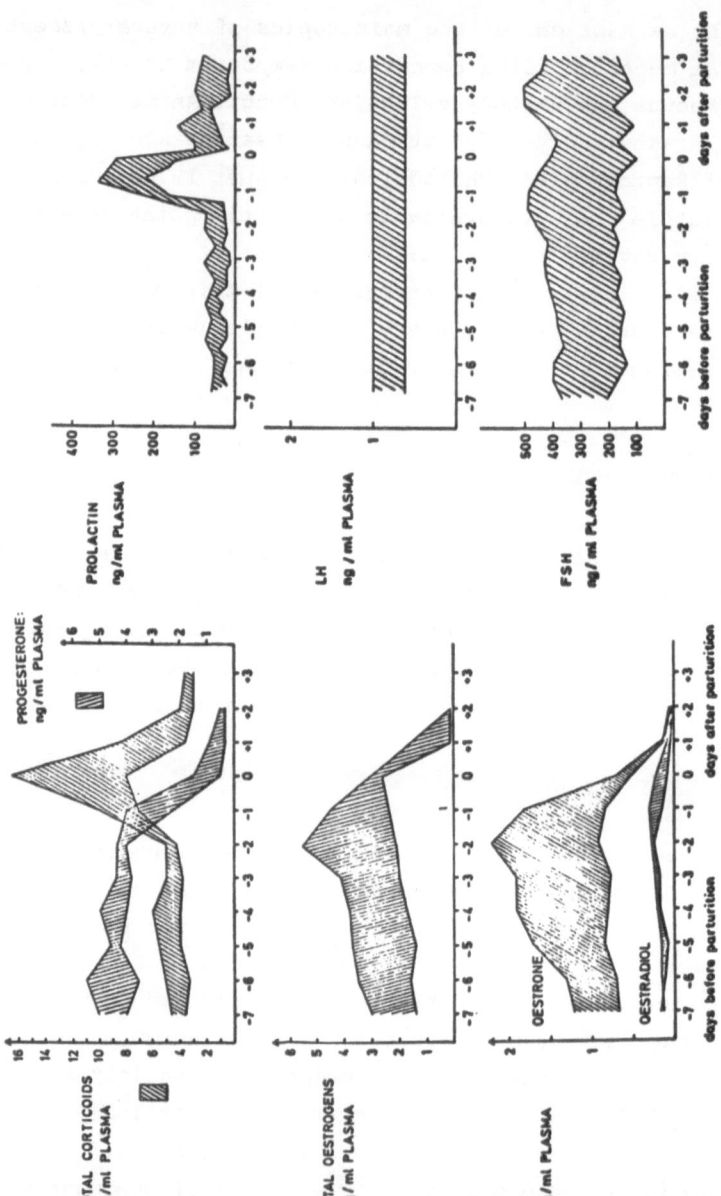

Fig. 1. Scheme of hormonal changes occurring around the time of parturition in jugular vein plasma in the cow (shaded areas indicate range of values) (taken from Hoffman et al., 1977).

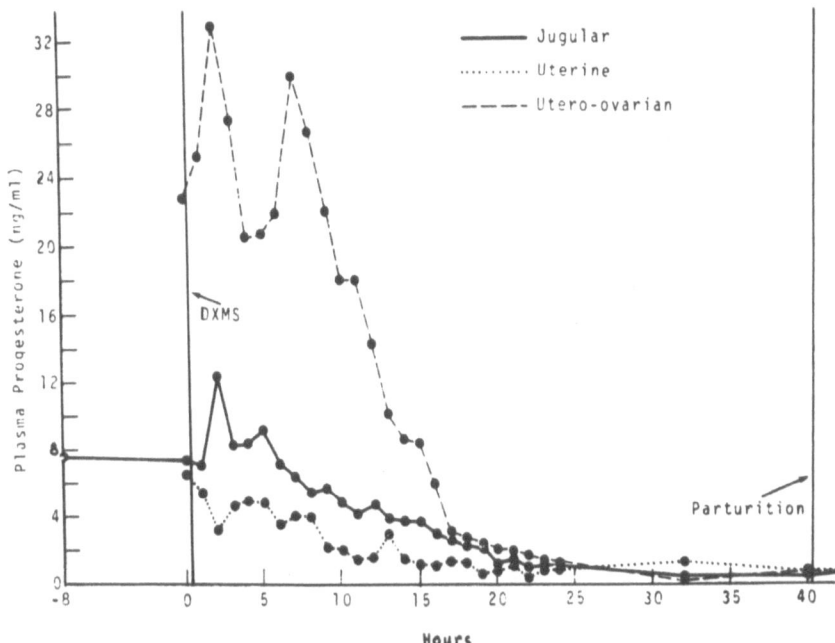

Fig. 2. Plasma progesterone (ng/ml) in jugular, utero-ovarian and uterine veins of four cows treated with dexamethasone (DXMS) (taken from Evans and Wagner, 1976).

of free corticosterone. No differences between the sampling sites (jugular versus uterine vein) were observed (Hoffman et al., 1976), whereas progesterone concentrations were highest in the utero-ovarian outflow and lowest in the uterine vein, as is shown in Figure 2. Oestrogen levels increased in the last three months of gestation with peak levels just before or at parturition. In general levels were higher in the uterine vein than in the jugular vein, where, in contrast to the uterine vein, short-term changes in oestrogen production are little if at all reflected. The dominating oestrogen is oestrone and, except for oestradiol-17β, conjugated oestrogens by far exceed the free oestrogens. These observations are illustrated for one example in Table 1.

In contrast to prolactin, which exhibits a characteristic pre-partal peak (Schams and Karg, 1969), LH and FSH levels in peripheral plasma show no significant changes around parturition. Similarly bovine placental lactogen levels remained at the

TABLE 1

CONCENTRATIONS (ng/ml) OF FREE AND CONJUGATED OESTROGENS IN JUGULAR AND UTERINE VEIN PLASMA OF ONE COW

Oestrogen		Vein	Day of pregnancy			
			270	271	272	273[1])
Free	Oestrone	Jugular	1.41	1.52	1.65	1.11
		Uterine	1.48	3.78	5.12	6.10
	Oestradiol-17α	Jugular	0.05	0.13	0.11	0.11
		Uterine	0.17	0.16	0.19	0.26
	Oestradiol-17β	Jugular	0.02	0.05	0.07	0.09
		Uterine	nd[2])	nd[2])	0.03	0.12
Conjugated	Oestrone	Jugular	18.35	19.73	17.68	20.34
		Uterine	20.92	26.85	35.72	42.32
	Oestradiol-17λ	Jugular	4.10	6.36	5.50	4.18
		Uterine	4.90	7.39	4.48	—
	Oestradiol-17β	Jugular	0.87	0.24	0.29	0.26
		Uterine	0.05	0.69	0.41	0.32

[1]) Day of premature parturition
[2]) Not detectable

plateau observed during the last three months of gestation until parturition (Bolander et al., 1976).

In respect to prostaglandin F, Fairclough et al. (1975) did not observe consistent changes until 48 - 72 hours before term. The final measurable increase always coincided with the beginning of luteolysis and was especially obvious during the final phase of labour when progesterone had already reached basal levels. Edqvist et al. (1976), who determined the prostaglandin F metabolite 15-keto-13, 14-dihydro-PGF$_{2\alpha}$ made very similar observations. However, in some of their observations prostaglandin levels were already elevated before the immediate decline of progesterone.

Foetus

Though still more information is necessary, considerable knowledge has been gained recently about hormonal profiles in the foetus during the peri-partal phase (for review, see Hoffmann et al., 1976). Remarkable changes concern especially the corticoid levels which increased from around 10 ng/ml about 10 days before parturition to values up to 100 ng/ml at term (Comline et al., 1974; Hoffman et al., 1976; Hunter et al., 1977). We could further show that during this time the quotient cortisol/corticosterone changes significantly from 10 to about 70, indicating a positive shift from corticosterone to cortisol production. The corticoid levels so far established did not differ within the foetal compartment (sampling sites: umbilical artery and vein, saphenous vein), whereas oestrogens were higher in the umbilical vein than in the umbilical artery, as shown in Figure 3. This figure also demonstrates that conjugated oestrogens exceed the free ones and that in contrast to the dam oestradiol-17α is the dominant oestrogen. As also becomes obvious from Table 2 oestrogen concentrations increase significantly in the foetus towards the end of gestation and conjugated oestradiol-17α can reach concentrations up to 100 ng/ml. In contrast to the dam and to other species, progesterone is constantly low in the bovine foetus.

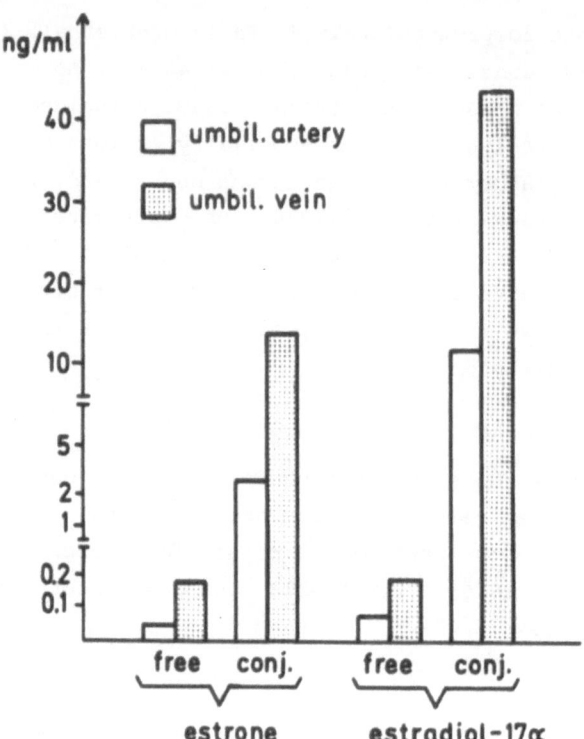

Fig. 3. Differences of oestrogen concentrations in umbilical circulation.
(Values based on five paired samples from four animals).

No special information is available so far about pituitary
hormones in the blood of the bovine foetus during the peri-
partal phase.

Role of the placenta

The situation described so far for the dam and the foetus
clearly demonstrates that the placenta acts as a barrier between
the maternal and foetal compartments. Figure 4 shows significant
differences in the hormone concentrations between the two
compartments. The dynamic changes illustrated in Figure 5
indicate very little, if any, transfer of steroids across the
placenta.

TABLE 2

CONCENTRATIONS (ng/ml) OF FREE AND CONJUGATED OESTROGENS IN UMBILICAL VEIN PLASMA OF ONE COW

Day of Pregnancy	Oestrone Free	Conj.	Oestradiol-17β Free	Conj.	Oestradiol-17α Free	Conj.
270	0.10	19.88	nd[2])	0.50	0.15	46.12
271	0.08	57.68	0.01	0.33	0.12	94.07
272	0.34	80.40	nd[2])	0.44	0.29	113.93
273[1])	0.55	27.97	0.03	0.29	2.09	101.97

[1]) Day of premature parturition

[2]) Not detectable

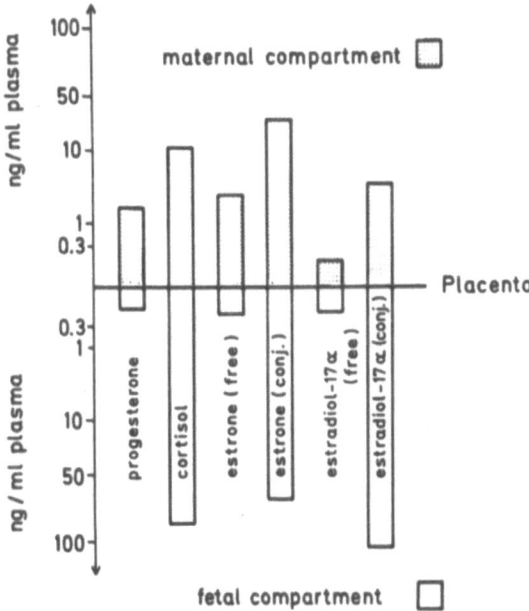

Fig. 4. Differences in steroid hormone levels in the foetal and maternal compartments (cow D, mean values from day 272/273 of gestation)

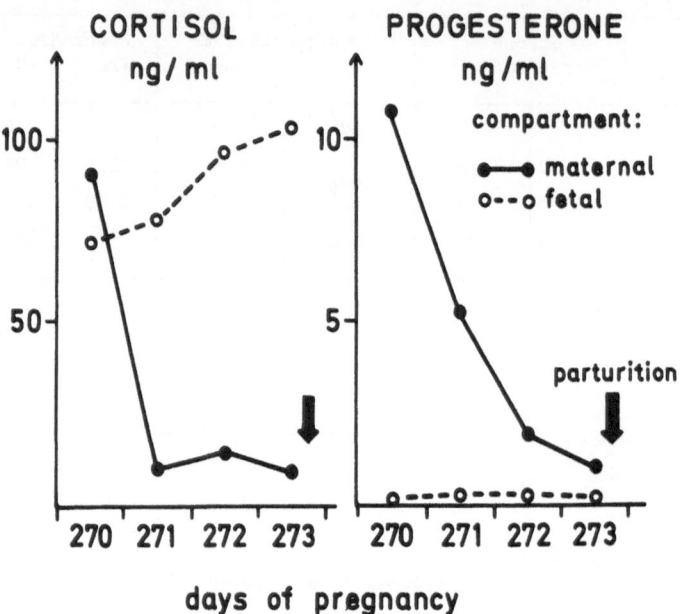

Fig. 5. Differences in dynamics of hormonal changes in plasma between
the foetal and maternal compartments, illustrated for cortisol
and progesterone (cow D)

A specific function must be attributed to the placenta in
respect to oestrogen production. The higher oestrogen
concentrations in the placental outflow (uterine vein and
umbilical vein compared with jugular vein and umbilical artery)
clearly point to the placenta as the main source of oestrogen
production in the pregnant cow. Based on this and the qualitative
differences observed in respect to oestrogens between the foetal
and maternal compartment (see Tables 1 and 2) it was postulated
that the placenta itself, as it synthesises oestrogens, also
regulates the type and amount secreted into each compartment
(Hoffmann et al., 1976).

3. HORMONAL FUNCTIONS AND INTERACTIONS IN RESPECT TO LUTEOLYSIS
   AND ONSET OF PARTURITION

In spite of the fact that the attempts to further
elucidate the hormonal control of parturition have yielded a
respectable amount of data, many questions concerning function
and mechanisms of action still can not be answered.

The only fact established is that in the cow proper
maintenance of gestation depends on a functioning, progesterone -
producing corpus luteum (Hoffmann and Karg, 1974; Wagner et al.,
1974). Elimination of corpus luteum function (mechanically or
biochemically, ie treatment with a high dose of corticoid or
prostaglandin F - like acting compounds) during the last 2
months of gestation will always lead to parturition. This
effect can be overcome by exogenous application of progesterone
(Jöchle et al., 1972). In respect to this very distinct role of
the corpus luteum, it is quite astonishing that some cows can
maintain pregnancy after elimination of the corpus luteum
between about 190 and 240 days of pregnancy, though dystocia
and other problems are observed, (Estergreen et al., 1967;
Edqvist et al., 1976). We have speculated that during this
period, when the onset of increased oestrogen production has
been observed (Grunert and Ahlers, 1969), the cow may develop
another mechanism to support pregnancy (Hoffmann et al., 1977).
The hypothesis is advanced that the precursor for the oestrogen
production in the placenta, which somehow has to be a gestagenic
compound, may be utilised locally in the uterus. Later on,
after adequate enzyme induction, this mechanism may no longer
be effective.

While the pregnancy - protecting effect of progesterone can
be demonstrated practically until the last day of gestation,
very little is known about the functional role of oestrogens.
Apparently in contrast to the human, the bovine foetus is not
involved in providing precursors for placental oestrogen
production. In experiments where premature parturition was
induced after ovariectomy (retained placenta was observed)

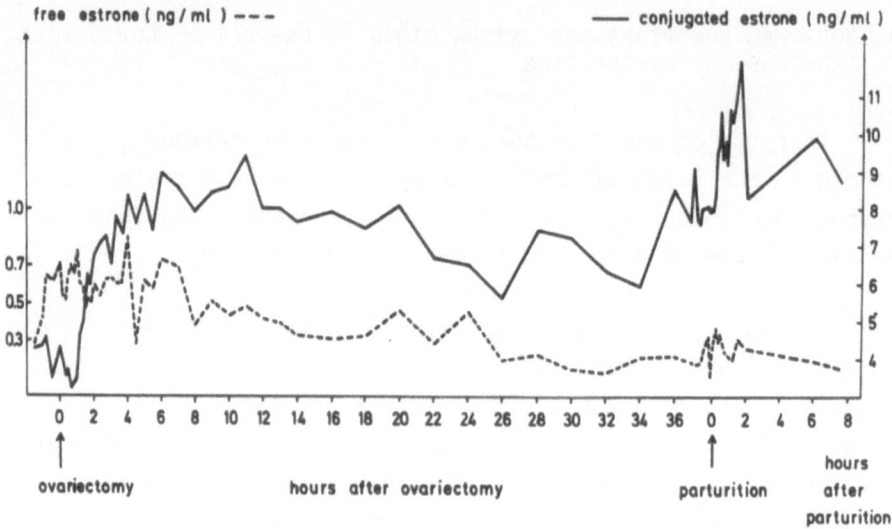

Fig. 6 Jugular plasma oestrone (ng/ml) (free and conjugated oestrone) of
an ovariectomised cow (cow Vl, 268th day of gestation)

oestrogens in the maternal circulation remained high or even
continued to increase after expulsion of the foetus (see
Figure 6). From these experiments it is also obvious that
this final increase in oestrogen production does not depend on
precursors supplied by the ovary. Thus an enhanced conversion
of progesterone into oestrogens, which was speculated to be one
of the mechanisms for luteolysis in the sheep (Flint 1976,
personal communication) can be excluded from further discussion.
In previous papers (Hoffmann et al., 1973; Hoffmann and Karg,
1974) we have discussed the role of oestrogens as a 'permissive'
one. Besides induction of uterine growth the action was seen
in relation to the preparation of the genital tract for
parturition and perhaps to allow certain (not yet known)
hormonal interactions. We still have no clear evidence for
further functional roles. However, it has to be assumed that
at least during the last weeks of gestation a vast surplus of
oestrogen is produced. The fact, that the biologically less
active oestradiol - 17α is secreted, predominantly in conjugated
form into the foetal compartment, may represent a detoxifying
mechanism. Nevertheless according to clinical observations,

calves can be born with symptoms of oestrogenic activity
(Grunert, 1976, personal communication). From observations in
the rat (Gemzell, 1952) we also have speculated (Hoffmann et al.,
1977) that in the foetus the oestrogens may indirectly stimulate
adrenal development.

The importance of an intact pituitary-adrenal axis in the
foetus for adequate initiation of parturition has been well
established, especially for some of the domestic species such
as the pig (Bosc, 1973; Bosc et al., 1974), sheep and goat
(Liggins, 1969; Liggins et al., 1973; Thorburn et al., 1972)
and the cow. In the cow genetically conditioned adrenal
hypoplasia of the foetus is combined with prolonged gestation
and large post-mature foetuses (Holm et al., 1961; Kennedy,
1971). On the other hand stimulation of the foetal adrenal
gland with corticotrophin will lead to premature parturition
(Comline et al., 1974), and it was assumed that this resembles
the changes of the activity of the foetal adrenal gland as
observed under normal physiological conditions (increased
output of corticoids, see above). However, this maturation
process of the foetal adrenal most likely does not provide
the direct trigger for luteolysis since it was shown that
progesterone production was not affected when corpora lutea
from late pregnant animals were incubated in the presence of
high concentrations of corticoid. Also the action of synthetic
corticoids in inducing parturition when given to the dam in
high dosages in the last three months of gestation is considered
to be more pharmacological than physiological in nature
(Hoffmann et al., 1977).

Much attention has recently been given to the action of
prostaglandin F, especially due to its luteolytic action after
administration to the dam. Yet most of the in vivo data
presented so far (Fairclough et al., 1975; Edqvist et al., 1976)
only indicate a coincidence of the pre-partal progesterone
decline with the first significantly elevated prostaglandin
levels. In our opinion this points towards the question "which
came first - the hen or the egg?" and a satisfactory answer

cannot be given at present. Though, especially in the sheep, good evidence has been obtained that prostaglandin $F_2\alpha$ may be the luteolytic compound (reviewed by Thorburn et al., 1977), the information available for the cow is still conflicting. For this species Rao (1976) has demonstrated that progesterone produced in the corpus luteum inhibits the binding of prostaglandin $F_2\alpha$ to luteal cell membranes containing the specific prostaglandin receptors. Similarly we could not demonstrate an effect of prostaglandin $F_2\alpha$ on progesterone synthesis in corpora lutea from late pregnant animals under in vitro conditions (Schmidt et al., 1977). The situation becomes even less clear when the observations of Flower (1977) in the rat that a removal of progesterone triggers the release of prostaglandin F, are transferred to the cow. Thus in the cow the search for the luteolytic compound or complex is still open. From the analytical data presented so far it cannot be questioned that the major prostaglandin F release occurs after luteolysis, possibly inducing contractility of the uterus. Finally some attention should be given to the pituitary hormones. The only hormone shown so far to exhibit drastic changes in peripheral plasma around parturition is prolactin. However, the peak observed seems to be rather a consequence of parturition (possibly a result of stress of labour) than an active mechanism involved (Karg and Schams, 1970, Hoffmann and Karg, 1974). LH and FSH levels in peripheral plasma apparently remain constant throughout the whole of gestation (Hoffmann et al., 1973, 1977). Especially for FSH this is quite remarkable since due to the elevated oestrogen levels during the last three months of gestation some feed back mechanisms with depressed FSH plasma levels as a consequence would be expected. This raises the question whether changes in receptor sensitivity have occurred during that period and whether the plasma levels determined are biologically significant. For LH it can be said that the constant peripheral plasma levels around parturition have provoked the interpretation that lack of LH, the luteotrophic hormone in the cow (Hoffmann et al., 1974), can not be made responsible for luteolysis. However, stimulation of LH release with GnRH shows a much smaller response around parturition than at other periods of time (Schams et al., unpublished data) which may point to an active role of LH in allowing luteolysis. The consequent question is whether the peripheral plasma levels at that time are biologically significant; by no means do they allow conclusions about LH concentration at the target organ and the situation concerning LH-receptor sites. In this respect an interesting hint was obtained by Hoffmann et al. (1974). They showed a possibly increased susceptibility of the corpus luteum to a lack of LH after an extended life span. Other than in cyclic animals progesterone levels already exhibited a rapid drop after one injection of LH antiserum in a hysterectomised heifer carrying a 158 - day - old corpus luteum; three injections completely eliminated corpus luteum function. All these observations call for new experimental approaches to further elucidate mechanisms involved in luteolysis. A new way to monitor corpus luteum function should also be included, since in peripheral plasma the actual stage of corpus luteum regression may be masked due to a depletion of body stores, like fat, where around parturition concentrations of up to 500 ng progesterone/g fat could be measured (Schmidt et al., 1977).

4. CONCLUSIONS IN RESPECT TO POSSIBLE CALVING DIFFICULTIES IN CATTLE BREEDING FROM THE ENDOCRINOLOGICAL POINT OF VIEW

In spite of the many data presented in section 2, more speculations and interpretation than discussions of known inter-actions were presented in section 3. The problems in choosing the right parameters and interpreting the data obtained, make it of course very difficult to establish relations to possible calving difficulties. Basically it seems that the two compartments (dam and foetus) go through a well balanced process of maturation, and parturition is initiated after the breakdown of this balance due to a signal from the foetus at an exact point of time. Evidence has been presented, that the maternal compartment can accept a signal for parturition long before physiological term. Thus, with unchanged receptor sensitivity in the maternal compartment, any hypo - or hyper - function of

the foetal pituitary - adrenal axis where this signal seems
to come from may lead to delayed or premature parturition with
all its consequences. This situation may also occur in a
'subclinical' way, when for example in crossbreeding the
balance of the two compartments is disrupted at an unphysiological
point of time due to the genetic information controlling the
activity of the foetal adrenal gland. At present we do not
see the possibility that the determined endocrine status of
the maternal compartment could cause calving difficulties.
This of course in no way excludes malfunctions in individual
animals which express themselves in distinct clinical symptoms.
The problem of retained placenta will not be discussed further
since it only affects the dam. In summary, much progress has
been made in recent years but further progress is necessary to
fully understand and control the mechanisms regulating
parturition in the cow.

ACKNOWLEDGEMENT

This work was supported by the Deutsche Forschungs-
gemeinschaft.

REFERENCES

Bolander, F.F., Ulberg, L.C. and Fellows, R.E. 1976. Circulating
placental lactogen levels in dairy and beef cattle. Endocrinology
99: 1273-1278.

Bosc, M.J. 1973. Modification de la durée de gestation de la truie par
administration d'ACTH aux foetus. Comptes Rendus de l'Académie des
Sciences, Paris 276: (Série D) 3183-3186.

Bosc, M.J. 1977. Physiology of parturition in domestic animals. Paper
for VIIIth International Congress on Animal Reproduction and Artificial
Insemination, Krakow, July 12-16, 1976.

Bosc, M.J.,du Mesnil du Buisson, F. and Locatelli, A. 1974. Mise en
évidence d'un contrôle foetal de la parturition chez la truie.
Interactions avec la fonction lutéale. Comptes Rendus de l'Académie
des Sciences, Paris. 278: (Série D) 1507-1510.

Comline, R.S., Hall, L.W., Lavelle, R.B., Nathanielsz, P.W. and Silver, M.
1974. Parturition in the cow: endocrine changes in animals with
chronically implanted catheters in the foetal and maternal
circulations. Journal of Endocrinology 63: 451-472.

Edqvist, L.E., Kindahl, H. and Stabenfeldt, G.H. 1976. On the role of
prostaglandins in bovine parturition. Proceedings VIIIth International
Congress on Animal Reproduction and Artificial Insemination, Krakow,
July 12-16, 1976. Vol. III, pp 357-360.

Estergreen, V.L. (Jr), Frost, O.L., Gomez, W.R., Erb, R.E. and Bullard, J.F.
1976. Effect of ovariectomy on pregnancy maintenance and parturition
in dairy cows. Journal of Dairy Science 50: 1293-1295.

Evans, L.E. and Wagner, W.C. 1976. Bovine plasma oestrogens, progesterone
and glucocorticoids during dexamethasone induced parturition. Acta
Endocrinologica 81: 385-397.

Fairclough, R.J., Hunter, J.T. and Welch, R.A.S. 1975. Peripheral plasma
progesterone and uteroovarian prostaglandin F concentration in the
cow around parturition. Prostaglandins 9: 901-914.

Flower, R.J. 1977. The role of prostaglandins in parturition with special
reference to the rat. In: 'The Foetus and Birth' Ciba Foundation
Symposium, No. 47.

Gemzell, C. 1952. Increase in formation and secretion of ACTH in rats
following administration of oestradiol-monobenzoate. Acta Endocrinologica 11: 221-228.

Grunert, E. and Ahlers, D. 1969. Harnöstrogenbestimmung beim Rind zur Diagnose intra-uterin abgestorbener Früchte. Deutsche Tierärztliche Wochenshrift 76: 501-504.

Hoffmann, B. and Karg, H. 1974. Endocrine balance of the cow at parturition. In:'Avortement et parturition provoqués' Rapporteurs généraux M.J. Bosc, R. Palmer et Cl. Sureau. Masson et Cie, Paris, pp. 123-138.

Hoffmann, B., Schams, D., Giménez, T., Ender, M.L., Herrmann, Ch. and Karg, H. 1973. Changes of progesterone, total oestrogens, corticosteroids, prolactin and LH in bovine peripheral plasma around parturition with special reference to the effect of exogenous corticoids and a prolactin inhibitor respectively. Acta Endocrinologica 73: 385-395.

Hoffmann, B., Schams, D., Bopp, R., Ender, M.L., Giménez, T. and Karg, H. 1974. Luteotrophic factors in the cow: evidence for LH rather than prolactin. Journal of Reproduction and Fertility 40: 77-85.

Hoffmann, B., Wagner, W.C. and Giménez, T. 1976. Free and conjugated steroids in maternal and foetal plasma in the cow near term. Biology of Reproduction 15: 126-133.

Hoffmann, B., Wagner, W.C., Rattenberger, E. and Schmidt, J. 1977. Endocrine relationships during late gestation and parturition in the cow. In: 'The Foetus and Birth'. Ciba Foundation Symposium No. 47: 107-125.

Holm, L.W., Parker, H.R. and Galligan, S.J. 1961. Adrenal insufficiency in postmature Holstein calves. American Journal of Obstetrics and Gynaecology 81: 1000-1008.

Hunter, J.T., Fairclough, R.J., Peterson, A.J. and Welch, R.A.S. 1977. Foetal and maternal hormonal changes preceding normal bovine parturition. Acta Endocrinologica 84: 653-663.

Jöchle, W., Esparza, H., Giménez, T. and Hidalgo, M.A. 1972. Inhibition of corticoid-induced parturition by progesterone in cattle: effect on delivery and calf viability. Journal of Reproduction and Fertility, 28: 407-412.

Karg, H. and Schams, D. 1970. Discussion on prolactin levels in bovine blood under different physiological conditions. In:'Lactation' Ed. I.R. Falconer, Butterworths, London pp 141-143.

Karg, H., Hoffmann, B. and Schams, D. 1970. Luteinising hormone, prolactin and progesterone relationships in vivo (data from the cow). Excerpta Medica International Congress Series No. 219; 691-698.

Kennedy, P.C. 1971. Interaction of foetal disease and the onset of labour in cattle and sheep. Federation Proceedings 30: 110-113.

Liggins, G.C. 1969. The foetal role in the initiation of parturition in
    the ewe. In: 'Foetal Anatomy'. Ciba Foundation Symposium, London,
    pp. 218-231.

Liggins, G.C., Fairclough, R.J., Grieves, S.A., Kendall, J.L. and Knox, B.S.
    1973. The mechanism of initiation of parturition in the ewe. Recent
    Progress in Hormone Research 29: 111-159.

Rao, Ch. V. 1976. Inhibition of ^{3}H prostaglandin $F_2\alpha$ binding to its
    receptors by progesterone. Steroids 27: 831-843.

Schams, D. and Karg, H. 1969. Untersuchungen über Prolaktin im Rinderblut
    mit einer radioimmunologischen Bestimmungsmethode. Zentralblatt für
    Veterinär-Medizin, Reihe A. 17: 193-212.

Schmidt, J., Hoffmann, B. and Rattenberger, E. 1977. Role of corpus
    luteum function in the cow as it relates to parturition. Acta
    Endocrinologica (in press).

Thorburn, G.D., Nicol, D.H. Basset, J.M., Shutt, D.A. and Cox, R.I. 1972.
    Parturition in the goat and sheep: changes in corticosteroids,
    progesterone, oestrogens and prostaglandin F. Journal of
    Reproducation and Fertility, Suppl. 16: 61-84.

Thorburn, G.D., Challis, J.R.C. and Currie, W.B. 1977. Control of
    parturition in domestic animals. Biology of Reproduction 16: 18-27.

Wagner, W.C., Thompson, F.N., Evans, L.E. and Molokwu, E.C.I. 1974.
    Hormonal mechanisms controlling parturition. Journal of Animal
    Science (Suppl. 1). 38: 39-57.

THE INFLUENCE OF THE SIRE ON THE OESTROGEN PRODUCTION OF THE BOVINE FOETUS-PLACENTAL UNIT

A. Osinga and W. Hazeleger

Department of Animal Husbandry, Agricultural University,
P.O. Box 338, Wageningen, The Netherlands.

ABSTRACT

Ease of parturition in heifers is limited by the size of the calf and by the available space in the birth canal during parturition. Variation in the ease of birth is mainly explained by the foetal genotype. Selection for ease of birth itself is not very promising for the parameter can not be measured easily and objectively. Selection for a lower birth weight is theoretically quite possible but it is not wanted because of the higher economic value of the heavier calf. Studies on the possibilities for selection for a wider birth canal (at parturition) seem therefore to be worthwhile. The space in the birth canal has been considered in the past to be a purely maternal character which can be improved by selection via progeny tests of sires (Ménissier et al., 1975). The foetal genotype also influences the ease of birth, however. This is probably mediated by oestrogens produced by the foeto-placental unit. High urinary oestrogens in late gestation coincide with easy births, even in the case of high birth weights. These oestrogens may exert a local preparative effect on the birth canal. This hypothesis requires to be studied and if it proves to be true a selection programme could be initiated for high levels of oestrogen in late gestation. There is a need for experimental models to test this hypothesis. It is also necessary to investigate what part of the birth canal (pelvis, cervix or vagina) is the most limiting in the primiparous cow and how this can be measured objectively.

INTRODUCTION

The basic mechanism of parturition, as recently described
by Thorburn et al. (1977), is mainly a hormonal process initiated
by the foetus by means of ACTH, glucocorticoids and oestrogens.
These foetal steroids induce the maternal release of prostaglandins and oxytocin and reduce maternal progesterone
production. The influence of varying quantities of these
hormones on the ease of calving is not yet understood. The
oestrogenic hormone is almost exclusively produced by the
foeto-placental unit in the late pregnant cow. It is produced
in very large quantities (over 30 mg per day - see Figure 1).
In this paper evidence will be presented for the influence of
the foetal genotype on the urinary excretion by the dam. Since
it is also known that the foetal genotype influences the
incidence of dystocia-related[1] stillbirths in heifers (Van
Dieten, 1963; Smidt et al., 1968; Pijnenburg, 1974) the
possible association between the amount of oestrogen produced
by the foeto-placental unit and the incidence of stillbirths
in heifers will be discussed.

URINARY OESTROGEN STUDIES

As described earlier (Osinga, 1970) urinary oestrogen was
estimated as oestrone and oestradiol-17α photometrically in the
survey and studies 1 and 2 (see Table 1). In the later studies
(3 - 6, Osinga, 1976) the oestrogens were determined fluori-
metrically, this being a more sensitive method. The oestrogen/
creatinine ratio (μg/g) proved to be a reliable criterion
(Osinga, 1970) for the comparison of the urinary oestrogen
excretion level between individual cows. The data were not
corrected for procedural losses. The urine samples were drawn
by metal catheter between 09.00 and 17.00 hr.

[1] Dystocia is defined as difficult, abnormal or prolonged birth.

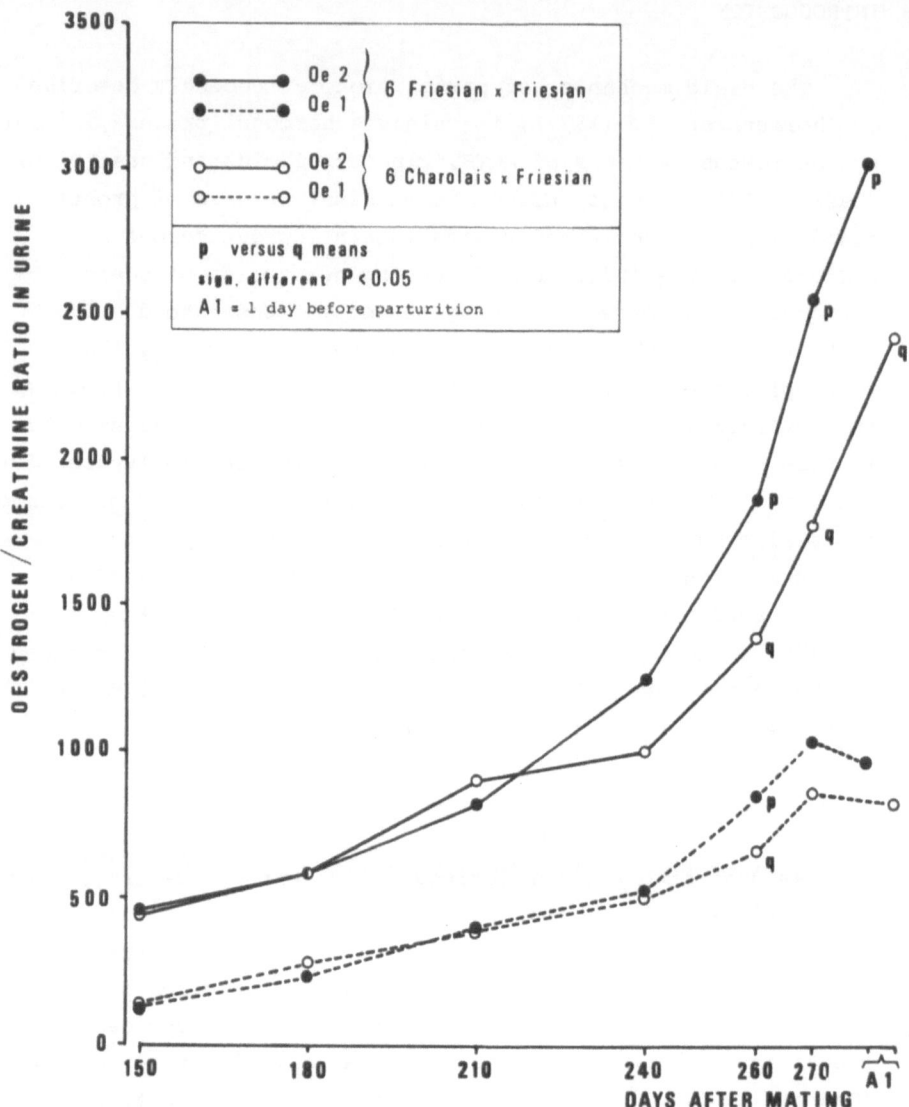

Fig. 1. Urinary oestrogen excretion rate in six Friesian identical twin
pairs, one of each twin being pregnant from a Charolais bull and
the other from a Friesian bull.

285

TABLE 1

SHORT DESCRIPTION OF OESTROGEN STUDIES

Study	Number of animals	Characteristics of animals	Sampling time
Survey	160	3 herds of identical twins	During 2 years
		All 3 main Dutch dairy breeds	2/3 of gestation period
		All twins bred by Friesian bulls	
		1 herd of regular cows	
Study 1	12	6 pairs Dutch Friesian twins	Day 150 to parturition
2	121	MRY-cows, 5 bulls. AI-"De Kempen"	Day 260
3	97	Dutch Friesian AI-Gouda	"
4	64	MRY + Belgian White[1] AI-Hasselt	"
5	50	Holstein x Dutch Friesian, 3 herds	"
6	103	Zebu and Dutch Friesian, Kenya	Day 270

[1] Belgian White are white animals from the Belgian Blue-White breed.

In the survey (Table 1) about 160 cows were sampled frequently during second and third three months of gestation. The oestrogen excretion increased with the stage of gestation (see Figure 1). A large variation was found to exist between animals and this variation was found to be associated with herd, month of calving and birth weight. The level was higher in twin pregnancies and in MRY cows pregnant by a Friesian bull than in cows of either breed carrying a purebred single foetus. After parturition the urinary oestrogen level dropped to almost zero within two days. This drop was also observed in a cow in which the foetus proved to be mummified.

The oestrogen excretion curves are quite similar in identical twin pairs if mated to the same sire and calving in the same season. The sex of the foetus did not affect oestrogen excretion by the dam. The correlation between birth weight and oestrogen level is positive, but low ($r = 0.08 - 0.10$).

In the first study (Table 1, Figure 1) six of the Friesian twin pairs, used in the survey study, were used again in the following pregnancy. One of each pair was now inseminated with Charolais semen and the other with Friesian semen (9 bulls used in total). The gestation period of the crosses was 6.5 days longer ($P<0.05$) and the birth weight 5 kg higher than those of the purebred calves. The total urinary oestrogen excretion ($Oe_2 + Oe_1$) from 260 days of gestation to parturition and the 17α-oestradiol alone (Oe_2) were significantly lower in the cows with the crossbred foetuses than in the dams with the pure Friesian foetuses (Figure 1). Oestrone excretion was significantly different at 260 days only.

Significant sire effects on oestrogen levels were found in study 2 (Tables 1 and 3) with five selected sires in the MRY breed. The sires were selected for the frequency of stillbirths in their offspring, born from heifers. These figures, obtained from the birth registration in their first year of insemination, are presented in Table 2, together with the mean estimated birth weight and gestation period of male calves, born out of multi-parous dams.

The practice of crossing Dutch dairy breeds with beef bulls (Charolais), Holstein bulls or bulls of faster growing dual-purpose breeds (MRY on Friesian cows or Belgian Whites on MRY cows) provided us with a good opportunity for further studies on the effect of foetal genotype on urinary oestrogen excretion by the dam in late gestation (Osinga, 1976).

Study 3 was performed with material from one AI district (Gouda). Analysis of variance showed a significant effect of foetal genotype on the urinary oestrogen level, though only the difference between the Charolais cross and pure Friesians was significant for oestradiol-17α while the oestrone excretion

was different between the double-muscled genotype and the MRY
x Friesian cross (P< 0.05; Tukey test according to de Jonge,
1963).

TABLE 2

STILLBIRTH FREQUENCY, MEAN BIRTH WEIGHT AND GESTATION LENGTH OF FIVE
SELECTED SIRES USED IN STUDY 2 IN THEIR FIRST YEAR OF SERVICE

Sire:	1	2	3	4	5
Stillbirth frequency in heifers (%)	3.0	5.5	9.1	20.1	33.4
Mean birth weight from multiparae (male, kg)	39.3	39.8	42.1	41.6	41.2
Mean gestation period from multiparae (male, days)	278.3	278.8	279.9	280.3	284.2

Study 4 concerned three foetal genotypes based on the
pure Belgian White, the pure MRY and the cross between Belgian
White bulls and MRY cows in the AI district of Hasselt in
Belgium where both breeds are kept commercially and the cross
is employed to improve beef characters in the dairy breed. The
cross is almost intermediate for oestrone and significantly
higher for oestradiol-17α than the pure paternal genotype
(Table 3).

Study 5 (Table 1) showed a higher oestrone excretion
(P< 0.05) in the dams, pregnant from a Holstein bull, than in
the pure Dutch genotype; oestradiol-17α was not affected.

Heterosis effects seem to occur in urinary oestrogen
excretion as is shown in study 6 (Table 3). This concerns the
purebred Zebu breeds Sahiwal and Boran of Kenya and their
crosses with Dutch Friesian bulls, all sampled at 270 days of
gestation. Dutch Friesian cows (imported) were also kept at
the same location and were sampled in the same season as the
Zebu cows. Dystocia and stillbirths are not common in these
Zebu breeds.

TABLE 3

URINARY OESTROGEN EXCRETION RATES IN COWS CARRYING FOETUSES OF DIFFERENT GENOTYPES

	Foetal genotype	Oe2 [a]	Oe1 [b]	n
Study 2				
5 MRY bulls x MRY cows	sire 1	1150 q	616 p	25
day 260	sire 2	1543 p	647 p	25
	sire 3	1297 pq	662 p	23
	sire 4	1098 q	486 p	25
	sire 5	1057 q	474 p	23
	F - value	3.56**	2.78*	
Study 3				
day 260	F x F	3383p	1120pq	33
	CH x F	2428q	921pq	27
	MRY x F	2911pq	1195p	17
	DM x F	2615pq	858q	20
	F - value	3.95**	3.30**	
Study 4				
day 260	MRY x MRY	2653p	1321p	22
	W x W	2147q	1013q	21
	W x MRY	2538p	1168pq	21
Study 5				
day 260	H x F	2804p	1354p	24
	F x F	2827p	1032q	26
Study 6				
day 270	F x F	3038p	1227p	35
	S x S	1468q	575q	18
	F x S	2903p	1199p	25
	B x B	1732q	658q	12
	F x B	2836p	1170p	13

(a) = μg oestradiol-17α per g creatinine (b) = μg oestrone per g creatinine

n = number of cows   ** : $P < 0.01$.   * $P < 0.05$

F x F = Friesian bull x Friesian cow   Ch = Charolais, MRY = Meuse - Rhine IJssel breed

DM = 1 double-muscled MRY bull

H = Holstein-Friesian   W = Belgian White   S = Sahiwal

B = Boran

p,q : different letters means difference significant $P < 0.05$.

DISCUSSION

Oestrogen and dystocia-related stillbirths in heifers

All these oestrogen studies provide evidence for the
influence of the sire or the sire's breed (foetal genotype) on
the urinary oestrogen excretion rate of the dam (path A in the
scheme).

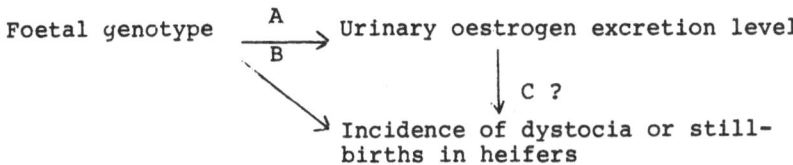

Foetal genotype $\xrightarrow[\text{B}]{\text{A}}$ Urinary oestrogen excretion level

$\downarrow$ C ?

Incidence of dystocia or stillbirths in heifers

Evidence for path B is provided as well (Pijnenburg, 1974).
It is much more difficult to collect evidence for the possible
association between the oestrogen level and the ease of birth
or dystocia-related stillbirth frequency (path C). The main
reason is that dystocia itself is difficult to measure
objectively. That is also the reason why some investigators
look for dystocia-related parameters, like pelvic area, as a
basis for selection instead of dystocia itself (Ménissier et al.,
1975). Scoring the ease of birth by a subjective score (eg 1
= easy, 2 = normal and 3 = difficult) seems to be the best
available method but it requires a large number of calvings to
obtain accurate means. Even the dystocia rate or mean score
for ease of birth in groups of 10 - 25 animals, as obtained in
the oestrogen studies, has only a very limited value and is
therefore omitted from Table 3. The dystocia scores can only
be evaluated within parities, sexes and seasons. Furthermore,
malpresentation of the calf may be a reason for difficult
births and this is expected to have no direct relationship with
the incompatibility of foetal size and available space in the
birth canal.

Some indirect evidence for path C may be obtained from the
low urinary oestrogen levels in the Charolais x Dutch Friesian
genotypes in studies 1 and 3. The Charolais crosses might have

been expected to show higher oestrogen excretion if judged only by the birth weight of the calves.

A number of publications (Milk Marketing Board, 1965; Vet. Clin. Obs. Unit, 1963; Bergström, 1973; Belic and Ménissier, 1968) indicate the occurrence of many more difficult births in dairy cows mated to a Charolais bull than with purebreeding. For the same reason it is not advisable to breed Friesian nulliparae with Charolais bulls.

The stillbirth frequencies of the five MRY sires (Table 2) were derived from 102 - 270 parturitions per sire in heifers and provide a rather accurate estimate of the incidence of dystocia caused by these sires. Evidence for path C may be derived from the low oestrogen levels in sire groups 4 and 5 since sires 4 and 5 gave rather high birth weights and a high incidence of stillbirths.

Offspring of a double-muscled bull will also frequently show dystocia, though literature data are scarce. A low oestrogen excretion is also found in this genotype. No literature data are available on dystocia in the Belgian White x MRY cross. The high oestrone excretion in the Dutch Friesian cows pregnant from a Holstein bull (study 5), may explain at least partly the fact that the dystocia frequency in Dutch Friesian cows is not very much increased if mated to Holstein bulls (Vos et al., 1976; Witt et al., 1971), even though the birth weight increased by 2 - 3 kg.

This indirect evidence for the relationship between urinary oestrogen excretion level and the incidence of dystocia-related stillbirths in heifers (path C) suggests that the oestrogenic hormones originating from the foetal compartment, exert a preparative effect on the birth canal, and that more oestrogens will prepare the mother better for parturition. This hypothesis requires further testing in experiments with groups of heifers where both high and low dystocia frequencies might be anticipated. For these studies it is essential to know what part of the birth canal (pelvic bones, cervix, vagina or vulva) is the most limiting one for an easy birth.

If association between oestrogen levels and dystocia in heifers is indeed confirmed in such studies, selection for high oestrogen levels in late gestation may be expected to contribute to the decrease in dystocia frequency and consequently to reduced stillbirth frequency in heifers.

REFERENCES

Belic, M. and Ménissier, F. 1968. Etude de quelques facteurs influençant
les difficultés du vêlage en croisement industriel. Annales de
Zootechnie 17: 107-142.

Bergström, P.L. 1973. Gebruikskruising voor vleesproduktie bij rundvee.
Rapport B-117. IVO-Zeist.

de Jonge, H. 1963. Inleiding tot de medische statistiek. Leiden.

Ménissier, F., Vissac, B. and Frebling, J. 1975. Optimum breeding plans
for beef cattle. Bulletin Technique du Département de Génétique
Animale, No.21.

Milk Marketing Board, 1965. 'The Charolais trials'. Report of the breeding
and production organisation No.15: 43.

Osinga, A. 1970. Oestrogen excretion by the pregnant bovine and its
relation with some characters of gestation and parturition.
Communications Agricultural University 70: 12.

Osinga, A. 1976. Differences in the level of urinary oestrogens in late
pregnancy between cattle breeds and breed crosses. Proceedings VIIIth
International Congress on Animal Reproduction and AI, Krakow, 1976,
Vol.III, pp 390-393.

Pijnenburg, W. 1974. Doodgeboorte bij MRY-Kalveren. Thesis, Wageningen.

Smidt, D., Dannenberg, K. and Weseloh, E. 1968. Die Bedeutung von
Geburtsvorgang und Laktation für die Fortpflanzungsleistung land-
wirtschaftlicher Nutztiere. Zeitschrift für Tierzüchtung und
Züchtungsbiologie 84: 236.

Thorburn, G.D., Challis, J.R.C. and Currie, W.B. 1977. Control of
parturition in domestic animals. Biology of Reproduction 16: 18-27.

Van Dieten, S.W.J. 1963. Mortaliteit van Kalveren bij de partus à terme
van MRY runderen. Dissertation, Utrecht.

Veterinary Clinical Observation Unit, 1963. The Charolais survey.
Veterinary Record 75: 1045.

Vos, H, Osinga, A. and Ogink, W.D. 1976. Kruisingsproeven met sperma van
Noord-Amerikaanse zwartbonte stieren in Nederland. 4. De vruchtbaar-
heid. Bedrijfsontwikkeling 7: 729-733.

Witt, M., Andreae, U., Kalweit, F.W., Pfeiderer, W.E., Röseler, W. and
Selhausen, D. 1971. Einfluss der Kreuzung von Vatertieren der
Holstein-Friesian-Rasse mit Deutschen Schwartzbunten Kühen auf Körper-
entwicklung, Milchleistung und Muskelbindungsvermögen. Schriftenreihe
Max-Planck Institut. Heft 56.

NEURAL CONTROL OF THE REPRODUCTIVE TRACT IN THE COW AS IT RELATES TO PARTURITION

M. Rüsse

Gynäkologische und Ambulatorische Tierklinik der Universität
München, Königinstraße 12, D-8000 Munich 22, West Germany

ABSTRACT

The female genital tract is innervated by the autonomic nervous system. Moreover, there are sensitive nerve fibres located in the uterine cervix and in the dorsal region of the vagina which regulate the release of oxytocin and the abdominal muscular pressure during labour.

Uterine smooth muscles are innervated by the sympathetic nervous system. The postganglionic sympathetic nerves innervating the myometrium originate from the last thoracic and the first lumbar segments of the spinal cord. They travel via the hypogastric nerve to the pelvic plexus, where the preganglionic nerves synapse with the so-called short postganglionic neurons supplying the uterine muscle. The smooth muscles cells of the uterus contain both α (excitatory) and β (inhibitory) adrenoceptors. Stimulation of the myometrial cells by the postganglionic nerve fibres is mediated by noradrenaline which acts mainly on α-adrenoceptors in the bovine uterus. Stimulation of the β-adrenoceptors is done by adrenaline. However, during pregnancy the noradrenaline storage granules in the postganglionic sympathetic neurons are emptied. This means that the sympathetic neurons can not influence the myometrium at parturition. However, the stimulating ability of the adrenoceptors present at the membrane of the myometrial cells remains unchanged. By this way adrenaline release from the adrenal gland caused by stimulation of the central sympathetic system may inhibit the myometrial contractions and uterine activity during labour. This mechanism plays an important role during the first stage of parturition. Moroever during the second stage the increased oxytocin release which is initiated by the foetus may be inhibited by this system, thus regulating intensity of labour.

The uterus of all mammals so far investigated is innervated by the autonomic nervous system. However, in the cow there is no special report about this innervation. So we have to look at the known situation in other animals and accept that it will be similar for the cow.

The sympathetic pathway innervating the uterus, cervix and vagina leaves the spinal column in the last thoracic and first and second lumbar vertebrae. The nerve fibres pass the sympathetic chain and the anterior mesenteric plexus and travel in the hypogastric nerves to the pelvic plexus. Each sympathetic nerve is composed of a preganglionic neuron and a postganglionic neuron. The cell body of the preganglionic neuron lies in the spinal cord, the cell body of the postganglionic neuron in the sympathetic chain or in one of the outlying sympathetic ganglia. The synapses of the sympathetic pathways to the uterine smooth muscle lie in the pelvic plexus in contrast to all other sympathetic innervation of organs and blood vessels. In this way the uterine smooth muscle is innervated only from so-called short sympathetic postganglionic neurons. This morphological difference leads to the conclusion that this innervation has some special importance, but not at the time of parturition.

The neurotransmitter of the postganglionic sympathetic nerve fibres is norepinephrine. The norepinephrine content of the postganglionic neurons between the uterine smooth muscle cells are demonstrable. Electron micrographs show that small vesicles of norepinephrine are stored in sympathetic nerve endings. Supposedly, nerve impulses cause short periods of increased permeability of the membrane at the fibre ending, thereby allowing sudden release of minute amounts of the neuro transmitter into the surrounding fluids. Ordinarily, the action of norepinephrine secreted into the uterine muscle lasts for only a few seconds, illustrating that its destruction or re-uptake in the nerve endings is rapid. However, norepinephrine and epinephrine secreted into the blood by the adrenal medulla are not removed or destroyed significantly until they diffuse into a tissue. Therefore, when secreted into the blood, these two neurotransmitters remain active for one or more minutes.

This is very important, because the norepinephrine granules in the short postganglionic neurons in the uterus disappear during pregnancy, showing that the influence of the sympathetic peripheral nerves in the uterus is eliminated at term. However, during the last third of pregnancy and during parturition the two neurotransmitters secreted from the adrenal medulla into the blood stream can reach the uterine muscle and stimulate the adrenoceptors of the smooth muscle cells. The amount of epinephrine synthesised in the adrenal medulla is dependent on the amount of corticosteroid synthesised in the adrenal cortex. An increase of corticosteroids increases the conversion of norepinephrine to epinephrine. To summarise these events, sympathetic stimulation before, during and after parturition will increase the output of epinephrine, while during the time before the increase of corticosteroid production in the adrenal cortex the adrenal medulla will synthesise more norepinephrine than epinephrine. During parturition we have to consider that there is no direct influence from the sympathetic innervation of the uterus. However the adrenergic neuro-transmitter norepinephrine and especially epinephrine from the adrenal medulla can influence uterine motility. Epinephrine stimulates the β-adrenoceptors of the smooth muscle cells causing a relaxation. During parturition this relaxing effect of epinephrine in the uterus is stronger than the action of oxytocin and seems to me the direct antoganist to this hormone during the second stage of parturition.

The parasympathetic system, like the sympathetic, has both preganglionic and postganglionic neurons, but, in general, the preganglionic fibres pass uninterrupted to the organ that is to be excited by parasympathetic impulses. In the ligaments of the uterus close to the cervix lie the pelvic ganglia of the parasympathetic system. The preganglionic fibres synapse with these, and then short postganglionic fibres leave the ganglia to spread in the organ. In the genital tract the parasympathetic fibres are only found in the cervix and its vicinity, but not in the myometrium.

The neurotransmitter in the postganglionic parasympathetic neuron is acetylcholine. The acetylcholine secreted by the parasympathetic nerve endings is destroyed by the enzyme cholinesterase, which is present in all the effector organs or the surrounding fluids. The acetylcholine liberated in the cervix and vagina persists for as long as several seconds after its release and therefore has a long period of action. Acetylcholine stimulates the smooth muscle cells. It seems to me that this effect helps to protect the tissue of the cervix from overdistension during labour. However, we have to consider that the uterine smooth muscle layers are not innervated by parasympathetic nerve fibres. Consequently, stimulation of the parasympathetic system has no direct influence on uterine motility before, during and after parturition. However activation of the parasympathetic nerves probably has something to do with the formation of the cervix after parturition.

PRELIMINARY OBSERVATIONS ON MYOMETRIAL ELECTRICAL ACTIVITY BEFORE, DURING AND AFTER PARTURITION IN THE COW

M.A.N. Taverne, G.C. van der Weyden and P. Fontijne.
Clinic for Veterinary Obstetrics, Gynaecology and Artificial Insemination, State University of Utrecht, Yalelaan 7, Uithof, Utrecht, The Netherlands.

ABSTRACT

In eight cows bipolar silver electrodes were implanted on the outside of the uterine wall at different locations, 9 - 30 days before calculated term. In addition, open-end catheters were placed between the uterine wall and the foetal membranes in seven of these animals in order to measure pressure fluctuations. In two cows parturition occurred within 4 days after the operation. In five cows parturition started around calculated term while in one cow parturition was induced with flumethason 30 days after operation. In three of the eight animals an uterine torsion had to be connected during the early state of parturition. During the last week of pregnancy, periods (6 - 30 minutes) of distinct electrical activity (groups of trains) occurred at irregular intervals (15 - 120 minutes) in the distended parts of the pregnant horn. The first sign of impending parturition was the increase in the duration of the individual trains between 24 and 12 hours before birth. Around the same time electrical activity became evident in the undistended parts of the uterus. Regular uterine contractions (tubo-cervical as well as cervico-tubal) occurred only after the expulsive stage had started. During the first hours postpartum the majority of the contractions progressed in a tubo-cervical direction.

Isoxsuprine blocked or partially decreased uterine activity during and after parturition. A gradual decrease in uterine peristalsis was observed during the first 3 days postpartum in animals with a retained placenta, the most frequent contractions occurring at the tubal end of the uterine horn.

Because uterine electrical activity was well correlated with pressure changes at all stages studied, our technique using implantation of electrodes appears to be useful in longitudinal endocrine studies of initiation of parturition, in the pathophysiology of parturition and in the in vivo evaluation of drugs which could influence myometrial activity.

INTRODUCTION

Among the different aspects of the physiology of calving, hormonal changes associated with the initiation of parturition have received the most attention during the last few years (Hoffmann and Karg, 1974; Thorburn, Challis and Currie, 1977). Yet it is not known how the observed endocrine changes are related in time with changes in myometrial activity. Zerobin (1970) inserted catheters between the uterine wall and the foetal membranes in two pregnant cows only 5 days before calculated term in order to measure pressure fluctuations. He found uterine activity increasing only a few hours before birth. Because only long-term studies of myometrial activity during pregnancy and parturition in the cow enable possible correlations with endocrine changes to be observed we applied an electro-physiological method, using bipolar electrodes implanted on the outside of the myometrium of pregnant cows. This technique allowed the recording of myometrial activity for several weeks during the last month of pregnancy and beyond parturition in sheep (Naaktgeboren et al., 1975) and the minipig (Taverne, 1976).

MATERIAL AND METHODS

During lateral laparotomy in 8 cows (Holstein-Friesian and Maas-Rijn-IJssel breeds) four bipolar silver electrodes, embedded in silastic sheet, were sutured on the myometrium at different locations of the uterus. The wires from the electrodes were embedded in silastic tubes and were led to the pelvic cavity. From here they emerged through the left sacrosciatic ligament. They were connected to a gold-lined 8-pin plug. The electrical activity was recorded and analysed according to the techniques described by Naaktgeboren et al. (1975). In seven of these eight cows recordings of intra-uterine pressure fluctuations were performed by insertion of two open-end catheters (outer diameter: 1.5 mm, inner diameter : 0.5 mm) between endometrium and chorion. The catheters also emerged through the left sacrosciatic ligament and were attached to Statham P23b pressure receptors which were connected with electromanometric amplifiers (EMT-311, Elema-

TABLE 1

EXPERIMENTAL ANIMALS

Cow number	Laparotomy (days before term[1])	Interval from laparotomy to parturition	Remarks
2	$\pm$ 30	30	Induced parturition.[2] Isoxsuprine treatment after delivery. Placenta delivered.
3	8	5	After connection of a uterine torsion a dead calf was delivered by foetotomy. Retained placenta.
4	9	7	Normal delivery after connection of a 180° uterine torsion. Isoxsuprine treatment after delivery. Retained placenta.
5	15	6	Normal parturition. Isoxsuprine treatment during and after delivery. Retained placenta.
6	15	11	Normal parturition. Isoxsuprine treatment during and after delivery. Placenta delivered
7	10	4	Normal parturition. Isoxsuprine treatment after delivery. retained placenta
8	10	2	Cervix failed to dilate completely. Dead calf delivered by foetotomy. Retained placenta.
9	9	9	Normal delivery after connection of a 360° uterine torsion. Isoxsuprine treatment during delivery. Placenta delivered.

1) Calculated term 280 days   2) Cortexilar (Syntex)

Schönander). The maximal number of six different signals (four electrodes and two catheters) were registered on a 6-channel recorder (Mingograf 81, Elema-Schönander) with a permanent paper speed of 2.5 mm per minute. Three of these signals could be registered simultaneously on a Cardiostat 3T-recorder (Siemens) with a much lower paper speed. The catheters were continually flushed during recording at a rate of 0.01 ml of physiological saline solution per minute. Recording sessions lasted for at least 2 hours each day while continual recordings were performed around the time of parturition. From three cows (no. 2, 3 and 4) daily blood samples were collected by vein puncture. Plasma progesterone concentrations were determined by radio-immunoassay as described by de Jong et al. (1974).

The day of pregnancy on which the laparotomy was performed, the interval between operation and calving, the course of parturition and experimental remarks are indicated for each animal in Table 1.

During this study, the influence of a single intramuscular injection of 10 - 20 ml duphaspasmin (11.58 mg isoxsuprinelactate per ml, Philips-Duphar, Holland) was studied during and/or after calving. The results have already been discussed elsewhere (Taverne et al., 1976).

RESULTS

Before parturition

In five cows (no.2, 4, 5, 6 and 9) we were able to record uterine activity during the last week of pregnancy. In cow no. 2 only the electrode at the tubal end of the pregnant horn functioned. No distinct electrical activity could be recorded at this place until about 24 h after the peripheral progesterone concentration had reached a value below 1 ng/ml after an intra-muscular injection of 10 mg flumethason (Figure 1A). In the other four cows myometrial activity in the undistended parts of the uterus was absent or only a few single spikes or short trains could be recorded while at the distended areas of the uterine

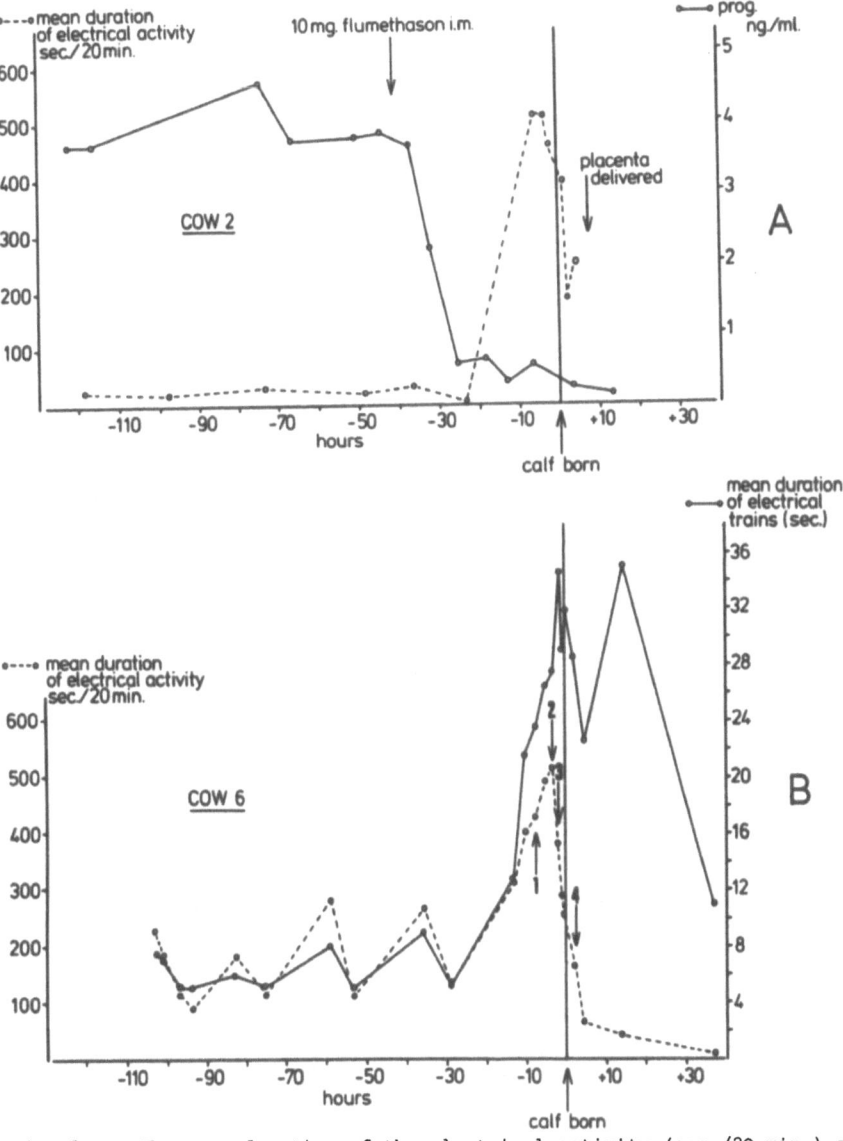

Fig. 1.A. The mean duration of the electrical activity (sec./20 min.) at the tubal end of the pregnant horn and the peripheral plasma progesterone concentration (ng/ml) before and after injection of flumethason.

Fig. 1.B. Changes of the mean duration of the electrical trains (sec.) and the mean duration of total electrical activity/ 20 min around normal parturition.

1. Only the external os of the cervix is still closed
2. The cow starts to strain
3. Isoxsuprine injection when the front legs become visible
4. Isoxsuprine injection postpartum.

wall distinct periods of electrical activity occurred at
irregular intervals (Figure 2A). These active periods lasted
for 6 - 30 minutes and alternated with quieter periods of up
to 120 minutes during which only local, individual spikes or
trains were registered. During the periods of activity, short
trains (5 - 11 seconds) occurred at a frequency of 25 - 60/20
min. At the start of a period of electrical activity the
pressure curves showed a gradual increase of the base line on
which high frequent pressure fluctuations were superimposed
(Figure 2B). At each recording site the electrical activity
was well synchronised with pressure increase. At the end of
the active period, the pressure curves again reached base-line
levels (Figure 2B).

Parturition

Between 24 and 12 hours before the calf was delivered, the
uterine electromyogram (EMG) showed the first changes
characteristic of impending parturition: the mean duration of
the individual trains, especially those between periods of
activity, increased from 5 - 11 seconds to 10 - 25 seconds
(Figure 1B). At this time the non-pregnant horn and the tubal
end of the pregnant horn showed the same type of activity as
the distended parts of the pregnant horn. In the three cows from
which blood samples had been collected, this significant change
occurred after the progesterone concentration had decreased to
levels of 1 ng/ml or less (Figure 1A). During these initial
changes no external signs indicated that the animal was in
parturition. Uterine activity was still similar to that recorded
earlier, that is, a periodic pattern with active phases of about
4 - 10 minutes alternated with short periods (4 - 10 minutes)
of relative inactivity. During dilatation of the cervix, as
diagnosed by vaginal exploration, uterine activity increased
gradually (Figure 1B). During and between the periods of
electrical activity (Figure 3A), individual, long-lasting
contractions became evident. The EMG was synchronised with
pressure fluctuations and contractions were sometimes propagated
in a peristaltic or antiperistaltic direction. In the 3 cows

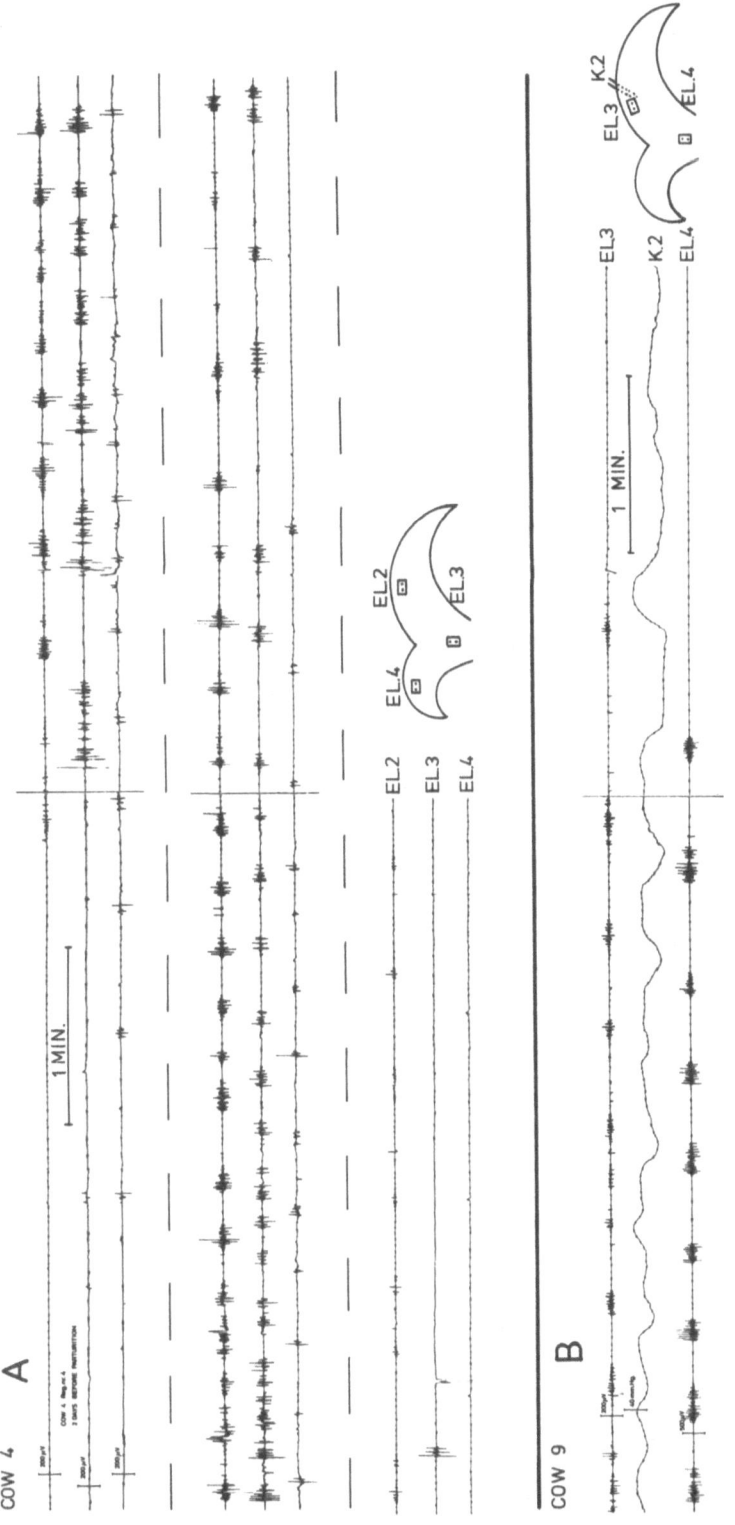

Fig. 2A. A 12 minute period of myometrial electrical activity two days before parturition. Note that electrical activity in the non-pregnant horn (lower line) is minimal.

Fig. 2B. Electrical activity and pressure fluctuations during the last minutes of an active period, two days before parturition.

304

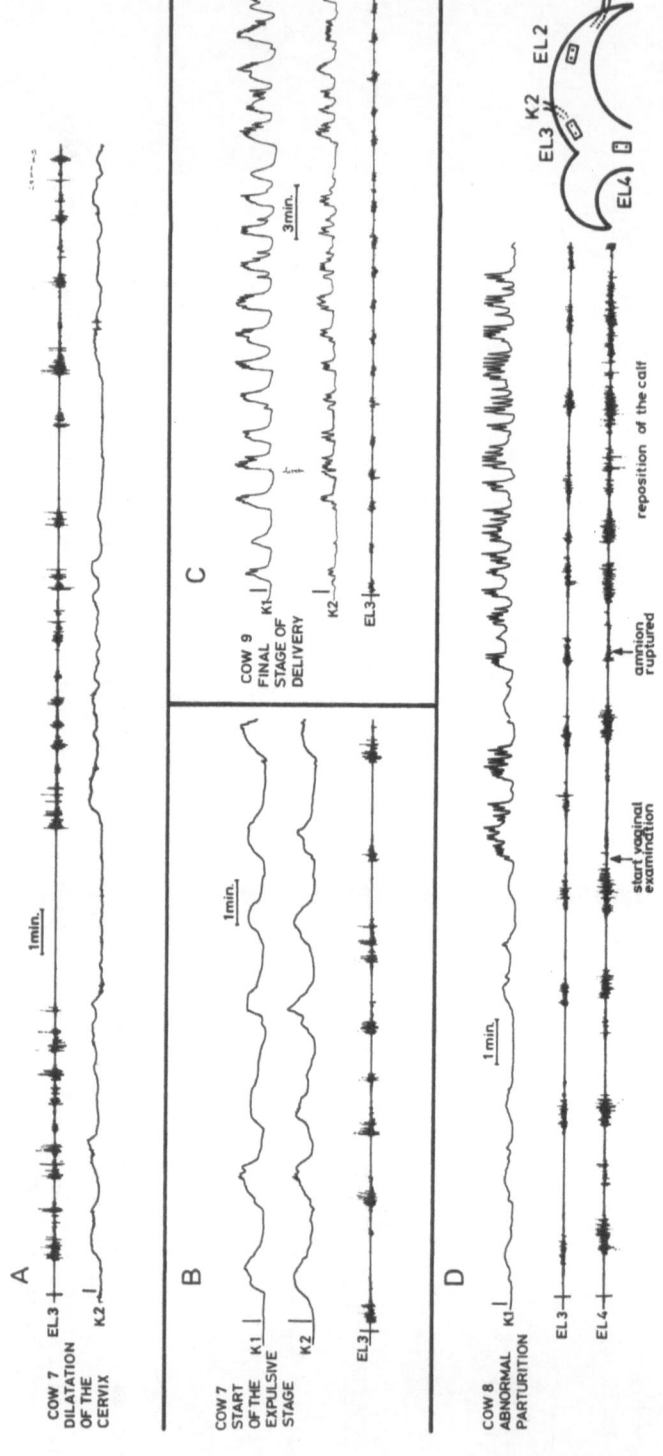

Fig. 3. Uterine electrical activity and intra-uterine pressure changes during dilatation of the cervix (A), the start (B) and the final stage of expulsion (C) in normal deliveries and during vaginal exploration of a cow with an incompletely dilated cervix (D).

with uterine torsions diagnosed at this stage, uterine activity
seemed to persist in the periodic pattern; individual, well
propagated, labour contractions had not become evident. However,
in all three of these cases the non-progressing parturition and
abnormal behaviour of the cow (rather than the abnormal uterine
activity) indicated abnormal parturition.

From the moment the cow started to strain, well propagated
contractions became more and more clear (Figure 3B). Nevertheless periods with an almost continuous electrical activity of
several minutes persisted until halfway through the expulsive
stage. In those animals treated with isoxsuprine during
delivery, uterine electrical activity significantly decreased
and pressure fluctuations completely vanished. (Sham injection
showed no effect). As a result labour was protracted in these
animals for at least 30 - 90 minutes.

During the final stage of expulsion almost every contraction
was accompanied by several straining movements (Figure 3C).
Besides peristaltic and antiperistaltic contractions, quite a
few contractions seemed to originate somewhere along the
pregnant horn, making their propagation difficult to determine.
Contraction frequency ranged from 8 to 15/20 min. during the
stage of expulsion while the mean duration of the electrical
trains ranged from 25 to 50 seconds. There seemed to be no
significant difference in myometrial activity between animals
calving normally and cows which delivered their calves after
correction of a uterine torsion. Only during vaginal exploration
of the abnormal parturient cows uterine activity increased
significantly (Figure 3D) concomitant with frequent straining
movements of the cow.

The postpartum period
Immediately after expulsion of the calf contraction
frequency decreased to 4 to 10/20 min. the mean duration of the
electrical trains ranging from 30 to 60 seconds. Very regular
peristaltic contractions, propagated along the entire length
of the horn, dominated during this stage (Figure 4). After

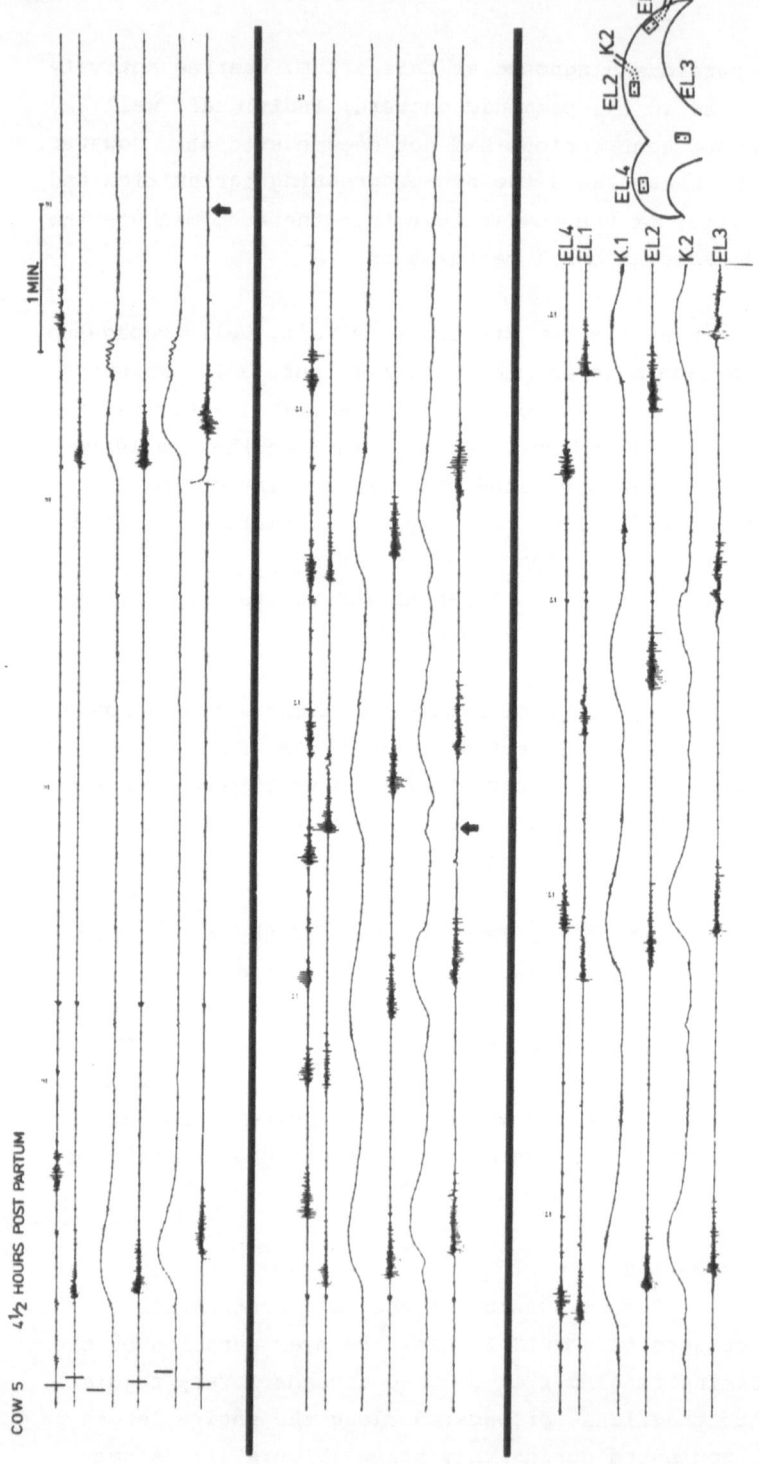

Fig. 4. Uterine activity during the postpartum period. The cow was milked during the time indicated by the two arrows.

injection of isoxsuprine 90 - 150 minutes postpartum, electrical activity and pressure changes disappeared almost completely.

In three of the eight cows the placenta was delivered within 7 h postpartum (Table 1). In the other five cows the placenta was retained and during the first 3 days postpartum in these animals a gradual increase in uterine peristalsis was observed, caused by a decreased propagation of contractions which originated at the tubal end of the pregnant horn. From the 4th day postpartum, only frequent short trains and small pressure fluctuations could be registered at the tubal end of the horn; other parts of the horn showing no activity at all.

Contraction frequency in one cow (no.6) with a normally delivered placenta was considerably lower during the first 2 days postpartum compared with that in the retention cows (Table 2).

TABLE 2

MEAN NUMBER OF CONTRACTIONS PER 20 MINUTES DURING THE FIRST 3 DAYS POSTPARTUM

Hours p.p.	Cow No.	Tubal end of the horn					Middle of the horn				
		3^x	5^x	6	7^x	8^x	3^x	5^x	6	7^x	8^x
12-27		10.3	6.4	3.8	6.0	4.9	-	6.1	3.4	6.5	4.7
36-54		-	6.5	3.4	6.5	5.4	-	2.6	0.6	3.5	4.3
60-77		-	6.5	0	7.8	-	-	0	0	3.2	-

^{x}cows with retained placenta.

DISCUSSION

Extensive abdominal surgery of pregnant cows in late gestation often causes abortion or premature calving within a few days, preceded by declining peripheral plasma progesterone concentrations (Vandeplassche et al., 1976). This seems to have taken place in at least 2 (no. 7 and 8) of our cows. The relatively high incidence of uterine torsions (three out of eight cows) may also have been a result of the operation; the uterus had to be rotated in most cows in order to implant electrodes and catheters at four different locations on the pregnant horn. In addition the electrode wires and catheters running through the abdominal cavity could have promoted the tendency to rotate. Because cows behaved quite normally after the operation and all showed the same type of uterine activity during the last days of pregnancy, it must be assumed that the uterine torsions developed during the initial stage of parturition. During implantation of only one or two electrodes, rotation of the uterus would not be necessary, surgery time would be decreased and therefore manipulation of the uterus should be limited during operations in cows.

The described qualitative and quantitative aspects of uterine activity during the last week of pregnancy are comparable with those of the pregnant sheep (Naaktgeboren et al., 1975). Alternating periods of myometrial activity and rest have also been reported for the pregnant miniature pig (Taverne, 1976). Because this type of uterine activity in these three species occurs clearly only in those parts of the uterus where a foetus or a part of the conceptus is located, one might speculate that this type of activity is caused by the continuous stretching of the myometrium by the uterine contents. The correlation between foetal movements and myometrial activity has yet to be revealed.

In cows as in the sheep, the first significant change in uterine EMG occurred only about 12 - 24 h before birth, although both qualitative and quantitative aspects of the changes in periparturient steroid concentrations are different in the cow and sheep (reviewed by Thorburn et al., 1977). However, in both species the ultimate results, ie a decreased progesterone concentration and a maximal oestrogen concentration, seem to coincide with the first obvious change in uterine activity. Periods of prolonged uterine activity with short intervals of rest sometimes persisted until the stage of expulsion, while

single, regular and totally propagated contractions seemed to
develop quite late during parturition. One might inquire
if this change in the pattern of uterine activity can be
correlated with changes in peripheral oxytocin concentrations,
the rupture of the foetal membranes and/or the entrance of the
calf into the pelvic cavity.

The occurrence of both peristaltic and antiperistaltic
contractions during the first and second stages of parturition
has been previously reported in cows (Zerobin and Spörri, 1972)
and in sheep (Naaktgeboren et al., 1975). Although their
relative importance is unknown, it is quite clear that during
parturition no definite part of the uterus functions as a
pacemaker, because contractions could originate at any place
on the pregnant horn. During the postpartum period, however,
the tubal end of the pregnant horn seemed to be the major site
for contractions to begin. Contraction frequency during and
immediately after delivery of the calf corresponds with that
reported by Rüsse (1965) and Zerobin et al. (1972). Although
contraction frequency in the puerperal cows with retained placenta
was greater in our study, the presence of contractions and the
decrease in peristalsis during the first 3 days postpartum
confirm the data of Zerobin et al. (1972). The difference in
uterine activity between cows with and without retained placenta
is marked although only a single observation (cow no. 6, Table 2)
was made. Probably a prolonged stretching of the uterine wall
by the presence of foetal membranes and/or elevated peripheral
oestrogen levels of cows with retained placenta (Agthe and
Kolm, 1975) promote uterine activity during the first days
postpartum.

Because the electromyographic technique used in this study
allows a longitudinal study of myometrial activity, it should
be combined with extensive endocrine studies in order to reveal
functional relationships. In addition it seems to be suited to
investigate pathophysiological aspects of parturition and puerperium and the in vivo effects of drugs that influence myometrial
cells.

REFERENCES

Agthe, O. and Kolm, H.P. 1975. Oestrogen and progesterone levels in the
blood plasma of cows with normal parturition or with a retained
placenta. Journal of Reproduction and Fertility 43: 163-166.

Hoffmann, B. and Karg, H. 1974. Endocrine balance of the cow at
parturition. In: "Avortement et parturition provoqués", edited by
M.J. Bosc, R. Palmer and Cl. Sureau, Masson et Cie., Paris, pp. 123-138.

Jong, F.H. de, Baird, D.T. and van der Molen, H.J. 1974. Ovarian
secretion rates of oestrogens, androgens and progesterone in normal
women and in women with persistent ovarian follicles. Acta
Endocrinologica 77: 575-587.

Naaktgeboren, C., Pool, C., van der Weyden, G.C., Taverne, M.A.M., Schoof,
A.G. and Kroon, C.H. 1975. Elektrophysiologische Untersuchungen
über die Uteruskontraktionen des Schafes während der Trächtigkeit
und der Geburt. Zeitschrift für Tierzüchtung und Züchtungsbiologie
92: 220-243.

Naaktgeboren, C., van der Weyden, G.C., Klopper, P.J., Kroon, C.H., Schoof,
A.G. and Taverne, M.A.M. 1973. Electrophysiological observations of
uterine motility during the oestrous cycle in sheep. Journal of
Reproduction and Fertility 33: 511-518.

Rüsse, M. 1965. Der Geburtsablauf beim Rind. Archiv für Experimentelle
Veterinärmedizin 19: 763-870.

Taverne, M.A.M., van der Weyden, G.C., Fontijne, P. and Dieleman, S.J. 1976.
Electrical activity of the myometrium and intra-uterine pressure
changes around parturition in the cow and the influence of a single
intramuscular injection of isoxsuprine. Proceedings of the VIIIth
International Congress on Animal Reproduction and Artificial
Insemination, Krakow, 1976, Volume III, 407-410.

Taverne, M.A.M. 1976. Die Physiologie der Geburt beim Zwergschwein.
Deutche Tierärtzliche Wochenschrift, 83, 527-528.

Thorburn, G.D., Challis, J.R.G. and Currie, W.B. 1977. Control of parturition
in domestic animals. Biology of Reproduction, 16, 18-27.

Vandeplassche, M., Corijn, M., Spincemaille, J., Bouters, R. and Bonte, P.
1976. Die Prophylaxe von Abortus und Frühgeburt beim Rind und Pferd.
Deutsche Tierärtzliche Wochenschrift, 83, 554-556.

Zerobin, K. 1970. Die Uterusbewegungen bei Kühen während der Geburt und der
    Nachgeburtsphase. Schweizer Archiv für Tierheilkunde, 112, 544-560.

Zerobin, K. and Spörri, H. 1972. Motility of the bovine and porcine uterus
    and fallopian tube. Advances in Veterinary Science and Comparative
    Medicine, 16, 303-354.

DISCUSSION

B. Hoffmann *(West Germany)*

Dr. Taverne, I appreciate your distinct statement;
parturition of course is not just a line. From our point of
view it starts with the onset of luteolysis as can be seen by
the decline of progesterone levels. We do not say parturition
is over when the umbilical cord is broken, but we take this
event as a marker point for our recordings.

M.A.N. Taverne *(The Netherlands)*

Well, as an extreme it could perhaps be said that the
preparatory state for parturition begins after Day 12 when
the corpus luteum of cycle changes into a corpus luteum of
pregnancy. What I wanted to indicate is that I don't like
studies where parturition is only indicated by one line.
Sometimes it is not even stated if the foetus was alive at
birth. Quite often parturition is described as expulsion of
the foetus, but it is not stated whether parturition was
normal, for example in duration. We have to be very careful in
all studies to define what we are talking about and possibly set
up some criteria showing that we are studying the physiology and
not the pathology of parturition.

H. Karg *(West Germany)*

I would like to make some kind of a conclusion and try to
bring the discussion of the first two papers dealing with the
endocrinological aspects into relation with the topic of our
seminar. What do we really want to know from the endocrinologist?
Firstly, what is the trigger mechanism for the initiation of
parturition, where - as you have seen - exact timing is most
important, and secondly, what triggers the endocrinological
events for relaxation? Concerning the trigger mechanisms and
having in mind what Drs. Hoffmann and Osinga said, there still
seems to be some kind of 'ping-pong' information, not knowing
which is first - 'ping' the foetal signal or 'pong' the maternal
sign. Dr. Hoffmann said that it is the foetal pituitary-
adrenal axis which the signal comes from and there is some

evidence for that from his analytical data obtained from foetal
and maternal blood samples collected between Days 260 and 270.
On the other hand Dr. Osinga showed that the rise of placental
oestrogens occurs much earlier. So Dr. Hoffmann, do you
exclude that already placental oestrogens may have a trigger
effect on the pituitary of the foetus? For example in the rat
oestrogens apparently have a stimulatory effect on adrenal
growth. The other question is, how does relaxation come about?
Is it a direct effect of oestrogens or do they rather act on
the production of relaxin?

B. Hoffmann

We think that the foetal and maternal compartments undergo
a well balanced process of maturation during pregnancy, starting
from the day of conception and constantly approaching parturition.
Therefore the real question which should be asked does not so
much relate to 'ping' or 'pong', but rather to the point, when
does the big 'bang' occur, that is when is the balance
between the two compartments interrupted and what is the cause
of this? In our opinion this 'noise' comes from the foetal
compartment.

M. Bosc (France)

The foetal pituitary-adrenal axis certainly plays an
important role in the processes for initiation and regulation
of parturition. From Dr. Hoffmann's paper it is obvious that
there is not only an increased corticoid output of the foetal
adrenal during the final phase of gestation but there is
also a shift from the production of corticosterone to cortisol.
This brings us to the question, what controls the foetal adrenal
gland in the cow? Would you speculate, Dr. Hoffmann, that there
is no negative feedback between the adrenals and the pituitary
in the foetus due to your observation, since there are no
increased plasma-corticoid levels in the foetus after surgery?

B. Hoffmann

This question is difficult to answer since we have no
data on foetal corticotrophin levels. However, from the data of

Comine et al., (1974), who infused ACTH into the bovine foetus
prior to term and who observed a gradual increase in foetal
corticoid levels within 5 days, it may be concluded that the
'block' itself is located in the adrenal. Perhaps the adequate
enzymes for corticoid synthesis are only provided just before
term and at least the shift in the production from corticosterone
to cortisol indicates that 17α-hydroxylase activity is
increased.

A. Osinga *(The Netherlands)*

I have three questions to Dr. Hoffmann. First, why didn't
you include data about foetal testosterone levels in your
paper? Those data could be of value particularly in relation
to the question why male calves cause dystocia more frequently
in primiparae than female calves do. It is known that the
foetal testosterone level is higher than the maternal level,
particularly in late gestation. The second question relates to
the metabolism of oestrogens where not only the placenta but
also the liver should be considered. We also think that the
placenta is not only a barrier but that we are dealing with a
foeto-placental unit, also in the cow, with the foetus playing
an active role for oestrogen synthesis. We derive this from
our crossbreeding studies. And my last question, could you
comment further on your observations on oestrogen levels after
parturition?

B. Hoffmann

Let me try to answer the last two questions first. The
liver undoubtedly plays an important role in oestrogen metabolism. However, this mainly concerns the oestrogens eliminated
with urine and faeces where, as you also have shown in your paper,
oestradiol-17α is the major metabolite. Yet, in our studies
with various sampling sites in the maternal and foetal
compartment we could demonstrate that the placental unit is not
only the major source of oestrogens in the late pregnant cow,
but also that the ratio of oestrogens released into the two
compartments is different. While conjugated oestrone was the
major oestrogen in the maternal compartment it was conjugated

oestradiol-17α in the foetal compartment. Secondly, in our
experiments with ovariectomised late-pregnant heifers it was
observed that - as measured on the maternal side - oestrogen
production was not affected by delivery of the foetus. We
therefore concluded that the foetus itself is not involved in
providing precursors for placental oestrogen production.
Oestrogen levels only decreased when part of the placenta,
which was retained, was removed. Similarly after normal
parturition peripheral plasma oestrogens only decreased just
prior to the delivery of the placenta. Thus, your observations,
Dr. Osinga, in relation to differences in oestrogen production
between breeds possibly are more related to the placenta itself.
I agree with you that the word 'barrier' in relation to
placental permeability may be rather strong since the steroid
molecule should be able to pass, but of course it may be
metabolised on its way through. So far we don't have any data
on this. To your first question on our testosterone measurements. I did not include the data because they were rather
scattered in the few animals examined. But so far we did not
see sex-related differences in the foetus, but perhaps Dr.
Oxender has some more information on that.

W. Oxender (USA)

Also in our studies on foetuses up to around Day 260 of
gestation we could not observe differences in testosterone
levels between male and female foetuses. In general andro-
stendione was higher in the foetus than in the dam.

M. Bosc

In your paper, Dr. Osinga, you mentioned the case of a
mummified foetus with an apparently unchanged oestrogen
production. Wouldn't that also demonstrate that the bovine
placenta alone is responsible for oestrogen production, not
needing precursors from the foetal adrenal gland as is the
case in the ewe?

A. Osinga

I meant to point out that this case was a comparison between

a pair of pregnant identical twins. At 180 days after mating
the urinary oestrogen level was about the same in both cows.
At 210 days it had increased normally in one cow but drama-
tically decreased in the other to about zero. A check of the
latter cows showed the presence of a mummified foetus.

E. Grunert (West Germany)

Is there any relationship between the weight of the foetal
membranes and the urinary oestrogen levels measured in the cow?

A. Osinga

It must be assumed that there is some kind of a positive
relation since there also seems to be a positive relation
between birth weight of the calf and weight of the foetal
membranes; as Dr. Bosc has already mentioned, Terqui and
Ménissier found a correlation coefficient of 0.8 between birth
weight and plasma oestrogen levels. In future studies we will
have a closer look at the relationships between plasma oestrogens,
urinary oestrogens and birth weight and, if possible, weight of
foetal membranes.

F. Pirchner (West Germany)

I have a more general question concerning the studies of
endocrine changes around parturition. Did anyone make use of
special cattle - genotypes with extreme short or long gestation
lengths?

M.A.N. Taverne

Actually the special observations made in pathological
situations of abnormally short or extended gestation lengths
in the sheep, goat and cow led to the more intensive studies
concerning the role of the foetus in initiation of parturition.
For example as early as 1961 it was observed by Holm et al.
in the USA, that calves with an adrenal insufficiency are born
postmature - we have discussed this already - and this of course
pointed toward a functional role of the foetal adrenal gland.

K. O'Farrell *(Ireland)*

Since foetal cortisol levels increase before parturition
and apparently do not cross the placental barrier, would you
Dr. Hoffmann like to speculate as to the mechanism of action of
exogenous corticoids in inducing parturition?

M. Bosc

I think this question will be discussed in more detail in
the next session. However, let me just say this. Treatment
with exogenous corticoid is always followed by a drop of
progesterone levels prior to calving. Furthermore, from our
observations in the sheep, we have developed a certain concept,
pointing towards two effects of the corticoid. Firstly the
exogenous corticoid acts as the endogenous cortisol does, but
secondly at the same time it inhibits the ACTH secretion of the
foetal pituitary. After this effect is abolished due to
metabolism of the exogenous corticoid, a rebound effect occurs,
leading to an increased ACTH release and thus cortisol production.
These corticoids then act at the placental and myometrial level.

B. Hoffmann

I have to agree that little is known about the mechanism
of action of exogenous corticoids in the cow as far as it
relates to induction of parturition. I want to remind you that
luteolysis can only be induced in the late pregnant but not in
the cyclic cow. We also could demonstrate that progesterone
synthesis of corpora lutea from late pregnant animals was not
affected under in vitro conditions when incubated with
corticoid. Thus the effect of luteolysis seen under in vivo
conditions is certainly an indirect one and it also should be
considered that the necessary dose of an exogenous corticoid
like flumethasone to induce parturition is equivalent to 30 000
mg cortisol in its biological activity and it therefore might
be justified to talk about a pharmacological action.

M. Bosc

Could you please comment on the fact that in some studies
the decline of progesterone prior to parturition is described as

a biphasic one, the first period generally occurring during
the last 20 - 30 days, the second one during the last two days.

B. Hoffmann

I think it is a matter of statistics and frequency of
blood sampling whether you talk about a 1-phase or 2-phase
decline. We have observed both kinds of pattern - animals with
a constantly high progesterone level until 30 - 40 hours prior
to parturition and animals where there was a tendency of declining progesterone levels during the last 30 days of gestation.

M. Bosc

You mentioned that LH is apparently the only luteotrophic
hormone in the cow. Yet in the pig du Mesnil du Buisson has
shown that there are three separate luteotrophic factors during
pregnancy - LH at the beginning, FSH at mid-pregnancy and
prolactin at the end of pregnancy. In regard to this point do
you have any information on the role of placental lactogen in
the cow?

B. Hoffmann

I have to restrict my answer to prolactin. Neither during
the cycle nor in the cow with a corpus luteum persistens could
a luteotrophic effect of prolactin be demonstrated by us.
Similarly no changes in parturition were observed when prolactin
release was blocked in the cow during the last 10 days of
gestation by giving the specific inhibitor CB-154 (Br-α-
argocryptin).

A. Osinga

I have two more questions to Dr. Hoffmann. The first point.
You suggested that the placenta acts as an effective barrier
between the foetal and maternal compartment. But since the
oestrogens are continuously produced in the placenta and partly
transferred to the foetus, an accumulation of oestrogens in
the foetal compartment would be expected unless they are broken
down which is not likely to occur since large quantities are
excreted as urinary oestrogens via the maternal compartment.

Since the amount of urinary oestrogens is dependent on the foetal genotype it would be assumed that these urinary oestrogens are of foetal origin. I expect your barrier theory to be valid for the corticosteroids but not for oestrogens. Wouldn't you accept the possibility that the corticoids produced in the foetal adrenal are metabolised in the placenta into oestrogens? These steroids must remain somewhere and I postulate that the oestrogens are transferred to the maternal compartment by diffusion, that they act locally on the uterus and birth canal and that they are inactivated as soon as they reach the peripheral plasma.

The second point I wanted to raise. You have demonstrated that the oestrogen concentrations remained high in peripheral plasma of the dam after expulsion of the foetus. Since you followed this for only eight hours this could be due to residual oestrogen precursors in the placenta being of foetal-adrenal origin. In my opinion your data do not absolutely prove that the foetus itself is not involved in the oestrogen synthesis. Placental perfusion studies should provide definite proof.

B. Hoffmann

This actually is a whole complex of questions and we have to come back to some of the points already discussed. For a better understanding of the discussion it should be said that the placenta consists of a maternal and a foetal part. We could show without any doubt that the placental effluent into the maternal compartment is different from the placental effluent into the foetal compartment. Previously we have discussed this as being due to differences in the biosynthetic capacity of the two placental compartments. It cannot be questioned that the oestrogens originating in the maternal part of the placenta are all excreted via the faeces and as urinary oestrogens. However, it is not known whether and to what extent oestrogens originating in the foetal part of the placenta are being transferred into the maternal compartment. The possibility exists that the free and conjugated 17α-

oestradiol appearing in the uterine vein is of foetal origin.
But this would only account for a minor percentage of the
oestrogens produced by the foetal trophoblast. It should also
be noted that there is a difference of oestrogen level in the
umbilical circulation indicating the the foetus is removing
substantial quantities of oestrogens from the circulation. At
term the whole foetal compartment is 'loaded' with oestrogens,
with especially high concentrations in the meconium and
amniotic fluid, and some newly born calves can even show
symptoms of heat. If the oestrogens produced by the placenta
depend on precursors supplied by the foetal adrenal gland it
could certainly not be attributed to the so far characterised
production of corticosterone or cortisol but only to a hitherto
unknown precursor which must be produced in large quantities,
beginning at around Day 200 of gestation when urinary
oestrogens begin to increase. I don't doubt that development
of the foetal part of the placenta, and thus also the qualitative
and quantitative attachment with the maternal part, depends on
the genotype of the foetus, but I doubt that the adrenal of the
foetus is providing precursors for placental oestrogen synthesis.
Of course from what is known in the human we speculated on
something similar in the cow, as you do, and this is why we
initially collected blood for only eight hours after parturition
in our ovariectomised cows. We could not see a drop of oestrogens
in these studies and we therefore extended our observation period
in the following experiment - I didn't get a slide ready - and
there a significant decline of peripheral plasma oestrogens in
the dam only occurred about 60 hours after parturition when
parts of the placenta were removed manually. So I think the
situation in the cow is different from other species.

A. Osinga

From the studies with induced parturition we know that
exogenous oestrogens are not as effective as endogenous
oestrogens, though in sheep parturition could be induced by
applying exogenous oestrogens. Perhaps Dr. Hoffmann has some
ideas about the different activity in the cow?

B. Hoffmann

To my knowledge parturition cannot be induced with oestrogens in the cow, as Dr. Bosc has also indicated. I relate this to the fact that in the cow the oestrogens around parturition rather play a more permissive role.

M. Bosc

There is no doubt that oestrogens are produced by the placenta, perhaps not in the same type of foeto-placental unit as seen in the human, and that they reflect the growth of the foetus. We have seen that oestrogen production is related to the genotype of the foetus, perhaps to calving difficulties, and it may be a valid parameter in other studies. There have been many studies in this field, but there is an almost total absence of work in relation to delivery of the placenta. We talk of it as a modern problem but people were already dealing with it 4 000 years ago, as is seen in ancient Egyptian paintings.

INDUCED PARTURITION

Chairman: H. Karg

BETAMETHASONE INDUCED CALVING: A COMPARISON BETWEEN INDUCED AND NON-INDUCED DAIRY COWS

K.J. O'Farrell,

Agricultural Institute, Moorepark Research Centre,
Fermoy, Co. Cork, Ireland.

ABSTRACT

A total of 104 British Friesian dairy cows were randomly assigned to induced and natural calving treatments. Fifty-one cows were injected intramuscularly with 40 mg of betamethasone between day 260 and 285 of pregnancy. Seventy-two % of these responded to one injection within 51.3 ± 36.1 hours. The remainder were retreated on the fourth day and calved within 40.8 ± 21.7 hours. Perinatal calf loss for treated and control groups was 7.5 and 7.2% respectively. Afterbirth retention occurred in 47% of treated and 4% of control cows***. The calving-to-service interval for treated and control cows was 64.5 ± 28.3 and 58.3 ± 18.2% days respectively. While the calving-to-conception inverval was 86.0 ± 32.1 and 73.1 ± 27.4 days respectively, seven treated and ten control cows failed to conceive subsequently. The calving-to-service and calving-to-conception intervals for treated cows which retained the afterbirth were 70.0 ± 32.8 and 93.3 ± 34.7 days respectively. The mean days pregnant at calving for treated cows was 274.5 ± 5.7 and calf birth weight was reduced on average by 0.30 kg per day advanced. Treatment did not lower Zinc Sulphate Turbidity levels in calves when compared with controls, nor did it influence the location length or yield of the cows.

1. INTRODUCTION

In Ireland calving data and calving spread are of particular importance to the summer milk producer. It has been calculated from survey data that for every day a cow calves before or after 23 March milk yield may be increased or decreased by 6.4 litres per day (Cunningham, 1972). In this survey it was found that 40% of cows calved after 31 March; therefore the national yield averages must be a reflection of this late calving date (Anonymous, 1976).

For the easy and efficent management of the dairy herd a compact calving period (3 months) is desirable (O'Farrell, 1975). Confining the calving spread of a herd to this period must necessarily involve advancing the calving date of the late calving cows. Advancing the calving date of these cows might also increase their lactation length and afford them a greater opportunity of being successfully inseminated within a limited breeding period.

Inducing calving with the short-acting corticosteroids may reduce milk production by 10 to 20% (O'Farrell and Langley, 1975; Welch, 1972). Beardsley et al (1974) found that milk production from induced cows was significantly lower than controls in the first 3 weeks after calving, but corrected averages for a 305-day lactation were not significantly different. Subsequent breeding performance does not appear to be adversely affected even for those animals which retain the afterbirth (O'Farrell and Langley, 1975; Wagner et al., 1974; Christiansen and Hansen, 1974). However control cows tended to have a higher conception rate to first and second service and a shorter mean calving to conception interval than treated cows (Beardsley et al., 1974).

The purpose of the present study was to compare the effects of induced calving on the welfare of calf and cow with natural calvings.

2. MATERIALS AND METHODS

2.1. Animals

A total of 104 spring-calving British Friesian dairy
cows were selected on the basis of similarity of lactation number,
expected calving date and previous lactation. Cows were then
randomly assigned to induced or natural calving treatments. One
of the cows assigned to the treatment group calved prematurely
and was excluded. The mean lactation number and lactation yield
were 4.25 ± 2.07 and 2940 ± 575 litres respectively for cows
allocated to induction and 3.90 ± 1.72 and 2929 ± 573 litres for
control cows.

2.2. Management

All animals were housed in cubicle-type sheds and had free
access to an outdoor silage face. Those cows which calved
before mid-March received 4.5 kg of concentrates per day in
addition to ad libitum silage. Cows calving after mid-March
were let out to pasture and received 0.9 kg of concentrates as
a carrier for calcined magnesite. Dams were left with their
calves for at least 12 hours and rejoined the herd after their
first milking.

2.3. Treatment

Fifty-one cows were injected intramuscularly with 40 mg
betamethasone[+] between day 260 and 285 (mean 271.4 ± 6.6) of
pregnancy. Those failing to show a response to treatment were
re-injected with a further 40 mg on the fourth day after initial
treatment. Treated cows were under constant observation until
calving.

2.4. Calves

Blood samples were taken 24 hours after birth from 39
calves born following induced parturition and from 23 full-term
control calves to assess their immune status. Serum from these

[+]Betsolan: Glaxo Laboratories Ltd.

samples was subjected to the zinc sulphate turbidity test. The detailed results of these tests are the subject of a separate paper (Langley and O'Farrell, 1976).

3. RESULTS

3.1. Response to treatment

A total of 37 (72%) cows responded to one 40 mg injection. The time of calving was not recorded for one of these. The mean treatment to calving interval was 51.3 ± 36.1 hours (Table 1). Three cows took over 96 hours to respond but were not retreated as no additional drug was available. They were 261, 264 and 272 days pregnant when injected and calved 191, 120 and 123 hours respectively after treatment. Cows treated earlier in pregnancy took longer to respond (Figure 1). Fourteen cows were retreated with a further 40 mg of betamethasone on the fourth day after initial treatment. The mean days pregnant at time of first treatment for these cows was 267.3 ± 4.8. The interval from second treatment to calving was 40.8 ± 21.7 hours. Control cows had a mean gestation length of 279.9 ± 6.2 days (Figure 2).

TABLE 1

GESTATION LENGTH AND RESPONSE TO TREATMENT FOR 51 COWS INJECTED WITH BETAMETHASONE. (MEANS $\pm$ SD)

	Injected once	Re-injected Day 4
Number of cows	37	14
Gestation length when treated (days)	273.1 ± 6.5	267.3 ± 4.8
Treatment to calving interval (h)	51.3 ± 36.1	40.8 ± 21.7
Gestation length at calving (days)	275.5 ± 5.6	272.3 ± 4.3
Number retained afterbirth	16 (43%)	8 (57%)
Number of calves weighed	33	14
Calf weight (kg)	38.6 ± 4.0	39.3 ± 5.5
Perinatal deaths	3 (twins)	1

3.2. Calf details

Induced. Fifty-three calves were born to the 51 treated cows. This included two sets of twins. One twin calf was

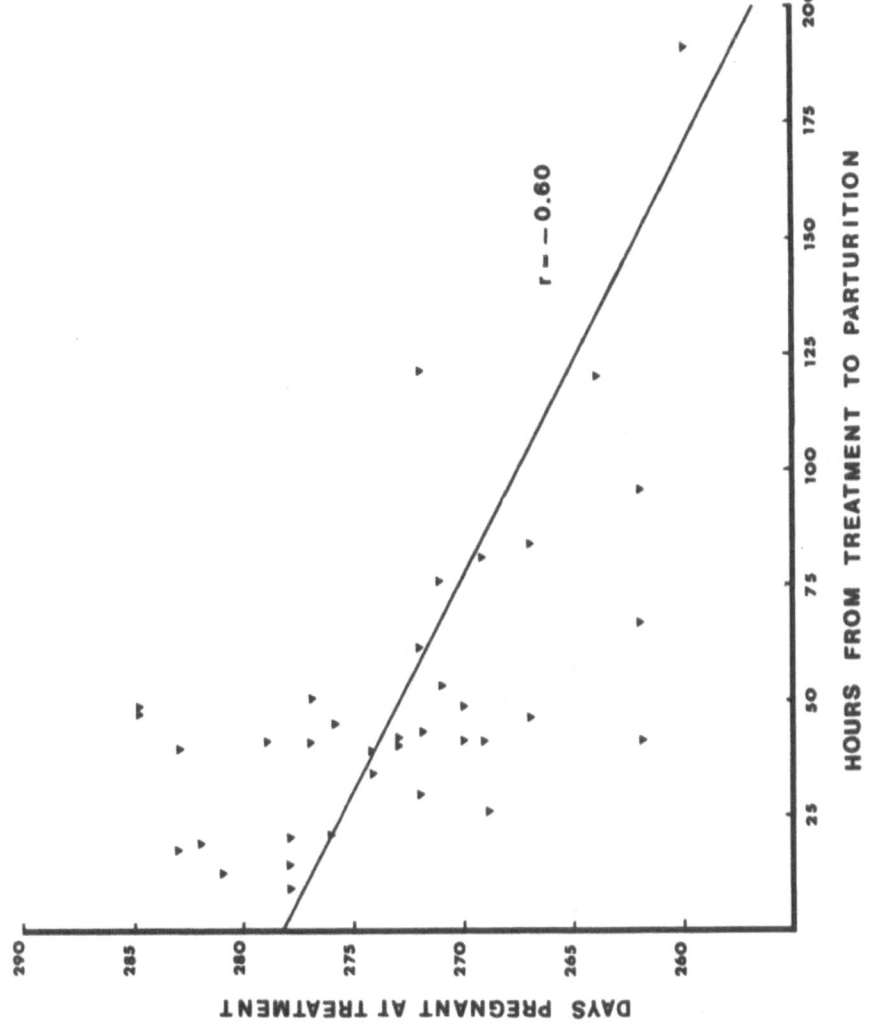

Fig. 1. Response to treatment.

stillborn and the co-twin also died within 4 hours of birth. Two more calves died within 48 hours of birth, bringing the total perinatal deaths to four. A total of four calves including one set of twins were not weighed. The average calf weight for 47 calves, excluding one set of twins, was 38.8 ± 4.4 kg. Calf weight was reduced on average by 0.3 kg per day advanced (Figure 2).

Control. A total of 55 calves which included three sets of twins were born to 52 cows. Four calves were stillborn including one set of twins. Seven calves were not weighed and the average calf weight for 48 calves was 37.7 ± 4.9 kg.

Immunity levels. The mean zinc sulphate turbidity units (ZST) for induced and control calves were 15.0 ± 12.1 and 18.3 ± 10.8 respectively. The number of calves with less than seven and less than ten ZST units were not significantly different (P>0.05), nor was there any relationship between these levels and the degree of prematurity (Langley and O'Farrell, 1976).

3.3. Afterbirth retention

The afterbirth was considered retained if still present 12 hours after calving. Twenty-four (47%) induced cows (P<0.001) and two control cows retained the membranes for longer than this period. The dependent membranes were then cut off approximately 10 cm below the vulva and no other treatment was given.

3.4. Culling post calving

Treated cows. Two cows which were lame and in poor physical condition were culled one week after calving. Another was culled due to low milk yield. Two cows died within three days of calving from milk fever and another which had given birth to twins died on the fourth day due to a severe pneumonia.

Control cows. One cow which had given birth to twins died four days after calving from an acute mastitis. A second cow which

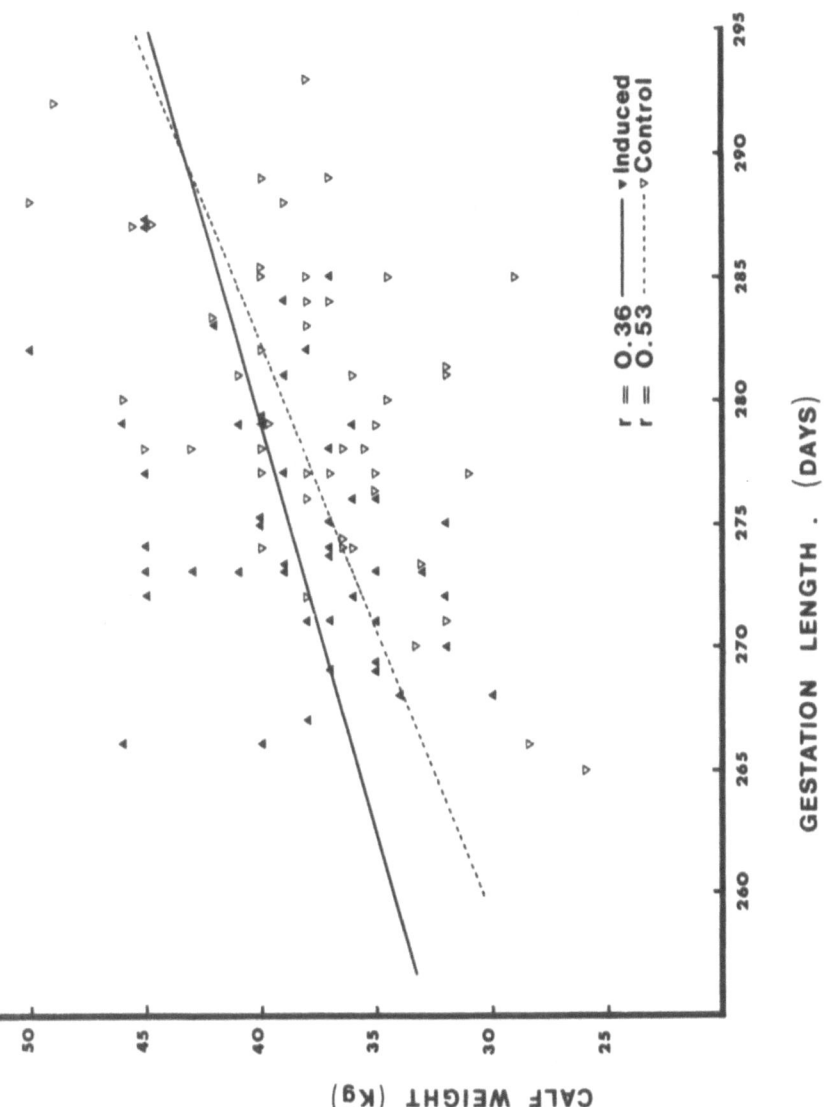

Fig. 2. Effect of gestation length on calf birth weight.

had a difficult calving developed a severe metritis and died.
A third cow contracted looping ill and died within a month of
calving.

3.5. Reproductive performance

The subsequent reproductive performance for the treated
and control cows is given in Table 2. There was no significant
difference in the mean calving to service or calving to
conception intervals for treated and control cows (P>0.05).
However treated cows which retained the afterbirth had a
significantly longer calving to conception interval (Table 3).

TABLE 2

REPRODUCTIVE PERFORMANCE OF INDUCED AND CONTROL COWS (MEANS ± SD)

	Induced	Control	
Number served	45	48	
Calving-to-service interval (days)	64.5 ± 28.3	58.3 ± 18.2	NS
Number conceived to first service	17	25	NS
Total in calf	38	38	
Calving to conception interval (days)	86.0 ± 32.1	73.1 ± 27.4	NS
Total number of services	81	78	
Services per conception	1.81	1.63	

TABLE 3

REPRODUCTIVE PERFORMANCE OF INDUCED COWS WHICH RETAINED THE AFTERBIRTH
(MEANS ± SD)

Number served	22	
Calving to service interval (days)	70.0 ± 32.8	(NS) *
Number conceived to first service	6	(NS)
Total in calf	19	
Calving to conception interval (days)	93.3 ± 34.7	(P<0.05)
Services per conception	2.21	(NS)

* Difference from control.

3.6. Lactation yield

There was no difference in lactation yield between treated and control cows (Table 4).

4.DISCUSSION

The mean treatment-to-response interval for those cows which calved to one and two 40 mg injections of betamethasone was similar to that previously reported (O'Farrell and Langley, 1976). Those cows requiring a second treatment on day four were treated earlier in pregnancy (Table 1) and, as would be expected, took longer to respond (O'Farrell and Langley, 1976; Welch, 1973). The spread in time from treatment to response with betamethasone is more variable than that reported by Beardsley et al. (1974) following dexamethasone administration. However in the latter instance all animals were treated on Day 273 of pregnancy.

The incidence of stillbirths and perinatal deaths for treatment and control groups was similar. Survey data (Cunningham et al., 1976) from dairy herds showed that 2.8% of calves were stillborn and another 1.3% died within 24 hours of birth. The incidence of stillbirths in this experiment was 1.8% and 7.2% for induced and non induced calves respectively. This is considerably lower than that reported following use of the long-acting corticosteroids (O'Farrell and Crowley, 1974; Welch et al., 1973). Only two treated cows were recorded as having required assistance at calving. One of these had twins. Beardsley et al. (1974) found that induced cows required more assistance at calving than controls, but this may have been due to the shorter response time to dexamethasone in that trial. The reduction in calf weight was 0.30 kg per day advanced and is similar to that found previously (O'Farrell and Langley, 1975).

The incidence of afterbirth retention was similar to that found by O'Farrell and Langley (1975) but on this occasion the time from treatment to parturition had no effect on retention rate. Treatment significantly increased the incidence of

TABLE 4

LACTATION LENGTH AND MILK YIELD FOR INDUCED AND CONTROL COWS

	Cows	Length (days)	SD	Yield (litres)	SD	Adjusted yield (litres)	SD
Induced							
Afterbirth retained	22	239.3	5.8	2743	104.6	2781	112.0
Not retained	23	237.6	5.7	2854	106.7	2912	109.6
Control							
Afterbirth retained	2	215.5	19.2	2960	354.7	3284	371.5
Not retained	45	247.7	4.1	2983	74.8	2920	78.3

retained afterbirths (P<0.001) (Table 1). The practice of cutting off the dependent membranes and periodically applying traction with a gloved hand to the remainder, appeared to be satisfactory, as none required additional therapy.

In calving-to-service and calving-to-conception intervals the treated cows did not differ significantly from controls. However, on average, the calving-to-conception interval for treated cows was 13 days longer. The calving-to-conception interval for treated cows which retained the afterbirth was, on average, 20 days longer (P<0.05). Moller et al. (1967) in a survey of retained afterbirths in New Zealand, found that those cases which were not given intra-uterine medication had normal fertility and those which were treated had reduced fertility. Beardsley et al. (1974) found that retention of the membranes was not detrimental to subsequent fertility even though the mean calving-to-conception interval was 12 days longer. In beef cows Moody and Han (1976) found that induced cows which retained the afterbirth had a lower pregnancy rate than controls, but the calving-to-conception interval for both groups was similar. In the present study the first service conception rate of induced cows which retained the afterbirth, was lower than that of controls but the difference was not significant (P>0.05).

There was no significant difference in lactation yield or length of induced and control cows (Table 4). The variations in yield merely reflected variations in lactation length. The variability was too great to allow detection of differences due to treatment (if these do exist) to be detected. Beardsley et al. (1974) found that while induced cows had a lower yield for the first three weeks of lactation, corrected yields for a 305-day lactation did not differ. The lower production in the first three weeks after induction has also been observed by the author(unpublished data).

5. CONCLUSIONS

The results of this experiment indicate that induced calving

with betamethasone did not have an adverse effect on cow
health, calf viability or subsequent milk yield. However afterbirth retention was significantly increased and this delayed
subsequent conception. Thus the advantage to be gained by
advancing calving may be lost due to the longer subsequent
calving-to-conception interval of the induced animals.

ACKNOWLEDGEMENTS

    The author wishes to thank Glaxo Laboratories Ltd., for
supplies of betamethasone, Messrs T. Condon and B. O'Rielly for
skilled technical assistance, the farm technicians for their
help and co-operation and Mr. P. Kearney for assistance in the
statistical analysis of the data.

REFERENCES

Anonymous 1976. EEC Dairy Facts and Figures p. 36. Milk Marketing Board,
    Thames Ditton, Surrey.

Beardsley, G.L., Muller, L.D., Owens, M.J., Ludens, F.C. and Tucker, W.L.
    1974. Initiation of parturition in dairy cows with dexamethasone. l.
    Cow response and performance. Journal of Dairy Science 57, 1061-1066.

Christiansen, I.J. and Hansen, L.H. 1974. Dexamethasone-induced parturition
    in cattle. British Veterinary Journal 120, 221-227.

Cunningham, E.P. 1972. Variation in dairy cow performance. Farm and Food
    Research 3, 133-134.

Cunningham, E.P., Shannon, M., Fallen, T.J. and O'Byrne, T.M. 1976. A survey
    of reproduction: calving and culling of cows in Irish dairy herds. Irish
    Journal of Agricultural Research 15, 177-183.

Langley, O.H. and O'Farrell, K.J. 1976. Immune status of dairy calves
    following induced parturition. The Veterinary Record 99, 187-188.

Moller, K., Newling, P.E., Robson, H.J., Jansen, G.J., Meursinge, J.A. and
    Cooper, M.G. 1967. Retained Foetal Membranes in Dairy Herds in
    the Huntley District. New Zealand Veterinary Journal 15, 7, 11.

Moody, E.L. and Han, D.K. 1976. Effects of induced calving on beef
    production. Journal of Animal Science 43, 298 (Abstr.).

O'Farrell, K.J. 1975. The role of management in dairy herd fertility. Irish
    Veterinary Journal 29, 118-124.

O'Farrell, K.J. and Crowley, J.P. 1974. Some observations on the use of two
    corticosteroid preparations for the induction of premature calving.
    The Veterinary Record 94, 364-366.

O'Farrell, K.J. and Langley, O.H. 1975. The induction of parturition in
    dairy cows with betamethasone. Irish Veterinary Journal 29, 151-155.

Wagner, W.C., Willham, R.L. and Evans, L.E. 1974. Controlled parturition in
    cattle. Journal of Animal Science 38, 485-489.

Welch, R.A.S. 1972. The effects of inducing early calving. Proceedings of
    the Ruakura Farmers Conference.

Welch, R.A.S., Newling, P.E. and Anderson, D. 1973. Induction of parturition
    in cattle with corticosteroids: an analysis of field trials. New
    Zealand Veterinary Journal 21, 103-108.

SOME RESULTS WITH INDUCED PARTURITION IN COWS AND HEIFERS

H.O. Gravert and E. Kordts

Institut für Milcherzeugung der Bundesanstalt für
Milchforschung, Kiel, Germany.

ABSTRACT

Parturitions were induced in 66 dairy cows and heifers 10 days prior to term. Corticosteroid (flumethasone) treatment resulted in parturition after 22 to 58 hours, prostaglandin treatment after 32 to 105 hours. In all cases of this early treatment placentas were retained while in heifers treated at a later stage only two placentas were retained. The average weight of the calves was reduced by about 5 kg and the frequency of calving difficulties was significantly lower.

1. INTRODUCTION

To avoid dystocia, induced parturition might be a practical aid to restrict gestation length and to reduce weight and size of calves. In our baby-calf programme which gives us an additional calf during the raising of replacement heifers we wanted to avoid calving difficulties in young heifers with an age at first calving of 17 to 20 months. In the beginning it was considered desirable to let the animals deliver approximately 10 days prior to term. Since the gestation length of German Friesians is about 280 days, treatment commenced on Day 267 to 270. Treatment was given to 23 early-bred heifers, additionally also to 22 normal-aged first-calf heifers (average calving age 22 to 35 months) and 21 cows with second to fifth calvings. Out of the 66 animals 50 were treated with corticosteroid (two 5 mg doses flumethasone, about 12 h apart; the first injection was intramuscular(im)or intravenous (iv)) and the other 16 with prostaglandin (PGF$_{2\alpha}$, 20 mg iv).

2. RESULTS

Flumethasone treatment induced parturition in all cases, with a mean interval between first treatment and parturition of 43 h (22 to 58 h). PGF$_{2\alpha}$ was effective in all heifers and in five of seven cows, with a mean interval of 57 h (32 to 105 h). Two cows did not respond to PGF$_{2\alpha}$ and calved 12 days after treatment. The placenta was retained in all animals in which parturition was induced. In early-bred heifers within our baby-calf programme this problem could be overcome by antibiotics since the heifers were dried off immediately after calving. In normal-aged heifers and cows the application of antibiotics was, however, restricted by its effect on milk quality.

The interval between parturition and new conception seemed to be extended by the treatment, especially with flumethasone. The average interval was 124 days after treatment with flumethasone, 94 days after treatment with prostaglandin and about 70 days for untreated animals which is about 80 days in

comparison with the above figures for induced parturition.
Due to a large variation (51 to 198 days) the differences were,
however, not significant.

As a consequence of these results we decided to change the
procedure and to treat only those animals which are expected to
show a delayed gestation. If there was no definite sign of
parturition on Day 278, one dose of 10 mg flumethasone (20 ml
Cortexilar) was injected im. Since frequency of calving
difficulties in cows was only 2 - 4% but, in heifers it was
8 - 10%, only heifers were treated.

Out of 50 early-bred heifers with an average age at first
calving of 18 months only 11 animals had to be treated. Only
two of these retained the placenta. Out of another 102 normal-
aged heifers only 14 were treated with no case of retentio
placentae. On average parturition occurred 32 h (9 to 50 h)
after injection.

It seems that with induced parturition the frequency of
dystocia in first-calving heifers can be reduced, especially
at a very low calving age. Without induced parturition the
average weight of calves of early-bred heifers was 37 kg, with
induced parturition the weight of these 'baby calves' was only
32 kg, the normal weight of German Friesian calves being 40 -
45 kg.

Up to 1976 altogether 180 young females were inseminated in
our experimental herd which resulted in 107 additional calves.
The average yield in the first lactation (after the second
calving) was 4 842 kg FCM with a calving age at the second
calving of 30 to 32 months.

INDUCTION OF PARTURITION IN THE BOVINE

R.W.J. Plenderleith

The Rase Veterinary Centre, Pasture Lane, Market Rasen,
Lincolnshire, England.

ABSTRACT

Much work has been carried out all over the world during the last eight years on the induction of parturition in the bovine. Understandably, a large proportion of this work has been aimed at the experimental understanding of the complex physiological problems associated with parturition. Until most of these problems are satisfactorily solved, the work will go on.

One of the problems which is encountered is the occasional failure of induction in large valuable pedigree animals. Perhaps a routine as suggested by the paper of Beal et al. (1976), where prostaglandin is given if induction with dexamethasone has not started witin 40 hours, could be considered.

This paper illustrates how induction has helped the practitioner in the problems which have occurred during the rapid modernisation of the beef and dairy industries of the UK. These problems, of course, occur in all EEC countries to a greater or lesser extent. In practical terms, induction will always be a useful procedure for managemental and medical problems, but keeping in mind its limitations, long-term strategy should be aimed at correcting the managemental malpractices which initiated the problems.

1. INTRODUCTION

This paper on induction of parturition in the bovine will discuss the practical application of the technique and the problems encountered in its use in a rapidly changing climate in the management of both dairy and beef herds in the United Kingdom in the last 10 - 20 years.

1.1. The dairy industry in the UK

A brief discussion of the dairy industry in the UK will be helpful in understanding how problems in parturition management have evolved. As in many countries, intensification of the industry has been radical over the last 25 years.

Figure 1 shows that the number of milk producers has fallen from 196 426 in 1950 to 71 197 in 1976. The number of dairy cows, however, has stayed remarkably constant and in 1976 was 3 238 000 compared with 3 165 000 in 1960.

The increase in herd size is amply illustrated in Tables 1 and 2.

TABLE 1

DAIRY HERD SIZE DISTRIBUTION IN ENGLAND AND WALES, 1969 AND 1975.

Herd size (cows)	Herds as % 1969	Herds as % 1975	Cows as % 1969	Cows as % 1975
3 - 9	13.7	9.7	2.6	1.3
10 - 19	24.1	16.0	10.7	5.0
20 - 29	20.9	15.8	15.5	8.4
30 - 39	13.9	13.9	14.5	10.4
40 - 49	9.1	11.3	12.4	10.9
50 - 69	10.2 (27.4)	14.8 (44.6)	18.2 (56.7)	18.9 (74.9)
70 - 99	5.3	10.8	13.4	19.4
100 - 199	2.5	6.6	9.9	18.4
200 +	0.3	1.1	2.8	7.3

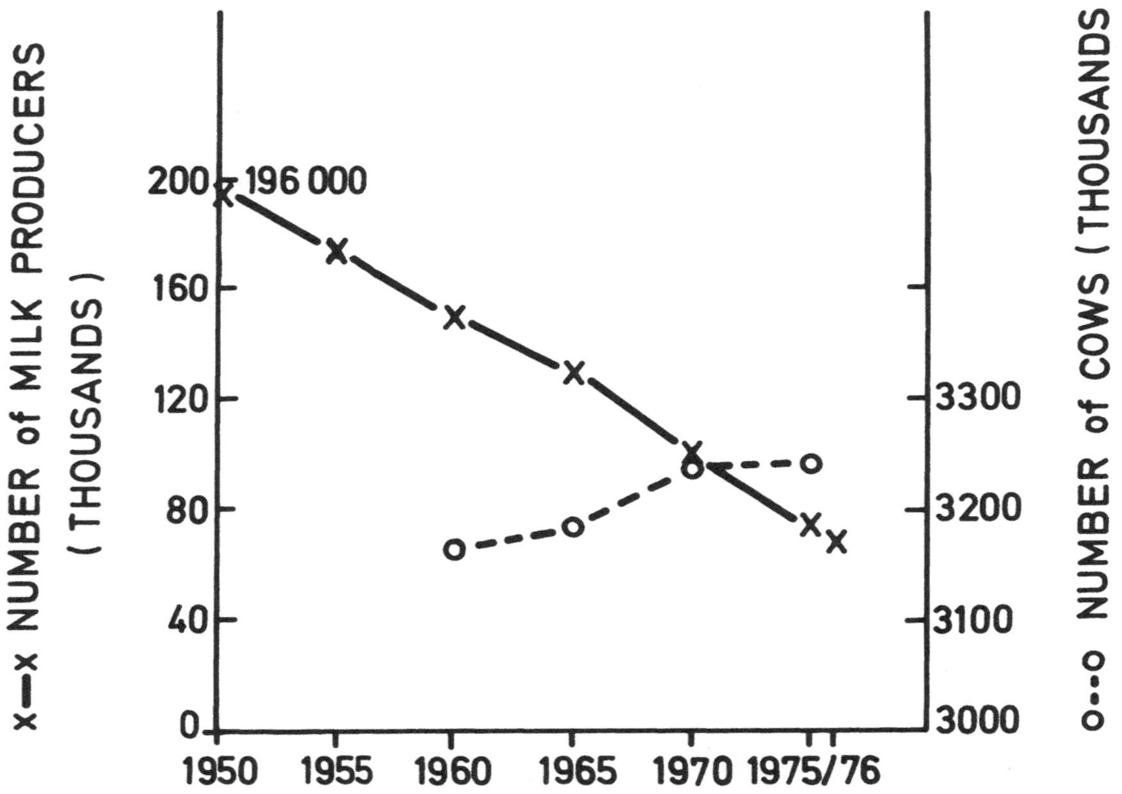

(SANASNOHT) SWOƆ ʇo ꓤƎᗺWUN o--o

3300
3200
3100
3000

1950 1955 1960 1965 1970 1975/76

200 196 000
160
120
80
40
0

(SANASNOHT)
ꓤƎƆUᗺOꓤꓷ ꓘ⅃IW ʇo ꓤƎᗺWUN x—x

REF. U.K. DAIRY FACTS and FIGURES 1976

Fig. 1.

343

TABLE 2

DAIRY HERD SIZE DISTRIBUTION IN SCOTLAND 1969 AND 1975

Herd size (cows)	Herds as %		Cows as %	
	1969	1975	1969	1975
40 +	63.3	77.7	81.3	90.8
80 +	14.7	29.5	30.2	50.3

The average milk yield per cow rose from 3 520 litres in 1965 to 4 270 litres in 1975/76.

The British Friesian is the predominant breed, as illustrated by the fact that, in 1975/76, 92.5% of dairy artificial insemination in England and Wales was with Friesian (including Canadian Holstein) semen.

Many larger herds serve groups of Friesian heifers according to age (ie 18 months old) to calve to suit their management system, eg spring grass, or for winter milk production.

The management often ignores basic concepts, for example, that the heifers should weigh at least 350 kg and that the sire used might be of a breed (such as Hereford or Aberdeen-Angus) to leave smaller calves than the purebred Friesian.

Because the heifers have not the good basic size at parturition and because the calves are large, dystocia occurs. The problem is presented to the practitioner after the first few animals have had dystocia, producing problems of Caesarean section, posterior paresis or ataxia, vaginal tears, metritis etc.

No accurate service data are available, unless the heifers have been inseminated either with or without prostaglandin synchronisation. Induction is used at an estimated 10 - 14 days pre-partum (by clinical signs), that is about Day 269; 20 mg betamethasone + 3.75 g of procaine penicillin + 3.75 g dihydrostreptomycin sulphate are given. The antibiotic is repeated after parturition. The calf's navel is dressed with oxytetracycline spray and the calf injected with 1.25 g penicillin + 1.25 g streptomycin. Induction usually results in a trouble-free calving within 72 hours. A 5 to 10% induction failure rate is expected.

The farmer happily tolerates the high rate of retained placenta (up to 80%) for a healthy heifer in the milking parlour. Efforts to reduce retained placenta have been made by using oestrogen with the inducing steroid. Grunert et al. (1975) found that 10 mg oestradiol had no effect in reducing the incidence (66.7%). When he used 120 mg diethylstilboestrol (60 mg intravenous + 60 mg intramuscular) plus 100 ml calcium iv he reduced the incidence to 23.5%. As the latter would not be practical under field conditions, further work is necessary in this area, especially in the light of conflicting reports from America (Garverick et al., 1974; Kesler et al., 1976). Christiansen and Hansen (1974) quote 2 - 25% as being accepted 'normal' levels of retained placenta.

1.2. The beef industry in the UK

The two main changes in the suckler herd in the last decade which have produced the need for induction of parturition are: the introduction of many European breeds to Britain, and their export as breeding animals to the New World.

These breeds were imported for crossbreeding with indigenous breeds to produce larger calves out of our traditional dams. Many breeds were and still are used, but the Charolais and Simmental are the main survivors.

Producers required larger calves, with good food conversion rates and producing the correct type of carcass for retailing. Again, management factors were not thoroughly investigated and severe dystocia, due to oversized foetuses, occurred.

The effect of the sire breed for crossing on calf weight can be seen in Table 3. Many problems arose in heifers and young cows.

TABLE 3

MEAT AND LIVESTOCK COMMISSION ANNUAL REPORT, 1974

Type of herd		Lowland	Upland	Hill
Sire breed		Calf birth weight		
		kg	kg	kg
Charolais		43.8	40.9	37.3
South Devon		42.7	40.5	36.4
Hereford	} most	36.4	34.5	31.8
Aberdeen Angus	} widely used	33.2	30.9	29.1

The use of artificial insemination with these sires also produced problems for the Milk Marketing Board, as little information was available from the source of the animals. This meant a great amount of field data had to be collected to see what breeds and individual bulls were most likely to cause dystocia.

TABLE 4

BREED DIFFERENCES OF SIRES ON FRIESIAN COWS (MILK MARKETING BOARD, 1975)

	Bulls	No. of progeny	Dystocia %	Mortality %
Simmental (Team 1)	11	4 507	3.5	4.5
Hereford (Team 5)	27	4 726	0.9	2.7
Charolais (Team 3)	10	2 311	3.6	4.7

Table 5 illustrates the difference between three bulls. Extremes are taken to emphasise the point.

TABLE 5

THREE SIMMENTAL BULLS (MILK MARKETING BOARD, 1975)

Bull	No. of progeny	Dystocia %	Mortality %
Harald	348	7.5	5.5
Caesar	238	6.7	9.2
Paul	380	1.3	2.6

It is important to remember in passing that the Charolais
and Simmental gestation periods are considered to be 290 days
while that of the British Friesian is 283 days and the Aberdeen-
Angus 278 days. Crossbred animals take mid-parent values.

The difficulties in crossbreeding of the early 1970's have
largely been overcome by a more cautious approach by the breeder
and by the careful selection of bulls for AI by the Milk
Marketing Board.

The other area which produced problems was the great surge
of interest in European breeds in the United States, Canada,
Australia and New Zealand. The United Kingdom, because of its
island state and its unique position vis-à-vis foot-and-mouth
disease, became the only acceptable source of these animals.

The massive prices (now somewhat less) led to many
pressures on the veterinary surgeons in practice to produce
live calves. Many practical lessons were learned in relation
to induction of parturition.

Meat and Livestock Commission figures just published (1977)
illustrate the problems that the Charolais and Simmental have
produced in pedigree herds.

TABLE 6

CALVING DIFFICULTIES IN UK PEDIGREE HERDS

Breed	No. of calvings	Dystocia %
Aberdeen-Angus	1 012	0.7
Welsh Black	1 700	1.2
Lincoln Red	1 420	1.3
Hereford	3 749	3.7
South Devon	800	5.6
Charolais	913	11.3
Simmental	541	11.8

From practical experience, the figures for Hereford and Simmental seem too high. It could be that, because they were in pedigree herds, the animals were allowed to become too fat for calving.

The direct effect of birth weight is seen in the breed differences in Table 7.

TABLE 7

Birth weight (kg)	Angus Total %	Dystocia %	Hereford Total %	Dystocia %	Charolais Total %	Dystocia %
Under 35	94	2.2	45	3.9	9	6.3
35 - 40	5	-	44	4.2	27	6.2
40 - 45	1	-	7	10.9	35	9.1
45 - 50	-	-	2	18.1	18	11.5
Over 50	-	-	1	18.2	12	23.4

Induction in heavy cattle must be carried out with an adequate dose of steroid, eg 35 mg of betamethasone.

2. MANAGEMENT OF PARTURITION

Even accepting the breed differences and the different market situations that exist in New Zealand and other countries, we would find it intolerable to induce in such a way as to accept such high calf mortality. (About 50% of UK beef comes from Friesian dairy calves). In New Zealand, it is estimated that 100 000 cows per annum are induced to calve for grass availability to get optimal milk production. These cows are pregnant from 150 days plus, and even when induced in the last 40 days of gestation, 30% of calves are stillborn (Widdows, 1974).

As veterinary surgeons, we have a moral duty to emphasise that the problem should be attacked from the other angle of using good herd records, adequate nutrition, good oestrus detection and prompt treatment of 'difficult' cows to ensure a logical approach to reducing the calving-to-conception interval. In this way, induction could be kept to a minimum.

Batches of heifers, of course, could be synchronised using prostaglandin and inseminated at the suitable time for correct calving.

2.1.Which steroid should be used?
Under field conditions,the use of long-acting corticosteroids shows no advantages. The high rate of predictability using short-acting steroids (Terblanche et al.,1976) makes them preferable in the UK when induction is carried out at 260 days plus. Any reduction in retained placenta claimed (Tervit, 1976) is of doubtful value as many workers have shown it has no effect on future reproductive performance (Wagner et al., 1974).

The fact that long-acting steroids last for about 17 days means that they are present in the calf, the cow and its milk post-partum, which may be a disadvantage in public health terms.

It would also appear from published work that long-acting steroids may affect the availability and absorption of immunoglobulins. Bailey et al. (1973) using long-acting steroids found that colostral immunoglobulin concentration was negatively related to the response time for the drug, but positively related to the number of days premature.

Calf serum immunoglobulins showed a significant relationship to (a) response time, (b) days premature and (c) colostral immunoglobulins. Eight out of thirteen premature calves were hypogammaglobulinaemic against one out of fourteen controls. The possible contributing factors were (a) lower immunoglobulins concentration in the dams' colostrum, (b) the lethargic attitude to suckling and therefore less volume ingested by premature calves and (c) the steroid may have the effect of premature closure of immunoglobulin absorption in the calf as happens in rodents. Becker and Guttman (1972) have demonstrated an effect of dexamethasone on immunoglobulin synthesis. Thus the longer the cow is exposed to the drug during the pre-calving period, the lower should be the state of immunoglobulin synthesis.

Hoerlein and Jones (1977) using short-acting flumethasone, found no difference in the immunoglobulin levels of calves born from induced cows compared to controls in two small trials. They stated that the total immunoglobulin was less in induced cows simply due to the fact that less colostrum was present, ie volume but not concentration was lower. An important point made in this paper that is relevant to the setting up of colostrum banks for calves was the statement that 'the rapid decrease of immunoglobulin in colostrum collected at successive milking points shows the necessity of saving the first colostrum after parturition.' The setting up of colostrum banks is very useful. Two litre packs of deep-frozen colostrum can be used for at least 6 months. Muller et al. (1975) also found little difference in immunoglobulin levels in induced and full-term calves when short-acting corticosteroids were used.

To sum up, the present evidence is in favour of short acting corticosteroids in relation to (a) predictability of time of response, (b) preservation of immunoglobulin status and (c) public health safety.

2.2. Medical termination of pregnancy

Hydrops. The treatment of this condition has always been extremely difficult. Christiansen and Hansen (1974) quote seven cases. Five were successfully induced within four days. One cow died. Van de Plassche et al. (1974) gave one or two injections of 20 - 40 mg dexamethasone to 21 cows with hydrops in their 7th to 9th month of pregnancy with a successful outcome in 17 cases.

The author has used 30 mg betamethasone successfully in two cases, a live calf (a requisite for steroid success) being born in both cases within four days.

Traumatic reticulitis. Ficarelli et al.(1970) successfully induced five cows diagnosed as suffering from acute traumatic reticulitis with flumethasone at 260 days or more of pregnancy. Three cases showed rapid clinical improvement, one slow improvement and one not at all. Three chronic reticulitis cases produced two live calves and one dead calf, and all dams were clinically cured.

Macerated or mummified foetus. These do not fulfil the criteria for corticosteroid induction; prostaglandin should be used.

The author has attempted induction without success in cases of pregnancy toxaemia.

Fractures. Cases where a valuable calf could be salvaged without too much distress to the dam might be considered.

REFERENCES

Bailey, L.F. et al. 1973. The use of dexamethasone trimethylacetate to
advance parturition in dairy cows. Australian Veterinary Journal
49, 567-73.

Beal, W.E. et al. 1976. Induction of parturition with $PGF_{2\alpha}$ following
dexamethasone. Journal of Animal Science 42, 1564.

Becker, M.J. and Guttman, H.N. 1972. Cell Immunology. 5, 122.

Christiansen, I.J. and Hansen, L.H. 1974. Dexamethasone-induced parturition
in cattle. British Veterinary Journal 130, 221-229.

Ficarelli et al. 1970. Nuova Veterinaria 46, 340-343.

Garverick, H.A. et al. 1974. Use of estrogen with dexamethasone for inducing
parturition in beef cattle. Journal of Animal Science 38, 584-590.

Grunert, E., Ahlers, D. and Jöchle, W. 1975. Effects of a high dose of
diethylstilbestrol on the delivery of the placenta after corticoid-
induced parturition in cattle. Theriogenology 3, 249-258.

Hoerlein, A.B. and Jones, D.L. 1972. Bovine immunoglobulins following
induced parturition. Journal of the American Veterinary Medical
Association 170, 325-326.

Kesler, D.J. et al. 1976. Concentrations of hormones in blood and milk
during and after induction of parturition in beef cattle with
dexamethasone and estradiol-17β.Journal of Animal Science 42, 918-926.

Muller, L.D. et al. 1975. Calf response to the initiation of parturition
in dairy cows with dexamethasone or dexamethasone with estradiol
benzoate. Journal of Animal Science 41, 1711-1716.

Terblanche, H.M. et al. 1976. Induced parturition in cattle. 1. Clinical
studies. Journal of the South African Veterinary Association 47,
113-115.

Tervit, H.R. 1976. Techniques and hazards of embryo manipulation and
induction of parturition. New Zealand Veterinary Journal 24, 74-79.

Van de Plassche, M., Bonters, R., Spincemailli, F. and Boute, P. 1974.
Induction of parturition in cases of pathological gestation in cattle.
Theriogenology 1, 115-121.

Wagner, W.C.,Willham, R.L. and Evans, L.E. 1974. Controlled parturition in
cattle. Journal of Animal Science 38, 485-489.

Widdows, F.E. 1974. Use of dexamethasone trimethylacetate in the induction
of parturition in cattle. New South Wales Veterinary Procs. 10, 46-48.

USE OF PROSTAGLANDINS FOR INDUCTION OF PARTURITION IN THE COW

M.J. Bosc

INRA - Station de Physiologie de la Reproduction,
Nouzilly 37380 Monnaie, France.

ABSTRACT

In the cow, the corpus luteum is considered to be the main source of production of progesterone, especially at the end of pregnancy. As has been shown by several authors, $PGF_{2\alpha}$ and its analogues have luteolytic properties and the end of the progesterone inhibition on the uterine motility results in abortion or calving with a high rate of success. The interval between treatment and calving depends on the compound. Thus with two 16-aryloxy-prostaglandins this mean interval is respecively 29.3 h ($\pm$ 12.1 h) and 41.6 h ($\pm$ 15.3 h). The main secondary effect which follows an induced calving with prostaglandins is the high incidence of placenta retention as is also observed with the use of corticosteroids.

1. INTRODUCTION

Birth is the result of uterine activity and of the preparation of the cervical canal. In the cow, as in many domestic mammals, the uterine activity begins and develops during the last day of gestation (Döcke, 1962; Gillette and Holm, 1963; Zerobin and Spörri, 1972). This uterine activity is under hormonal regulation; it depends upon the antagonsim of an inhibiting factor, progesterone (Csapo, 1956), and of stimulating ones (Csapo, 1977). Oxytocin and oestrogens are traditionally considered as stimulating; they would act by $PGF_{2\alpha}$ in the myometrial cell (Csapo, 1977).

The recent studies undertaken in the ewe, the goat and the sow have shown that these hormonal factors are under the control of the foetal system; hypothalamus, hypophysis and adrenals (Liggins et al., 1973; Currie et al., 1973; Fevre et al., 1975). In the cow, cases of abnormal prolonged gestation suggest that the foetus plays a determining role for its own birth (Kennedy, 1971). ACTH administration to the foetus by intramuscular injection (Bosc, 1974; Welch et al., 1973) or by perfusion (Comline et al., 1974) is followed by premature calving. Moreover, as in the ewe and the goat, the increase of foetal cortisol precedes those of oestrogens and of $PGF_{2\alpha}$ (Fairclough et al., 1975) and the final drop of the maternal level of progesterone (Comline et al., 1974). The use of corticosteroids for the induction of parturition takes into account these properties (Jöchle, 1973, Bosc, 1974).

2. CONTROL OF CALVING WITH PROSTAGLANDINS

2.1. The fall in the production of progesterone

The other possibility to induce calving is to remove the inhibition due to progesterone. Normally in the cow the level of progesterone in maternal blood plasma decreases gradually during the last thirty days of pregnancy and then falls more rapidly in the two days before parturition (Stabenfeldt, 1974). Inversely, pregnancy can be artificially prolonged by high doses of progestagen (Nellor, 1963) or of progesterone (Jöchle et al., 1972; Bosc, 1974).

In the cow, progesterone is essentially produced by the corpus luteum (CL) during pregnancy. Removal of the CL before Day 200 of gestation or after Day 260 leads to abortion (Estergreen et al., 1967; Hoffmann et al., 1977; McDonald et al., 1953). However, during the seventh or the eighth month of gestation, enucleation of the CL is compatible with continuation of pregnancy for 4 to 74 days (Estergreen et al., 1967). As has been suggested, gestation can be maintained with low doses of progesterone (Tanabe, 1966; Thorburn et al., 1977) and it is possible that an extra-ovarian source of progesterone exists.

Prostaglandins are well known as luteolytic agents, particularly during the oestrous cycle of the cow. At the end of pregnancy, their secretion is simultaneous with the fall in progesterone (Fairclough et al., 1975), although it starts earlier (Edqvist et al., 1975). The administration of $PGF_{2\alpha}$ or of an analogue of $PGF_{2\alpha}$ (A-PGF) is also followed by a fall in the maternal level of progesterone (Lamond et al., 1973; Aboul-Fadle et al., 1975; Bosc et al., 1975; Henricks et al., 1977).

2.2. Induction of abortion

The luteolytic effect of prostaglandins during pregnancy results in abortion (Lamond et al., 1973; Aboul-Fadle et al., 1975; Sloan, 1976).

The results obtained by different authors at various stages of pregnancy are shown in Table 1. $PGF_{2\alpha}$ or analogues are very efficient at the beginning of pregnancy (Lauderdale, 1974; Aboul-Fadle et al., 1975); at mid-pregnancy, some differences between analogues appear (Lauderdale, 1974; Cooper et al., 1976). In pathological cases, such as prolonged gestation, or when the cow bears a mummified foetus, the A-PGF seems an adequate tool for termination of pregnancy; in case of macerated foetus, their efficiency is not evident (Van de Plassche et al., 1974).

TABLE 1
INDUCTION OF ABORTION WITH PROSTAGLANDINS

Compound	Administration			Abortion		Reference
	Stage of pregnancy	Mode	Dose (mg)	Success	Range (h)	
A-PGF[1]	Beginning	im	45	100%	(?)	Lauderdale, 1974
A-PGF[2]	"	im	0.5	4/4	(24 - 72)	Aboul-Fadle et al., 1975
A-PGF[1]	Mid-pregnancy	im	45	<66%	(?)	Lauderdale, 1974
A-PGF[2]	"	im	0.5	82/83	(?)	Cooper et al., 1976
PGF	"	Foetal fluids	15 - 30	6/9	(24 - 96)	Sloan, 1976
A-PGF[2]	Mummification	im	0.5	15/16	(?)	Cooper et al., 1976
A-PGF[2]	"	iu	10.5 - 45	3/5	(72 - 96)	Van de Plassche et al., 1974
A-PGF[2]	Macerated foetus	iu	22.5 - 35	0/2		Van de Plassche et al., 1974
A-PGF[2]	Prolongation	im	0.5	2/2		Aboul-Fadle et al., 1975

A-PGF: Analogue of $PGF_{2\alpha}$

(1) - (2):Two different compounds

im - iu: Intramuscularly - Intra uterine.

2.3. Induction of calving

Recent studies have reported the use of natural prostaglandins or of their analogues for the induction of calving. Thus, Zerobin et al. (1973) have administered intravenously or 'in utero' $PGE_{2\alpha}$ or $PGF_{2\alpha}$ at various doses (5 to 50 mg); they note an uterine activity rapidly after the PG administration and calving 2 - 3 days after. Kordts and Jöchle (1975) have made a similar study with 20 mg of $PGF_{2\alpha}$ intravenously. Spears et al. (1974), Bosc et al. (1975), Henricks et al. (1977) have used analogues of $PGF_{2\alpha}$ at different doses varying with the compounds.

Table 2 presents the results obtained with two 16-aryloxyprostaglandins (Binder et al., 1974). With the first compound, the mean interval from treatment to calving is 29.3 h ($\pm$ 12.1). With the second, this interval is higher for the reason that parturition was not induced in one case with the lowest dose (0.175 mg); calvings were induced 41.6 h ($\pm$ 15.3) after A-PGF injection in the other cows. In this experiment, the two A-PGF give a mean interval smaller than with dexamethasone, but the variability seems to be greater than with corticosteroids.

The published results indicate a very high rate of success. Following Henricks et al. (1977), a minimum level of endogenous oestrogen is required before $PGF_{2\alpha}$ is effective. They noted an oestrogen level higher in the responding animals than in the non-responding animals at the time of $PGF_{2\alpha}$ injection. This permissive action of oestrogen has also been noted with corticosteroids (Evans, 1973).

The combination dexamethasone (16 mg) and A-PGF has been made (Bosc et al., 1975). In this case, calving occurred in a shorter time with A-PGF alone than with dexamethasone (Table 2). With the two, a smaller spread of births is noted than with either compound alone.

TABLE 2

INDUCTION OF CALVING WITH AN ANALOGUE OF PROSTANGLANDIN $F_{2\alpha}$ AND/OR DEXAMETHASONE

(Sire) cow	Treatment			N	Interval treatment/ calving (h)	Dystocia	
	Day	Product	Dose (mg)			Placenta- retention	Malpresentation
(Ch) Nor (Nor) Nor	208 ± 1	A-PGF[1]	4 to 10	8	29.30 ± 12.10	6/8	3/8
(Ch) Ch	281 ± 1	DX + A-PGF[1]	16 2x2	6	38.30 ± 6.40	3/6	0/6
(Ch) Nor	278 - 279	DX	16	9	32.58 ± 9.06	4/9	0/9
(Nor) Nor	277 ± 1.3	A-PGF[2]	0.175 to 1	11	47.10 ± 22.42	8/11	1/11

Ch = Charolais; Nor = Normand.

A-PGF = Analogue of $PGF_{2\alpha}$ ICI compound (1) 79939; (2) 80996.

DX = Dexamethasone phosphate

Control cows, duration of pregnancy: (Nor) Nor: 284.2 (± 6)
(Ch) Nor: 284.5 (± 3)
(Ch) Ch: 288.1 (± 4)

2.4. Induction with oestrogens

At the end of pregnancy, the administration of stilboestrol
in the ewe (Liggins et al., 1973) or of oestradiol-17β (Currie
et al., 1973) in the goat is followed by an increase of the
secretion of $PGF_{2\alpha}$ in the utero-ovarian vein. In these two species,
oestradiol benzoate is a potent means for the induction of
parturition (Terqui and Delouis, 1975). From these results,
one can assume that oestrogens would be efficient also in the
cow. The results obtained with different doses of oestradiol
benzoate given intramuscularly a few days before the mean term
are reported in Figure 1. As it can be noted, oestradiol
benzoate is ineffective in inducing calving even at a dose
of 180 mg. In this trial, 24 Friesian cows were treated; their
gestation lengths were the same as those of the control (275.4
days ± 39 v 275.7 days ± 4.7). This does not confirm earlier
reports (Gronborg-Pedersen, 1969) and does not support an active
role for the exogenous oestrogens in regard to calving.

2.5. Secondary effects

The main secondary effect, after calvings induced with PGF,
is the high rate of placental retention as it has been pointed
out after the corticosteroids (Jöchle, 1973; Bosc, 1974). The
addition of oestradiol-17β at a dose of 400 μg does not reduce
significantly their number (Henricks et al., 1977). This has
also been noted with corticosteroids (Bosc, 1973) even with
large doses of 90 mg (Bosc, unpublished). However, by
comparison, the addition of high dose of diethylstilboestrol
results in a lower rate of the non-delivery of the placenta
(Grunert et al., 1975).

Henricks et al. (1977) note a high rate of dystocia of
presentation. In fact, the calf is oriented several weeks
earlier (Naaktgeboren, 1963) and induction cannot explain his
observation.

As it has often been reported, birth weight is lower in
induced cows than in controls without there being a significant
effect on calf viability. Udder development was affected in the

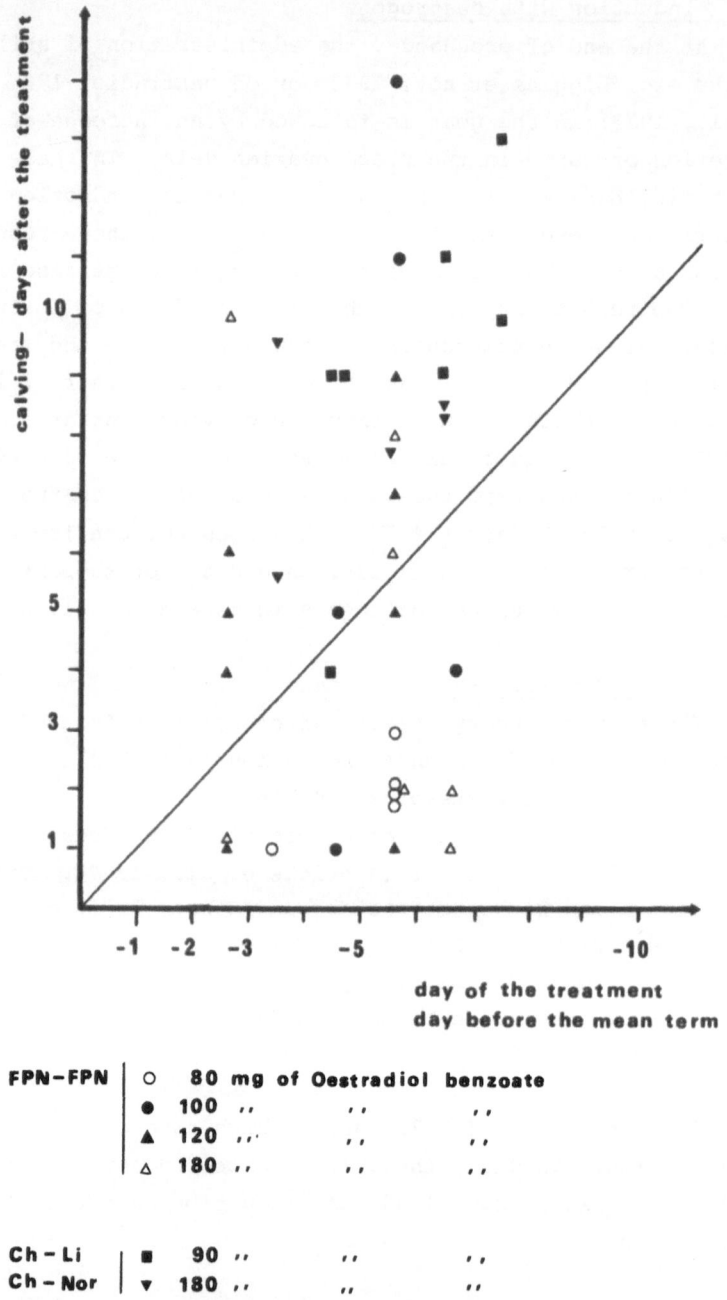

Fig. 1. Effect of oestradiol benzoate on duration of pregnancy in the cow.

experiments of Henricks et al. (1977), but no adverse effects were observed by Zerobin et al. (1973) on lactation and post-partum fertility.

3. CONCLUSION

Prostaglandin $F_{2\alpha}$ or its analogues are effective in inducing parturition in the bovine species. They can be used to induce abortion, especially in some pathological conditions. All the secondary effects have not yet been well defined. The incidence of retained placenta is high; it is not reduced by combination of A-PGF with oestrogens or corticosteroids. It has been suggested that $PGF_{2\alpha}$ or A-PGF induces parturition in the cow by their luteolytic effects and then progesterone production falls before birth in the induced animals. In the non-responding animals, this does not occur (Henricks et al., 1977). Further investigations are needed for a better understanding of their mode of action and of their possible interaction with endogenous oestrogens.

362

REFERENCES

Aboul-Fadle, W., Grunert, E., Schulte, B. and Matas, V. 1975. Progesteron-
    bestimmung im Blutplasma von tragenden Kühen vor und nach Applikation
    von Östrogenen und Prostaglandin. Zuchthygiene 10, 169-176.

Binder, D., Bowler, J., Brown, E.D., Crossley, N.S., Hutton, J., Senior,
    M., Slater, L., Wilkinson, P. and Wright, N.C.A. 1974. 16-Aryloxyprostag-
    landins: a new class of potent luteolytic agent. Prostaglandins 6,
    87-90.

Bosc, M.J. 1973. Données actuelles sur le contrôle de la parturition chez
    la brebis et la vache. Recueil de Médecine Vétérinaire 149, 1463-1480.

Bosc, M.J. 1974. La parturition provoquée chez les mammifères domestiques.
    In 'Avortement et Parturition Provoqués'. Bosc, Palmer and Sureau Ed.,
    Masson Cie, Paris, pp 1-24.

Bosc, M.J., Fevre, J. and Vaslet de Fontaubert, Y. 1975. A comparison of
    induction of parturition with dexamethasone or with an analogue of
    prostaglandin $F_{2\alpha}$ (A-PGF) in cattle. Theriogenology 5, 187-190.

Comline, R.S., Hall, L.W., Lavelle, R.B., Nathanielsz, P.W. and Silver, M.
    1974. Parturition in the cow: endocrine changes in animals with
    chronically implanted catheters in the foetal and maternal
    circulations. Journal of Endocrinology 63, 451.

Cooper, M.J., Jackson, P.S. and Norman, J.A. 1976. Utilisation thérapeutique
    des prostaglandines chez les bovins. Economie et Médecine animales
    17, 209-213.

Csapo, A.I., 1956. Defense mechanism of pregnancy. In 'Progesterone and
    the defense mechanism of pregnancy'. Ciba Foundation Study Group no.
    9. Little Brown & Co., Boston.

Csapo, A.I. 1977. The see-saw theory of parturition. Ciba Foundation
    Symposium 47, Elsevier Excerpta Medica, North Holland, Amsterdam, pp.
    159-210.

Currie, W.B., Wong, M.S.F., Cow, R.I. and Thorburn, G.D. 1973. Spontaneous
    or dexamethasone induced parturition in the sheep and goat: changes
    in plasma concentrations in maternal prostaglandin F and foetal
    oestrogen sulphate. Memoirs of the Society for Endocrinology
    20, 95-118. Endocrine Factors in Labour. Ed. A. Klopper and J. Gardner.

Döcke, F. 1962. Untersuchungen zur Uteruskontraktilität beim Rind. Archiv
    für experimentelle Veterinärmedizin 16, 1205-1307.

Edqvist, L.E., Kindahl, H. and Stabenfeldt, G.H. 1976. On the role of prostaglandins in bovine parturition. VIIIth International Congress on Animal Reproduction and Artificial Insemination, Krakow, 1, 76.

Estergreen, V.L., Frost, O.L., Gomes, W.R., Erb, R.E. and Bullard, J.F. 1967. Effect of ovariectomy on pregnancy maintenance and parturition in dairy cows. Journal of Dairy Science 50, 1293-1295.

Evans, L.E. 1973. Physio-pathologic studies on induced parturition in the cow (Bos taurus). Ph.D. Thesis, Iowa State University, Ames, Iowa.

Fairclough, R.J., Hunter, J.T. and Welch, R.A.S. 1975. Peripheral plasma progesterone and utero-ovarian prostaglandin F concentration in the cow around parturition. Prostaglandins 9, 901-914.

Fevre, J., Terqui, M. and Bosc, M.J. 1975. Mécanismes de la naissance chez la truie. Equilibres hormonaux avant et pendant le part. Journées de la Recherche Porcine en France, ITP - INRA Ed., Paris, pp 393-398.

Gillette, D.D. and Holm, L. 1963. Pre-partum to post-partum uterine and abdominal contractions in cows. American Journal of Physiology 204, 115-1121.

Gronborg-Pedersen, H. 1969. Partus provocatus hos koen. Nordisk Veterinär-medecine 21, 591-597.

Grunert, E., Ahlers, D. and Jöchle, W. 1975. Effects of high dose of diethylstilbestrol on the delivery of the placenta after corticoid induced parturition in cattle. Theriogenology 3, 249-258.

Henricks, D.M., Rawlings, N.C., Ellicott, A.R., Dickey, J.F. and Hill, J.R. 1977. Use of prostaglandin $F_{2\alpha}$ to induce parturition in beef heifers. Journal of Animal Science 44, 438-441.

Henricks, D.M., Rawlings, N.C. and Ellicott, A.R. 1977. Plasma hormone levels in beef heifers during prostaglandin-induced parturition. Theriogenology 7, 17-28.

Hoffmann, B., Wagner, W.C., Rattenberger, E. and Schmidt, J. 1977. Endocrine relationships during late gestation and parturition in the cow. Ciba Foundation Symposium 47 'The foetus and birth'. Elsevier Excerpta Medica North Holland, Amsterdam, pp. 107-125.

Jöchle, W., Esparza, M., Gimenez, T. and Hidalgo, M.A. 1972. Inhibition of corticoid induced parturition by progesterone in cattle; effect on delivery and calf viability. Journal of Reproduction and Fertility 28, 407-412.

Jöchle, W. 1973. Corticosteroid induced parturition in domestic animals. Annual Review of Pharmacology 13, 33-55.

Kennedy, P.C. 1971. Interaction of foetal disease and the onset of labour
in cattle and sheep. Federation Proceedings 30, 110-113.

Kordts, E. and Jöchle, W., 1975. Induced parturition in dairy cattle: a
comparison of a corticoid (fluomethasone) and a prostaglandin ($PGF_{2\alpha}$)
in different age groups. Theriogenology 3, 171-178.

Lamond, D.R., Tomlinson, R.V., Drost, M., Henricks, D.M. and Jöchle, W.
1973. Studies of prostaglandins in the cow. Prostaglandins 4, 269-284.

Liggins, G.C., Fairclough, R.J., Grieves, S.A., Kendall, J.L. and Knox, B.S.
1973. The mechanism of initiation of parturition in the ewe. Recent
Progress in Hormone Research 29, 111-159.

Lauderdale, J.W. 1974. The use of prostaglandins in cattle. Annales de
Biologie animale, Biochimie, Biophysique 15, 419-425.

McDonald, L.E., McNutt, S.H. and Nichols, R.E. 1953. On the essentiality
of the bovine corpus luteum of pregnancy. American Journal of
Veterinary Research 14, 539-541.

Naaktgeboren, C. 1963. Untersuchungen über die Geburt der Säugetiere.
Thesis University of Amsterdam.

Nellor, J.E. 1963. Induced delayed paruturition in swine and cattle.
Physiologist 6, 244.

Sloan, C.A. 1976. Comparison of dexamethasone trimethylacetate and
prostaglandin $F_{2\alpha}$. Journal of Reproduction and Fertility 46, 529.

Spears, L.L., Bercovitz, A.B., Reynolds, W.L., Kreider, J.L. and Godke,
A.A. 1974. Induction of parturition in beef cattle with oestradiol
and $PGF_{2\alpha}$. Journal of Animal Science 39, 227.

Stabenfeldt, G.H. 1974. The role of progesterone in parturition: premature,
normal, prolonged gestation. In 'Avortement et parturition provoqueés'.
Bosc. Palmer, Sureau Ed., Masson and Cie, Paris, pp. 97-122.

Tanabe, T.Y. 1966. Essentiality of the corpus luteum for maintenance of
pregnancy in dairy cows. Journal of Dairy Science 49, 731.

Terqui, M. and Delouis, C. 1975. Les oestrogènes au cours de la gestation
et de la parturition chez la brebis. lères Journées de la
Recherche Ovine et Caprine, Paris INRA - ITOVIC ed.; 2, 332-341.

Thorburn, G.D., Challis, J.R. and Currie, W.B. 1977. Control of parturition
in domestic animals. Biology of Reproduction 16, 18-27.

Van de Plassche, M., Bonters, R., Spincemailli, F. and Boute, P. 1974.
Induction of parturition in cases of pathological gestation in cattle.
Theriogenology 1, 115-121.

Welch, R.A.S., Frost, O.L. and Bergman, M. 1973. The effect of administering
ACTH directly to the foetal calf. New Zealand Medical Journal 24, 365.

Zerobin, K. and Spörri, H. 1972. Motility of the bovine and porcine uterus
and Fallopian tubes. Advances in Veterinary Science and Comparative
Science $\underline{16}$, 303-354.

Zerobin, K., Jöchle, W. and Steingruber, Ch. 1973. Termination of pregnancy
with prostaglandins $E_{2\alpha}$ ($PGE_{2\alpha}$) and $F_{2\alpha}$ ($PGF_{2\alpha}$) in cattle.
Prostaglandins $\underline{4}$, 891-901.

DISCUSSION

H. Karg *(West Germany)*

The basic question discussed during this session was whether there are efficient biochemical ways to advance calving in order to avoid calving difficulties or whether a Caesarean section at term should be done. We have heard from several papers this morning that the endocrinological events at parturition follow an exact countdown, and that induced parturition means to acquire control of all these processes. Perhaps we can divide the discussion and first talk about the methods of inducing parturition and then about the problem of retained placenta. There seems to be no doubt that short-acting corticoids are better than long-acting corticoids. Dr. Bosc indicated that oestradiol benzoate was not an active compound for induction of parturition in the cow. Perhaps it should be kept in mind that it is not a natural compound.

M. Bosc *(France)*

I agree that oestradiol benzoate, like diethylstilboestrol, is not a natural oestrogen. But I would remind you that it is very efficient for inducing parturition in the ewe and the goat. The dose recommended by Terqui and Delouis is 15 mg. We tried oestradiol benzoate in the cow in much higher doses (between 30 and 180 mg) without any success.

R.D. Politiek *(Netherlands)*

The question which actually should be raised right at the beginning of the discussion is whether there is a need at all for a widespread application of induced parturition or whether it should be limited to certain special problems only.

R.W.J. Plenderleith *(UK)*

I have tried to express my concern about the uncritical application of induced parturition. The method certainly has its limitations; perhaps it should be applied only by veterinarians in special clinical cases or for certain problems of cross-breeding, eg when using a Simmental or Charolais bull on a Friesian heifer.

H.O. Gravert *(West Germany)*

Under the special conditions of our baby calf programme induced parturition has proved to be an adequate biotechnical procedure.

H. Karg

Perhaps Dr. O'Farrell could also comment on the question raised by Dr. Politiek?

K. O'Farrell *(Ireland)*

We think that in certain management situations - for example late calving dairy cows, or heifers in calf to heavy beef bulls - there is a need for induced calving to reduce calving difficulty. Furthermore, in Ireland, as in New Zealand, coinciding the onset of lactation with the beginning of the grazing season is of high economic importance.

G. Averdunk *(West Germany)*

The use of induced parturition, as mentioned by Mr. Plenderleith with exotic breeds in Britain, is of deep concern to the geneticist. All recording systems and sophisticated analyses are worthless if the incidence of difficult calving cannot be expressed because of artificial induction of calving. Some of the results with exotic breeds in the United States and especially in Australia indicate that natural selection must already have been bypassed with the measures indicated above. Modern multiplication methods, for example egg transfer, and the high value of the transplanted embryos in small cows, may even exacerbate the problem. Responsibility for intensifying beef production rests on all disciplines of animal production and interdisciplinary cooperation seems to be urgently needed, to avoid counter-selection in the elite herds of a hierarchical breeding system. Is there a possibility that information from a local verterinarian can be brought into a genetic information system without disturbing the business basis of the practitioner?

R.W.J. Plenderleith

Induced parturition can only serve as an interim method to solve the problems. In the long term there must be another solution. I agree with Dr. Averdunk that there are big problems related to the genetic aspects of breeding and selection.

M.A.N. Taverne *(Netherlands)*

Jöchle and Zerobin (1973) administered PGF_2a or PGE intravenously or directly into the uterus of pregnant cows and recorded uterine contractions immediately after application. Contractions subsided, however, within a few hours and only after luteolysis had taken place did labour contractions start to develop. Would Dr. Bosc speculate on a similar mechanism of action of exogenous corticoids when administered to the late pregnant cow?

M. Bosc

Actually you should be in a better position to answer that question than I am. I have to repeat our theory, that exogenous corticoids act via a negative feedback on the foetal pituitary, bringing about a rebound effect with more ACTH and cortisol produced which then induce parturition.

M.A.N. Taverne

We have a recording from one cow where parturition was induced with corticoids. No immediate effect of flumethasone on myometrial activity could be seen. Uterine activity increased only after the peripheral progresterone plasma concentrations had reached levels below 1 mg/ml.

H. Karg

Another problem concerning induced parturition is its influence on lactation and hence milk yield. Did I understand correctly from your paper, Mr. Plenderleith, that you could not observe any effects at all?

R.W.J. Plenderleith

No, in general milk yield for the whole lactation period does not differ, but there is a distinct difference in the course of the lactation curve during the first 100-day period.

A. Osinga *(Netherlands)*

According to the study of Remmen (1976) the 100 days lactation yield of heifers in which parturition was induced with flumethasone was 300 kg less than that of the control heifers.

H. Karg

But this is probably due to the slow start of lactation after induction of parturition in heifers and can be compensated afterwards.

A. Osinga

Only partly, since the 300 kg milk cannot be produced in 10 days, which was the average period by which gestation was shortened.

K. O'Farrell

Following induced calving with the short-acting corticoids, udder development may take some time. We have observed that for the first 3 to 4 weeks after induction milk yield is lower than normal but after this it picks up rapidly so that the overall lactation yield is not different. This has also been noted by workers in the USA.

H. Karg

Perhaps we can now switch to the second part of the discussion and talk about the problem of retained placenta. I think I am in agreement with the clinicians that delivery of the placenta is the final part of the parturition process and should occur within a short period after the calf is born.

E. Grunert *(West Germany)*

From the clinical point of view we have to differentiate five stages in the calving cow:
1. Preparatory stage.
2. Stage of dilation of the cervix (duration: 6 - 14 h) (finishes with the rupture of the allantois sac).
3. Stage of dilation of the birth canal (duration: 2 - 6 h)
4. Stage of the expulsion (duration: 10 - 15 min)
5. Afterbirth stage (duration: 8 h).

H. Karg

The question is, how should a retained placenta be treated: is it advisable to try to remove it manually, or should it be left untreated?

K. O'Farrell

In the absence of overt illness we generally leave the placenta alone. If it has not been passed within 12 hours of calving we cut off the dependent membranes approximately 10 cm below the vulva.

R.W.J. Plenderleith

I agree with Dr. O'Farrell. Under practical conditions we try to avoid touching the placenta for several reasons. The use of antibiotics not only brings up the problem of passage into the milk, but also the process of introducing unnecessary antibiotics into the uterus may alter the uterine environment and cause other problems.

H. Karg

Are there no other possible ways of influencing the situation of a retained placenta? For example, doesn't the process of milking itself, via the induced release of oxytocin, have a stimulatory effect on the uterine contractility by which expulsion of the placenta could be enhanced?

M.A.N. Taverne

In cows with retained placenta the uterus shows spontaneous contractility for some days longer than in cows without retention. The influence of milking could not be recorded with our techniques for more than 2 to 3 days post partum, probably because of the absence of any oestrogenic influence.

E. Grunert

Rectal palpation and massage of the genital tract during the puerperium can also stimulate uterine contractility.

R.W.J. Plenderleith

According to our observations in beef herds involution of the uterus is also greatly enhanced by suckling.

M.A.N. Taverne

I agree with both Dr. Grunert and Mr. Plenderleith and their observations on the effect of oxytocin. However, many drugs can induce or influence uterine contractions during the puerperium. But these effects depend on the ovarian activity postpartum.

K. Sejrsen *(Denmark)*

As I see it, induced parturition and also Caesarean sections are unacceptable in routine practice. On the other hand induced parturition and Caesareans obviously play an important role in handling difficult cases. However, the problem is to identify the difficult cases prior to parturition. As we have heard, birth weight has a big influence on calving performance. Therefore a prepartum indicator of the birth weight would be helpful in order to decide whether it will be necessary to induce parturition or perform a Caesarean. As far as I know the concentration of placental lactogen is related to the size of the placenta, which in turn is related to the birth weight. Could a measurement of placental lactogen give an indirect measurement of the birth weight prior to calving? And could this measurement be used to decide whether induced parturition should be applied or not?

M. Bosc

I cannot say whether it is possible or not, since I have read only one article about placental lactogen in the cow. Apparently the concentration of placental lactogen in the peripheral plasma of the dam increases during gestation up to a certain level and remains constant thereafter while the calf is still growing. However, I said yesterday that the level of oestrogens in peripheral plasma on day 220 of gestation is related to the weight of the calf at birth. May I raise a final question. Is there any information on the efficiency of β-inimetics in controlling the time of explusion of the calf?

M.A.N. Taverne

These compounds are widely used in human medicine, for example in cases of a risk of premature labour or in cases of reanimation of the baby during labour.

M. Rüsse *(West Germany)*

There are some experiences in cattle with a tocolytic compound with the code name NAB (Boehringer, Ingelheim) which can block labour for some time even when it has already started.

NUTRITION AND MANAGEMENT OF THE DAM IN RELATION TO CALVING PROBLEMS, PART 1

Chairman: R. Bar-Anan

EFFECT OF REARING INTENSITY AND AGE AT
CALVING ON CALVING PERFORMANCE*

J. Brolund Larsen and K. Sejrsen
National Institute of Animal Science
25 Roligdhedsvej, 1958 Copenhagen V, Denmark.

ABSTRACT

A brief review of the effect of rearing intensity and age at calving on calving performance is given partly based on the report from the EEC seminar, 'The early calving of heifers and its impact on beef production'. Normally heifers have 2 to 4 times higher calf mortality than older cows. The reason may be more calving difficulties resulting from disproportion between the size of the dam and of the calf. It has been found that 50% of the variation in calving difficulties in heifers was explained by birth weight of the calf. The relationship between birth weight and still-birth is not linear indicating that some of the light calves are weak due to malnutrition or too early delivery. The calving performance is also related to the weight of the calf in relation to that of the mother.

The rearing intensity seems to affect the birth weight of the calf, so that heifers reared on high intensity have heavier calves than heifers reared on low intensity. The birth weight in relation to the mother, how-ever, will often, due to lower weight of the mother, be highest at low intensity. This may be the explanation for the tendency to more difficult calvings of heifers reared on a low plane of nutrition.

The birth weight decreases with a decrease in age at calving. However, when reared on the same plane of nutrition, the relative size of the calf often will be highest at the lowest age. This may partly be the explanation of the increase in calving difficulties with decreasing calving age. Other factors must also be involved, since there are more calving problems of younger than older heifers even if they are reared to the same weight at calving.

* Conclusions derived from EEC Seminar, 'The early calving of heifers and its impact on beef production'.

INTRODUCTION

According to the information given at the first EEC
seminar on nutrition and management in beef production, the
average age at first calving in the different breeds varies
from 24 to 34 months. However, in a number of countries there
are signs of a decrease in the calving age. This happens
especially in advanced herds in connection with an intensification of production. In Denmark the average calving age in the
late sixties was 29 months for the large dual-purpose breeds,
with only 12 - 15% of the heifers calving below 27 months
(Elleby and Mygind-Rasmussen, 1971). However, the latest
statistics show a trend towards lower age at first calving, with
about 25% of the heifers calving at 25 months of age or earlier
(Mygind-Rasmussen, 1975).

The advantages of early calving and the possibilities of
increasing the number of calves for beef production by lowering
the age at first calving are often mentioned. Gravert (1976)
found that full use of early calving could increase the total
beef production by about 36%. Larsen and Sejrsen (1975) found
that a decrease in the age at first calving from 30 to 18 months
would increase the number of animals for slaughter per year by
24%, when the number of animals in the herd is kept constant.

The main reasons for early calving not being commonly used
in practice are the increased risk of calving difficulties and
the increase in calf mortality. Danish field data (Elleby and
Mygind-Rasmussen, 1971) show that heifers calving about the
average calving age had lower calf mortality than either younger
or older heifers. This may be caused more by the development
of the heifers rather than the actual age. The results, however,
are in good agreement with other investigations (Dreyer, 1965;
Philipsson, 1976).

The frequency of dystocia and stillbirth reported in most
experiments is 2 - 4 times higher for heifers than for older
cows (Philipsson, 1976). The factors affecting calving

performance are often divided into cow factors (parity, small pelvic opening) and calf factors (sex, birth weight and malpresentation). However, environmental factors such as season of calving may also be of importance. Disproportion between dam and calf is a major factor leading to calving difficulties. Ménissier (1975) has found that 50% of the variation in calving difficulties was explained by birth weight of the calf. The relationship between birth weight and still-birth frequency is not linear (Philipsson, 1976) indicating that some of the light calves are weak due to mulnutrition or too early delivery.

In the last 50 years several experiments have been carried out. The experiments can be divided into three different models in relation to age, rearing intensity and body weight at calving (Henningson, 1970). This is an advantage since there exists a complete interaction between these factors and production. The models may help to explain these interactions.

	Age	Body weight	Rearing intensity
Model 1	equal	different	different
Model 2	different	different	equal
Model 3	different	equal	different

Most experiments up to 1960 were carried through according to Model 1 with normal calving age and different intensity (Steensberg, 1940; Hansson et al., 1967; Reid et al., 1964, among others). The reason was that the feed expenses were very important for the whole production and the question of the optimal rearing intensity was mainly the optimal amount of concentrates (grain) for a given amount of roughage, since there was a big difference between the concentrate and roughage prices. The calving age was almost fixed and there was generally no interest in a lowering of the age at calving.

In the sixties there arose an interest in lowering the age at calving because it was considered that it would be of enonomic advantage to decrease the 'unproductive' part of the cow's life. Experiments according to Model 2 are reported by many authors (Swanson, 1961; Wickersham and Schultz, 1963; Hibbs and Conrad, 1965; Witt et al., 1971). The results showed a lower·yield from heifers of lower age and weight at calving. However, a Danish investigation showed positive influence of body weight on milk yield at a constant age (Nielsen, 1962). These results seemed to indicate that normal milk yield could be achieved with early calving, if the heifers were reared to the same weight at calving. Experiments according to this concept (Model 3) were carried out in Denmark (Larsen et al., 1975).

The calving performance as well as the feeding intensity and the weight of the cow and the calf are not registered in field data. Therefore the information that can be derived from field data concerning the effect of age at calving and feeding intensity is limited. The information, however, can be supplemented with results from experiments, where the animals have been controlled individually. Unfortunately in this material information is often missing concerning calving performance, calf mortality, birth weight etc - possibly because calving performance was not earlier regarded as a serious problem. Another drawback in the experimental data is the limited number of animals per group.

The material mentioned is used together with results, comments and conclusions from the seminar in Holte. However, the material is so comprehensive that everything cannot be shown here; therefore only typical experiments are used to describe the different situations.

EQUAL AGE, DIFFERENT BODY WEIGHT AND DIFFERENT REARING INTENSITY (Model 1)

Experiments according to this model were carried out in Denmark (Steensberg, 1940; Steensberg and Østergaard, 1946). However, especially well known are the Swedish experiments by

Hansson et al. (1967) with monozygous twins, where the feeding
level was varied between 40 and 140% of normal. The heifers on
a very low feeding level had puberty delayed by 5 - 6 months.
Similar results were obtained in American experiments (Reid et
al., 1964). In the American experiments the calving results were
recorded and the results are shown in Table 1.

Birth weight increased with increasing feeding intensity
in the rearing period. The weight of the dam, however, was
increased relatively more, and therefore the weight of the
calves in relation to that of the mothers was highest (9.46%)
in the group on the low plane of nutrition. This must be the
explanation for the fact that twice as many of the calves
required assistance at parturition in this group. This is in
close agreement with the conclusion from English results with
low-fed early-calving heifers (Roy et al., 1975). These authors
concluded that 'for successful parturition the weight of a
Friesian dam should probably not be less than 460 kg'.

In an American experiment (Swanson and Hinton, 1964) two
groups of heifers calving at 2 years of age were reared on
normal intensity and 25% below. Dystocia was common in the
group fed on a low plane of nutrition. This was mainly caused
by high weight of the calf in relation to the dam since very
low feeding only had a small effect on the weight of the foetus.
The conclusion was that a very low plane of nutrition should not
be used, since it leads to more calving difficulties, even if
there is a slight reduction in the birth weight. Furthermore
there was a reduction in the subsequent milk production. The
feeding intensity has an obvious effect on the viability of the
calves up to 1 month of age.

DIFFERENT AGE, DIFFERENT BODY WEIGHT AND EQUAL REARING INTENSITY
(Model 2).

At the beginning of this century American cattle scientists
showed that well fed heifers exhibited their first oestrus
as early as 6 months of age. It was therefore natural that

TABLE 1

DIFFERENT LEVEL OF NUTRITION FOR HEIFERS (Reid et al., 1964)

Group	Number of animals	Intensity %	Calving age months	Dams weight kg	Birth weight kg	Birth weight Dam's weight	Calving difficulty %
Low	31	62	32.0	384	36.0	9.46	48
Normal	34	100	28.5	482	38.6	8.00	27
High	33	146	27.9	548	41.2	7.51	25

several experiments were made to investigate the possibilities
of lowering the age at first calving, when this became of
interest. One of the most well known experiments was referred
to at the seminar by Gravert (1975): 'In experiments with
monozygous twins by Witt et al. (1971) each pair was split
(Table 2) one heifer calving at 2 years old and one at 3 years
old. Though there was a weight difference of the dams of 110
kg (21%) the calf weight only differed by 3.1 kg (9%). The
difference could partly be explained by a shorter gestation
length (3 days). It seems to be important that there were no
more calving difficulties with calving at 2 years than at 3
years'.

In an American experiment (Wickersham and Schultz, 1963)
with Holstein-Friesian heifers reared on 114% of Morrison's
standard there seemed to be more calving difficulties in the
youngest age-group. The difficulties seemed to be attributed
to a tendency for the extremities of the calf to lock in the
pelvic girdle. The heifers had an average body weight of 486
kg after calving, which is generally accepted as a satisfactory
weight for Holstein heifers at first calving. Also the average
height of withers was normal. This means that body weight at
calving cannot be used as the sole criterion for determining
a safe calving, since early calving heifers with sufficient
body weight are not necessarily large enough to deliver a full-
size calf successfully.

In an Italian experiment with Holstein-Friesians
(Romita, 1975), three groups of heifers were bred to Limousin
bulls at 9, 14 and 19 months of age (Table 2). The birth weight
increased with increasing age, while the birth weight in
relation to the dam decreased. There was similar ease of calving
in the groups, and the mortality was low. In another trial
Friesian heifers were bred early to bulls of the Marchigiana
Piemontese, Charolais and Limousin breeds (Table 2). The
calving difficulties, except for Limousin were about 50%
higher than normally obtained with purebred Friesians. Three
groups of a local Italian breed, the Maremmana, whose heifers

TABLE 2

EQUAL INTENSITY – DIFFERENT CALVING AGE (Model 2)

Group	Witt et al 1971 A	B	Schultz 1969 Early	Medium	Late	Romita (1) 1975 PP	P	N
Breed	GF	GF	HF	HF	HF	HF	HF	HF
Number of heifers	21	21	7	11	10	18	18	17
Intensity	100	100	114	114	114			
Calving age, months	37	25	20.3	24.2	27.9	18	23.5	28.6
Dam weight after parturition, kg	520	411	486	557	613	405	475	537
Birth weight, kg	34	31	39.0	38.0	37.6	33.9	37.2	38.5
Birth weight x 100 / Dam weight	6.54	7.54	8.02	6.82	6.13	8.37	7.83	7.18
calving difficulty 1	5	6				10	11	13
2	13	13				3	5	2
3	3	1	} 88%	55%	56%	4	2	–
4						–	–	–
5	1	1				1	–	1
Average						1.83	1.83	1.52
Dead calves						1		2

TABLE 2 (CONT.)

EQUAL INTENSITY – DIFFERENT CALVING AGE (Model 2)

Group	Romita (2) 1975				Romita (3) 1975			Roy et al. 1975	
Breed	MxF	PxF	CxF	LxF	M	CxM	LxM	I	D
	HF	HF	HF	HF	M	M	M	F	F
Number of heifers	9	5	5	12	15	31	15	5	24
Calving age, months	21.4	20.4	20.1	22.2	23.5	23.3	23.2	15.8	18.6
Dam weight after partutition, kg	435	438	435	409	438	532	429	383	437
Birth weight, kg	38.3	39.9	41.9	33.3	29.8	36.8	33.1	32.1	35.9
Birth weight x 100 / Dam weight	8.80	9.10	9.63	8.15	6.80	6.92	7.74	8.7	8.1
Calving difficulty 1	4	2	1	7	11	2	2		
Calving difficulty 2	–	1	1	1	2	3	2		
Calving difficulty 3	1	–	1	2	2	4	2		
Calving difficulty 4	2	2	2	2	–	–	–		
Calving difficulty 5	2	–	1	–	–	–	–		
Average									
Dead calves	3	–	1	1		4.0		60	
Mortality %						19.0		6.3	

usually are bred the first time at 24 months of age or later, were bred at one year of age with bulls from its own breed, Charolais and Limousin respectively (Table 2). The Charolais crosses had the highest birth weight and mortality rate.

Roy et al. (1975) divided a group of early calving heifers in two, according to whether they were born during increasing (I) or decreasing (D) daylength (Table 2). They mention that opinions differ about the relative importance of calf weight and dam weight in causing dystocia. In the I-heifers the over-riding factor was the low live weight of the dams at parturition. Whereas within the D-group, dystocia was a reflection of excessive calf weight of the heifer. In the I-heifers the incidence of dead calves was very high even though the relative weight of the calves was only slightly higher. As mentioned earlier, Roy et al. (1975) estimate that a Friesian heifer should weigh not less than 460 kg. They also mention that the early calving system requires a much higher standard of management than conventional systems.

DIFFERENT AGE, EQUAL BODY WEIGHT AND DIFFERENT REARING INTENSITY (Model 3)

The positive results from the earliest Israeli experiments concerning early calving started several experiments throughout the world. The Israeli experiments have shown that the modern dual-purpose breeds had capacity for rapid growth and early breeding. Furthermore the concentrates (grain) were cheap in relation to roughage, but other economic factors made it profitable to breed earlier, for example the invested capital would return earlier. The experiments in Israel, however, had also shown that the early calving was not without problems (Amir and Kali, 1975). Amir mentioned that there was a high priority of the embryo for nutrition and that the weight of the newborn, even from young heifers, was only slightly reduced. Dystocia therefore is often the most important obstacle to the introduction of early calving. Research with the objectives of reducing or overcoming dystocia is of the highest priority.

That the size of the calf in relation to the size of the dam
is of importance can be seen from results of one of the Israeli
experiments

	Calving performance	
	Normal	Difficult
No. of animals	35	10
Age at calving	19.2	17.9
Live weight, kg	432	432
Weight of calf, kg	34.6	39.5
% of dam's weight	8.1	9.7
Gestation period, days	279	278

Dystocia can partly be overcome by selecting sires known
for small offspring or from small breeds such as Jersey or
Aberdeen-Angus. This practice has been used in Israel (Amir,
1974) quite successfully and in Danish work (Larsen et al., 1975),
where the calving age was decreased from 30 to 24 and 20 months
respectively (Table 3 (2), (3), (4)). In experiment 2 two-thirds
of the heifers were inseminated with Jersey semen. The lower
age at calving resulted in lower weight at calving of the heifer
and in lower birth weight. The results indicate that early
calving in connection with heavy feeding (exp. 2 group C) leads
to more difficult calvings and higher mortality rate. In
experiment 3 the number for this, however, is connected with
the bull selected. The bull was later shown to give an
abnormally high number of dead calves.

Swanson (1975) said about the calving problems: "Research
has shown that these difficulties are mainly due to the relative
size of the calf at birth, but also are affected by age and
fatness of the heifer. Young heifers have more calving trouble,
especially if they are fat at parturition." Arnett et al. (1971)
in a comparison of 12 twin pairs of beef type heifers showed that
obesity resulted in six times as many parturitions requiring
assistance and twice as many calves lost at or soon after birth
compared to moderately lean heifers.

TABLE 3

DIFFERENT INTENSITY - DIFFERENT CALVING AGE

	Little 1975			Gravert 1975				Larsen et al. (1) 1975	
Group	1	2	3	70/71	71/72	72/73	73/74	B	A
Breed	F+A	F+A	F+A	GF	GF	GF	GF	RDM	RDM
Number of heifers	35	35	40	7	4	6	22	17	17
Intensity	high	high	normal					100	108
Calving age, months	18	27	27	19	20	19	18	30	24
Dam's weight after parturition, kg				423	389	348	341	516	474
Birth weight, kg								37.0	35.4
Birth weight x 100 / Dam's weight								7.2	7.5
Calving difficulty 1									
2									
3									
4	} 9			} 1	2	–	5		
5	} 5								
Average									
Dead calves								5	6
Mortality %								29	35

TABLE 3 (CONT.)

DIFFERENT INTENSITY – DIFFERENT CALVING AGE

	Larsen et al. (2) 1975			Larsen et al. (3) 1975				Sejrsen and Larsen, 1977			
Group	B	C	C_1	N	S	K	U	1	2	3	4
Breed	RDM	RDM	RDM	RDM	RDM	RDM	RDM	SDM	SDM	SDM	SDM
Number of heifers	28	28	13	11	10	11	11	9	10	12	10
Intensity	100	125	116					143	127	109	100
Calving age, months	30	20	19	20	20	21	22	20	20	21	24
Dam's weight after parturition, kg	550	472	512	450	419	426	448	488	496	470	482
Birth weight, kg	34.4	32.8	31.2	38.1	32.2	34.5	33.4	31.9	35.1	32.6	34.3
Birth weight x 100 / Dam's weight	6.3	7.1	6.1	8.5	7.7	8.1	7.5	6.6	7.1	7.0	7.2
Calving difficulty 1	5	5	5	–			1	4	5	8	3
Calving difficulty 2	12	8	7	–	5	4	3	3	4	2	2
Calving difficulty 3	8	8	–	5	3	4	4	2	–	2	2
Calving difficulty 4	3	7	1	6	1	3	2	–	–		
Calving difficulty 5	–	–	–	–	1	–	1	–	1		1
Average											
Dead calves	4	8	1	6	5	2	4	1			
Mortality %	14	29	8	55	50	18	36				

For overcoming calving difficulties the most promising
development has been the use of corticosteroids for the
induction of parturition, as mentioned by Amir and Kali (1975).
By this method gestation is shortened and birth weight reduced.
So far results have been variable but successful. Similar
results were obtained by Gravert (1975) and Swanson (1975) said:
"The method has been accompanied by a very high incidence of
retained placenta" (Beardsley et al., 1974).

REFERENCES

Amir, S. 1974. Early breeding of dairy heifers - prospects and limitations.
25th Annual Meeting of EAAP, Copenhagen, 27 pp.

Amir, S. and Kali, J. 1975. Some topics for future research. In: 'The early
calving of heifers and its impact on beef production',ed. J.C. Tayler,
Commission of the European Communities, Luxembourg, pp. 274-280.

Arnett, D.W., Holland, G.L. and Totusek, R. 1971. Some effects of obesity
in beef females. Journal of Animal Science, 33, 1129-1136.

Beardsely, G.L., Muller, L.D., Owens, M.J., Lundens, F.C. and Tucker, W.L.
1974. Initiation of parturition in dairy cows with Dexamethazone.
I. Cow response and performance. Journal of Dairy Science, 57, 1061-
1066.

Dreyer, D. 1965. Geburtsablauf und Kälberverlust, untersucht an Nachkommen
ostfriesischen Besamungsbullen in Testbetrieben. Diss. Göttingen.
156 pp.

Elleby, F. and Mygind-Rasmussen, V. 1971. Kaelvningsstatistik. 4. medd. fra
LPH-udvalget. Århus. 107 pp.

Gravert, H.O. 1975. Practical and economic aspects of early calving in
dairy herds. In: 'The early calving of heifers and its impact on beef
production',ed. J.C. Tayler, Commission of the European Communities,
Luxembourg, pp.220-224.

Gravert, H.O. 1976. The early calving of heifers and its impact on beef
production. 27th Annual Meeting, European Association for Animal
Production, Zürich.

Hansson, A., Brännäng, E. and Liljedahl, L.E. 1967. Studies on monozygous
cattle twins. XIX. The interaction of heredity and intensity of
rearing with regard to growth and milk yield in dairy cattle.
Lantbrukshögskolans Annaler, 33, 643-693.

Henningsson, T. 1970. Studies on monozygous cattle twins. XXII. The effect
of rearing intensity, age and body weight at first calving on milk
production. Lantbrukshögskolans Annaler, 36, 419-429.

Hibbs, J.W. and Conrad, H.R. 1965. Breeding heifers early. Ohio Report
50(S), 72, 73 and 77. Ohio Agricultural Research and Development
Centre, Wooster, Ohio, USA.

Larsen, J.B., Foldager, J. and Sejrsen, K. 1975. A comparison between different feeding intensities using various (concentrate/roughage/grass) rations for rearing dairy heifers to calve at an early age. In: 'The early calving of heifers and its impact on beef production', ed.J.C. Tayler, Commission of the European Communities, Luxembourg, pp. 250-262.

Larsen, J.B. and Sejrsen, K. 1975. The Danish cattle population and beef production. In: 'The early calving of heifers and its impact on beef production',ed J.C. Tayler, Commission of the European Communities, Luxembourg, pp. 2-9.

Little, W. 1975. Early calving of dairy heifers in the United Kingdom. In: 'The early calving of heifers and its impact on beef production',ed. J.C. Tayler, Commission of the European Communities, Luxembourg, pp. 214-219.

Ménissier, F. 1975. Genetic aspects related to use of beef breeds. In: 'The early calving of heifers and its impact on beef production',ed. J.C. Tayler, Commission of the European Communities, Luxembourg, pp. 81-122.

Mygind-Rasmussen, V. 1975. Personal communication.

Nielsen, E. 1962. Afkomprøver med tyre 1945-60. 333. beretning fra Forsøgslaboratoriet, København. 148 pp.

Philipsson, J. 1976. Calving performance and calf mortality. Livestock Production Science, 3, 319-331.

Reid, J.T., Loosli, J.K., Trimberger, G.W., Turk, K.L., Asdell, S.A. and Smith, S.E. 1964. Causes and prevention of reproductive failures. IV. Effect of plane of nutrition during early life on growth, reproduction, production, health and longevity of Holstein cows. Cornell University Agricultural Experimental Station, Bulletin 987, Ithaca, New York, 31 pp.

Romita, A. 1975. Research results from Italy. In: 'The early calving of heifers and its impact on beef production', ed. J.C. Tayler, Commission of the European Communities, Luxembourg, pp. 191-204.

Roy, J.H.B., Gillies, C.M. and Shotton, S.M. 1975. Factors affecting first oestrus in cattle and their effects on early breeding In: 'The early calving of heifers and its impact on beef production', ed. J.C. Tayler, Commission of the European Communities, Luxembourg, pp. 128-142.

Schultz, L.K. 1969. Relationship of rearing rate of dairy heifers to mature performance. Journal of Dairy Science, 52, 1321-1329.

Sejrsen, K. and Larsen, J.B. 1977. Effect of silage:concentrate ratio on feed intake, growth rate and subsequent milk yield of early calving heifers. Livestock Production Science, in press.

Steensberg, V. 1940. Foderenhedsmængdens indflydelse på ungkvægets vækst. 189. beretning fra Landøkonomisk Forsøgslaboratorium, Copenhagen.

Steensberg, V. and Østergaard, P.S. 1946. Foderenhedsmængdens indflydelse på ungkvægets vækst II. 216. beretning fra Forsøgslaboratoriet, Copenhagen.

Swanson, E.W. 1961. Milk production and growth of identical twin heifers calving for the first time at two and three years of age. Journal of Dairy Science, 44, 2027-2034.

Swanson, E.W. 1975. Future research on problems of increasing meat production by early calving. In: 'The early calving of heifers and its impact on beef production', ed. J.C. Tayler, Commission of the European Communities, Luxembourg, pp. 281-288.

Swanson, E.W. and Hinton, S.A. 1964. Effect of seriously restricted growth upon lactation. Journal of Dairy Science, 47, 267-272.

Wickersham, E.W. and Schultz, L.H. 1963. Influence of age at first breeding on growth, reproduction and production of well-fed Holstein heifers. Journal of Dairy Science, 46, 544.

Witt, M., Andrea, U. and Röschler, W. 1971. Einfluss des Erstkalbealters auf Verlauf der Kalbung, die Milchleistung und die weitere Körperentwicklung von einigen Zwillingskühen. 2. Teil. Zeitschrift für. Tierzüchtung und Züchtungsbiologie, 88, 32-46.

PRE-CALVING MANAGEMENT AND FEEDING OF THE BEEF COW
IN RELATION TO CALVING PROBLEMS AND VIABILITY OF THE CALF

B.G. Lowman

The Edinburgh School of Agriculture, West Mains Road,
Edinburgh EH9 3JG, Scotland.

ABSTRACT

*On-farm recording for suckler herds in the UK indicates that 4% of calf
mortality results from difficulties at parturition (Barnes and Kilkenny, 1976);
a more detailed survey (Allen, 1977) demonstrated the importance of dam and
sire breed on calving difficulties in both pedigree and commercial beef herds.*

*Calf birth weight is a major factor determining the incidence of
calving difficulties. Birth weight is influenced by gestation length, sex
of calf and both short and long-term changes in cow nutrition. Extreme
reductions in the level of nutrition immediately before calving have fallen
into disrepute in commercial herds due to their effect on calf survival.
Longer term influences have not been quantified, but do affect birth
weights in commercial practice within the UK as is clearly demonstrated by
effects of season and locality of calving.*

*Managing beef herds in terms of annual changes in body condition is
now being advised as an aid to reducing calving difficulties. In addition,
emphasis is being placed on the importance of adequate supervision to reduce
calving problems which can only be economically achieved in herds with a
compact calving period. In herds with a widespread calving period adequate
supervision can rarely be justified in economic terms and correct
rationing of individual cows becomes difficult.*

*While the obvious financial cost of calf mortality at parturition is
reduced in commercial herds by fostering bought-in calves, the more severe
hidden costs, in terms of disease risk and reduced reproductive performance,
have not been quantified.*

1. THE PROBLEM

There are about 1.9 million beef cows in the United Kingdom supplying approximately half of the calves available for beef production. A survey of 800 of these herds in 1976 involving 19 091 calvings, demonstrated an overall level of dystocia of 5.5% where dystocia was classified as involving <u>mechanical</u> assistance, ie calving ropes, jacks, pulleys, veterinary assistance etc. (Kilkenny and Stollard, 1976).

1.1. Influence of breed of sire and dam

The major factor influencing the level of dystocia was the breed of sire used (Table 1). Calvings by Charolais, Simmental, South Devon and Limousin sires resulted in significantly (P<0.01) higher levels of dystocia and calf deaths than Hereford sires.

TABLE 1

DYSTOCIA AND CALF MORTALITY IN BEEF HERDS BY BREED OF SIRE (Kilkenny and Stollard, 1976)

	Dystocia %	Calf mortality %
Charolais	10.2	4.6
Simmental	10.1	4.4
South Devon	9.6	4.2
Limousin	8.1	3.8
Lincoln Red	7.0	3.2
Devon	6.7	3.6
Sussex	4.7	2.1
Hereford	4.3	1.6
Aberdeen-Angus	2.5	

Similar differences occurred due to breed of dam, but the results were largely confounded with the breed of sire used on certain dam crosses. Thus the majority of Hereford x Friesian cows were mated to large breed sires whilst Angus x Friesian, Hereford cross and Blue-Grey cows were mainly crossed with

Hereford sires (Table 2).

TABLE 2

DYSTOCIA IN LOWGROUND BEEF HERDS BY BREED OF DAM (Barnes and Kilkenny, 1976)

	Dystocia %
Charolais cross	13.4
Hereford x Friesian	8.4
Hereford cross	6.9
Angus cross	2.0
Blue-Grey	1.4

This survey also demonstrated the effect of parity of the dam on the incidence of dystocia, which was 8.3% for three-year-old cows (mainly calving for the first time) compared with 2.9% dystocia in cows 4-6-years-old and only 1.7% thereafter (Allen, 1977). These results confirm those of Laster et al. (1973) at Clay Center in the USA and those reported by the Simmental and Limousin Steering Committee (1977) in the UK. These surveys justify the recommendation that the use of heavy sire breeds should be restricted to cow matings only and preferably to those crosses of cows which characteristically have low levels of dystocia.

The effects of a reduction in the incidence of dystocia and calf mortality in crossbred compared with purebred cows is of particular importance in the UK beef herd which is based, in general, on a crossbred cow mated to a third breed of bull.

1.2. Influence of birth weight

The major single factor influencing the incidence of dystocia in beef herds is calf birth weight. This has been shown to be highly correlated with calving difficulties by Crowley (1965), Milk Marketing Board (1960) and Kilkenny and Stollard (1976). Similar studies in Edinburgh have confirmed this relationship (Table 3).

TABLE 3

INFLUENCE OF BIRTH WEIGHT ON THE INCIDENCE OF DYSTOCIA

Calf birth weight (kg)	< 30	31-35	36.40	41-45	46-50	>50
Dystocia (%)	0	13.0	6.1	21.4	23.8	45.0

In these studies involving 236 calvings an assisted calving was classified as dystocia: ie where one or more people assisted the calving with or without the use of mechanical assistance.

2. FACTORS INFLUENCING BIRTH WEIGHT

2.1. Breed

Breed has a major impact on calf birth weight (MLC data sheets, 1977) and hence dystocia (Allen, 1977), but equally important is the choice of animal within a breed. This has been clearly demonstrated for individual sires within a breed (MMB, 1966) and it is generally accepted that differences between individual cows within a breed or cross are associated at least in part with variation in pelvic size.

2.2. Parity

Birth weight is also influenced by parity of the dam (Lampo and Willem, 1965), but is not reflected in a corresponding increase in dystocia (Table 4) possibly due to the greater pelvic size and dilation at calving in mature animals.

TABLE 4

EFFECT OF AGE OF DAM ON CALVING DIFFICULTIES IN PEDIGREE BEEF HERDS
(Barnes and Kilkenny, 1976)

Age of dam (years)	Dystocia %	Calf mortality %
<3	8.3	2.1
4 - 6	2.9	0.7
7 - 9	1.7	0.9
> 10	1.7	0.7

2.3. Gestation length

Within breeds, birth weight is correlated with gestation length. The greater birth weight of the heavy beef breeds is also associated with longer gestation lengths (Preston and Willis, 1970).

Several workers have demonstrated a positive correlation (r = 0.24 to 0.42) between gestation length and birth weight (Lampo and Willem, 1965; De Fries et al., 1959). We have recorded a correlation of 0.56 (P<0.01) between gestation length and birth weight and associated with this a 0.45 kg increase in birth weight for each additional day of gestation. As birth weight is positively correlated with both the incidence of dystocia and gestation length, it is probable that gestation length will be positively correlated with the incidence of dystocia (Table 5) as has been shown in our investigations.

TABLE 5

THE INFLUENCE OF GESTATION LENGTH ON THE INCIDENCE OF DYSTOCIA

Gestation length	Number of calvings	Incidence of dystocia %
< 280	37	0
281 - 285	29	6.8
286 - 290	51	9.8
291 - 295	29	34.5
296 - 300	13	30.7
301 - 305	2	50
(Mean 286 days)		

The implications of this are that provided the date of effective mating and the average gestation length expected are known, this information can be successfully used to concentrate supervision at calving on those cows which have gone past their expected date of calving.

In this investigation 91% of all assisted calvings occurred in cows with longer than average gestation lengths. Recording

dates of service does, therefore, enable the suckled calf
producer to concentrate supervision at calving on those cows
most likely to suffer dystocia, since approximately 25% of all
cows with longer than average gestation lengths are likely to
experience this problem.

2.4. Calf sex

Sex of calf also influences gestation length and birth
weight (Anderson and Plum, 1965), resulting in almost double
the incidence of dystocia in bull calvings compared with
heifer calvings. However, Allen (1977) suggested that the
higher incidence of dystocia for bull calves was greater than
could be explained by their heavier birth weights and may also
have been influenced by their thicker conformation.

2.5. Nutrition in late pregnancy

Major differences during late pregnancy in feed intake
in general, and energy intake in particular have been shown
by several workers to affect calf birth weight and levels of
dystocia (Hight, 1968; Tudor, 1972). However, extremely low
levels of nutrition during late lactation, which are often
required to show significant differences in calf birth weight,
(Smithson et al., 1966) are often associated with a depression
in both early and total lactation performance, especially when
lactation occurs during a period of inadequate nutrition.
This has been especially true when this extreme 'starvation
during late pregnancy' technique has been applied commercially.
Although dystocia problems have been reduced, calf mortality
has been increased due to poor colostrum availability coupled
with a delayed lactation. The effect of colostrum quantity
and quality on early calf survival has been clearly demonstrated
in both pail-fed and suckled calves (Fisher et al., 1968; Logan
et al., 1974).

Commercial producers have suggested from their own
experience that type of supplementary feeding during late
pregnancy may influence the incidence of dystocia via calf birth
weight. They suggest that cereal supplementation in late

pregnancy increases calf birth weight rather than cow condition in comparison with forage supplements. This alteration in the partition of energy may work through the different rumen fermentation associated with the two types of feedstuffs in relation to the differing requirements of the cow and foetus (Edwards, personal communication).

2.6. Season and environment

Longer term variations in plane of nutrition also influence birth weights. This area has received little scientific attention, but is demonstrated under UK conditions by season of calving in the dairy herd (Figure 1), spring calving cows being pregnant during a period of feed restriction compared with autumn calving cows which are pregnant during the summer grazing season when feed supplies are plentiful. The overall nutritional status of the herd (hill versus lowground herds) similarly influences birth weights and the incidence of dystocia in beef herds (Table 6).

TABLE 6

THE EFFECT OF ENVIRONMENT ON CALF BIRTH WEIGHT AND THE INCIDENCE OF DYSTOCIA FOR CALVES SIRED BY HEREFORD BULLS (Barnes and Kilkenny, 1976)

	Lowground	Upland	Hill
Birth weight (kg)	36	34	32
Dystocia (%)	4.3	3.4	1.4

2.7. Nutrition - long-term effects

From nutritional studies, feeding either 90, 125 or 175% of maintenance to autumn-calving Hereford x British Friesian cows during the first five months of lactation, we have examined the residual effect of the level of early lactation, ie early pregnancy nutrition, on subsequent calf birth weight. During the last 4 - 6 months of pregnancy the cows were at grass and cows previously fed only 90% of maintenance, had higher live-weight gains at grass than cows previously fed 175% of maintenance (Figure 2). Nevertheless, for 50 cows over a 3-year period there was a significant (P<.05) reduction in subsequent calf birth

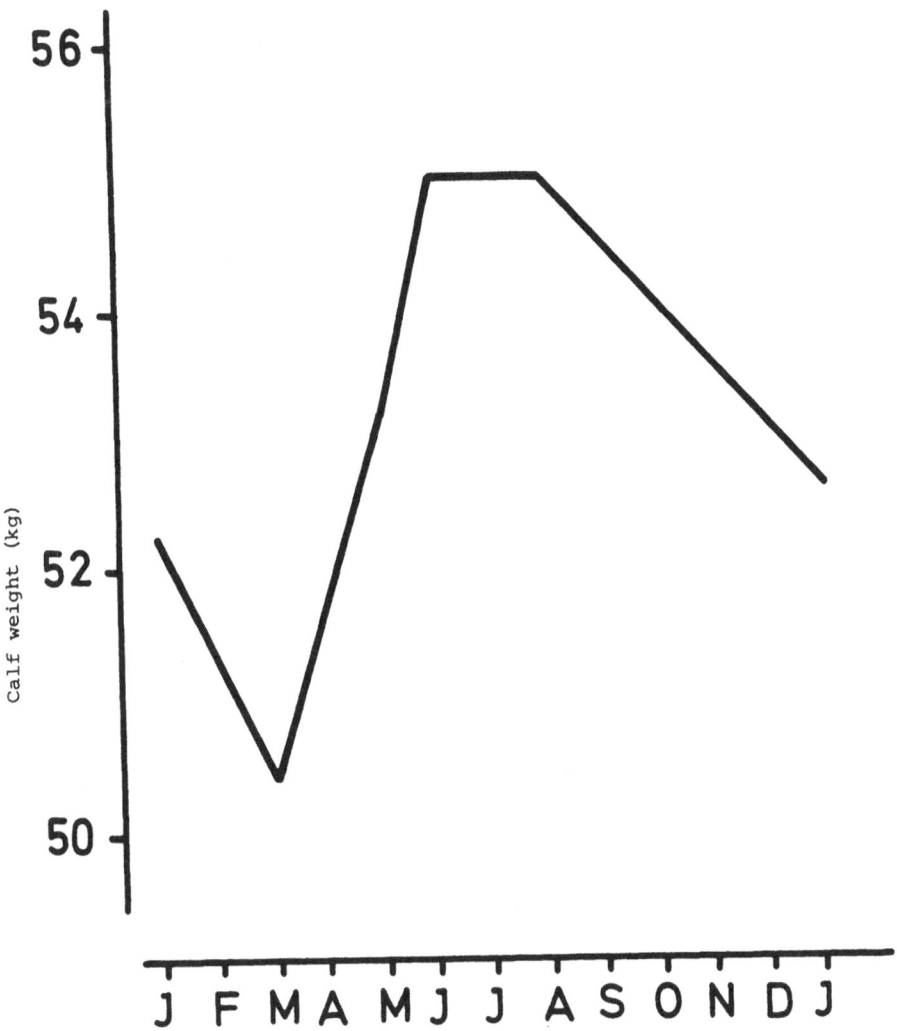

Source: MMB 1969

Fig. 1. Effect of season on 10 day calf weight for Hereford x Friesian
steer calves

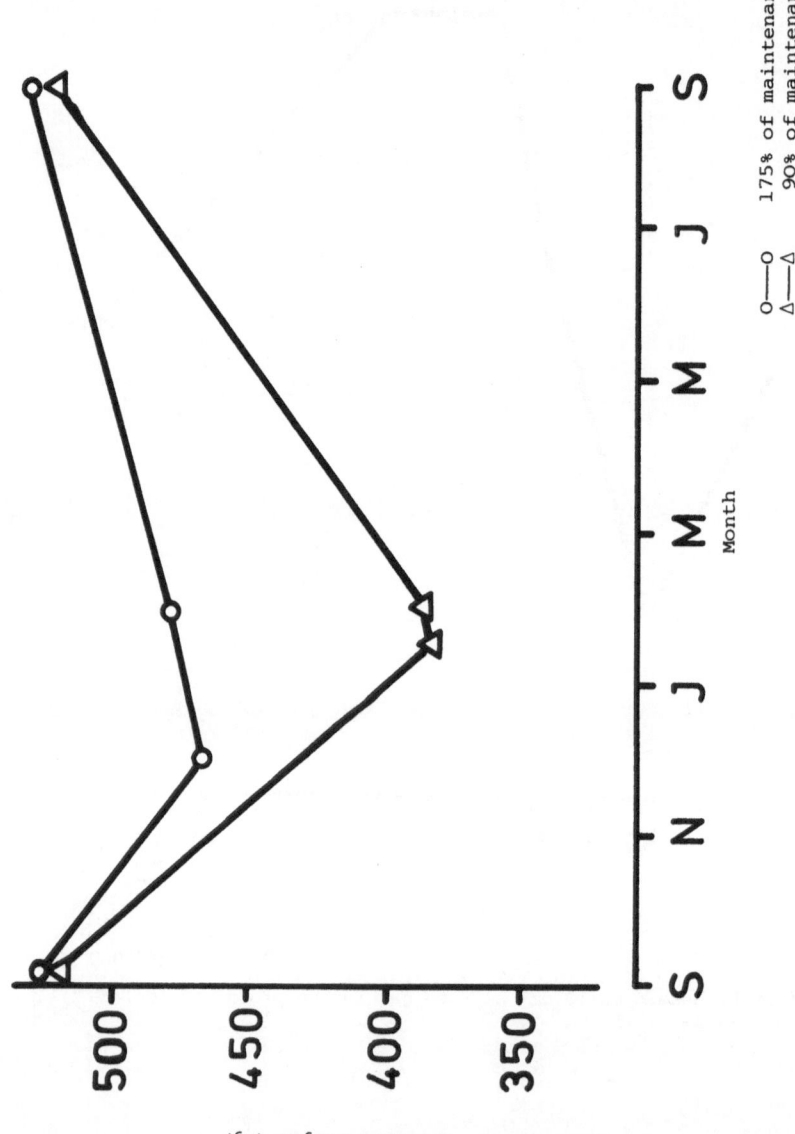

Fig. 2. Effect of two planes of nutrition during the first 150 days of lactation on cow live weight.

weight (1.37 kg, adjusted for sex) for cows receiving the low
treatment in comparison with cows receiving the high treatment.
This was associated with a non-significant difference in
gestation length which was 286 and 285 respectively for the
high and low treatments. This may partially account for the
1.37 kg difference in birth weight and agrees with the
hypothesis of Preston and Willis (1970), that high levels of
energy intake tend to prolong gestation. The corresponding
incidences of dystocia at the parturition following the three
early lactation/pregnancy treatments was 7.9, 19.1 and 23.4%
respectively for the 90, 125 and 175% of maintenance treatments.

2.8. Cow condition at calving

The importance of feed intake during pregnancy on the
incidence of dystocia may also act through the condition
(level of fatness) of the dam. Commercially it is accepted that
above a moderate level of condition the incidence of dystocia
increases with the increasing condition (fatness) of cows.
This may be partially attributable to an excessive fat
deposition around the pelvic canal in fat cows which reduces
the ease of parturition (Allen, 1977). This, coupled with
excessive fat deposition in muscle tissue decreasing muscular
efficiency in expelling the foetus - a feature of 'lazy' cows
which tire quickly at parturition, will increase the incidence
of dystocia and calf mortality.

In view of these considerations, a simple subjective
condition scoring system is now being recommended by the
advisory services in the UK for use as a management aid to the
achievement of a yearly cycle of condition change in beef cows.
The target condition scores at calving aim to prevent cows being
excessively fat (above a condition score of 3) to reduce the
incidence of dystocia or from being excessively lean (below a
condition score of 2) to ensure early calf survival.

3. PRACTICAL IMPLICATIONS

Although feeding and management of the cow throughout the

year to obtain the correct target condition score at calving
coupled with good supervision at calving are important management factors to reduce the incidence of dystocia and early
calf mortality, few beef herds can economically implement these
techniques. Excessively long calving periods currently
accepted in UK beef herds prevent this (MLC, 1974), (Table 7).

TABLE 7

DISTRIBUTION OF RECORDED HERDS BY CALVING SPREAD

Calving spread (days)	Proportion of herds %
< 90	9
90 - 119	12
120 - 149	32
150 - 179	30
180 - 209	10
>210	7

Over three-quarters of recorded beef herds in the UK calve
over a period in excess of four months but are managed as one
herd, few farmers have facilities for managing sub-herds based
on date of calving. In this situation correct feeding and
management becomes impossible (Lowman and Somerville, 1976) and
adequate supervision at calving uneconomic especially where
natural service is employed and dates of service are not
recorded.

In addition to these problems it would appear from our own
results that the incidence of dystocia is highest in late-calving
cows within a herd (Table 8).

It is possible that the longer dry period of the late
calving cows (Table 8) may be a factor influencing both birth
weight and gestation length. This relationship may also
partially explain the smaller seasonal variation in birth weight
recorded in commercial beef berds (Table 9) in the UK by
Barnes and Kilkenny (1976) in comparison with the UK dairy herd

(Figure 1) where there is no seasonal effect on the length of the dry period. Autumn calving beef cows, however, will on average have a dry period of only two months compared with four months for spring calving beef cows. If this trend for the length of the dry period to affect calf birth weight and gestation length can be confirmed it will increase still further the importance of maintaining a compact calving period.

TABLE 8

EFFECT OF DATE OF CALVING WITHIN THE CALVING PERIOD ON THE INCIDENCE OF DYSTOCIA

Month of calving	1	2	3	4
Dystocia (%)	10.6	21.0	38.6	31.8
Birth weight (kg)	42.1	40.8	43.3	46.4
Gestation length (days)	285.1	285.0	289.4	289.5

TABLE 9

EFFECT OF SEASON ON CALF BIRTH IN COMMERCIAL BEEF HERDS (MLC, 1977)

Month of birth	Birth weight (kg)
January	34.4
February	34.6
March	35.4
April	36.7
May	36.8
June	36.4
July	35.5
August	36.0
September	35.3
October	35.0
November	35.8
December	34.3

4. ECONOMIC IMPLICATIONS

An average 3.5% calf mortality in UK beef herds from dystocia and early calf mortality results in a loss of over 60 000 calves or a potential beef output valued at approximately £18 million to the UK beef market.

However, the cost to the individual producer is less clearly realised due to the practice of fostering purchased calves on cows which are lactating, but have lost their own calf at or around parturition. Whilst this limits the initial financial loss to the producer to the price of a replacement calf, such practices often incur greater hidden losses following the introduction of disease, generally *E. coli*, to the home bred calves. Septicaemia accounts for the majority of calf mortality in beef herds between birth and weaning and in many cases an outbreak of septicaemia can be traced back to a new strain of *E. coli* being introduced into the herd through a bought-in calf. However, whilst calf mortality following an outbreak of septicaemia is an obvious financial loss, the insidious effect on infected calves which survive, reducing calf weaning weight by as much as 15 - 20 kg in comparison with healthy calves, can be even greater.

Similarly, whilst the financial cost of cow mortality resulting from dystocia is obvious, the insidious effect of dystocia on subsequent fertility is not. Laster et al. (1973) have quantified the relationship between dystocia and delayed conception in beef herds in the USA and observed that cows which experienced dystocia suffered an increase of 5.8 days in subsequent calving interval. There are no comparable figures for UK conditions, but such an increase in calving interval will only reduce calf weaning weight by approximately 5 kg in UK herds. However, in herds with a compact calving period (two months) it is liable to increase the number of barren cows. Any delay in the interval between calving and conception in herds with a short mating period will result in more barren cows at that mating and also in more late calving cows in the subsequent calving period. Lowman and Somerville (1976) have shown these late calvers to be less fertile than early calving cows.

5. CONCLUSIONS

Dystocia and early calf mortality cause a large direct loss of potential beef from the national beef herd. To the individual producer the direct loss is less obvious but indirect losses in the form of reduced performance can be considerable. Apart from a careful selection of breeds and individuals within a breed, producers have only two areas in which they can limit the incidence and consequences of dystocia - adequate supervision at calving of cows with long gestation periods, and feed management to control cow condition prior to and at parturition. Both techniques can only be economically justified in commercial herds with a short calving period which in itself may be associated with a reduced incidence of dystocia.

REFERENCES

Allen, D.M. 1977. Ease of calving in beef herds. British Cattle Breeders
    Club Digest 32.

Anderson, H. and Plum, M. 1965. Gestation length and birth weight in cattle
    and buffaloes. Journal of Dairy Science 48, 1224.

Barnes, B. and Kilkenny, J.B. 1976. Calving survey in pedigree and
    commercial beef herds. MLC Summer Scholarship.

Crowley, J.P. 1965. The effect of Charolais bulls on calving performance.
    Irish Journal of Agricultural Research 4, 205.

Fisher, E.W., Selman, I.E., McEwan, A.D. and de la Fuente, G. 1968. Fifth
    International Meeting on Diseases of Cattle. Opatija, Yugoslavia,
    Reports and Contributions p. 27.

De Fries, J.C., Touchberry, R.W. and Hays, R.L. 1959. Heritability of the
    length of the gestation period in dairy cattle. Journal of Dairy
    Science 42, 598.

Hight, G.K. 1968. Plane of nutrition effects in late pregnancy and during
    lactation on beef cows and their calves to weaning. New Zealand
    Journal of Agricultural Research 11, 71.

Kilkenny, J.B. and Stollard, R. 1976. Calf birth weights in beef breeding
    herds and the relationship between birth weight, calf mortality and
    calving difficulties. Animal Production 22, 159.

Laster, D.B., Glimp, H.A., Cundiff, L.V. and Gregory, K.E. 1973. Factors
    affecting dystocia and the effects of dystocia on subsequent reproduction
    in beef cattle. Journal of Animal Science 36, 695.

Lampo, P. and Willem, A. 1965. Causes of variation in the birth weight of
    calves. Vlaams Diergeneeskunde Tijdschrift, 43, 79.

Logan, E.F., McBeath, D.G. and Lowman, B.G. 1974. Quantitative studies on
    serum immunoglobulin levels in suckled calves from birth to five
    weeks. Veterinary Record 94, 367.

Lowman, B.G. 1975. Method of condition scoring suckler cows - feed and
    management. British Cattle Breeders Club Digest 30, 30.

Lowman, B.G. and Somerville, S.H. 1976. Managing the beef herd. Meat and
    Livestock Commission National Beef Conference, Harrogate

Meat and Livestock Commission, 1974. Beef Improvement Services - Records
    Report No. 35 p. 34.

Meat and Livestock Commission.1977. Data sheets on Beef Production and
    Breeding. Data sheet 16. MLC Bletchley, Milton Keynes.

Milk Marketing Board. 1960. The incidence of difficult calving in Ayrshire
    and Friesian heifers. Production Division Report No. 10, 98. MMB,
    Thames Ditton, London.

Milk Marketing Board. 1966. Progeny testing dairy beef. Report of Breeding
    and Production Organisation No. 16, 54. MMB, Thames Ditton, London.

Milk Marketing Board. 1969. Calf weights. Report of Breeding and Production
    Organisation No. 20, 95. MMB, Thames Ditton, London.

Preston, T.R. and Willis, M.B. 1970. Intensive Beef Production. Pergamon
    Press, Oxford.

Simmental and Limousin Tests Steering Committee. 1977. In press MAFF, London.

Smithson, L., Ewing, S.A., Renbarger, B. 1966. Influence of level of
    nutrition of the dams on birth weight of beef calves. Journal of
    Animal Science 25, 909 Abs.

Tudor, G.D. 1972. The effect of pre- and post-natal nutrition on the growth
    of beef cattle. Australian Journal of Agricultural Research 23, 389.

PROBLEMS ASSOCIATED WITH THE CALVING AND NEONATAL PERIOD IN BEEF CATTLE

W. Oxender [1] and W. Adams [2]

1) Department of Large Animal Surgery and Medicine, Michigan State University, East Lansing, Michigan 48824, USA and

2) Department of Veterinary Clinical Studies, Western College of Veterinary Medicine, University of Saskatchewan, Saskatoon, Saskatchewan, Canada.

ABSTRACT

The mortality rate for beef calves is increased significantly when the cow requires assistance to complete parturition. Calf losses at or near the time of birth were four times greater from cows experiencing dystocia (20.4%) than for calves from cows with no dystocia (5.0%). Calf birth weight was a significant factor which increased calving difficulty an average of 2.3% for each kilogram increase in birth weight. Sire breed, dam breed, dam age and sex of the calf were significant factors contributing to dystocia.

Infectious disease of the digestive system are the major cause of early neonatal mortality in beef calves. Several enteropathogenic bacterial and viral agents have been identified. Enterotoxic Escherichia coli appears to be the most common single agent, but multiple infections appear to cause the highest mortality.

Serum immunoglobulin concentrations of calves appear to play an important role in calf survival and were increased significantly only in calves fed colostrum compared to other diets. Oral challenge with a septicaemia producing E. coli caused septicaemia and rapid death in all calves except those on the colostrum diet.

Calves with neonatal diarrhoea have severe dehydration and metabolic acidosis within hours. Successful clinical treatment requires fluid replacement and correction of the acidosis. Antibacterial therapy for these calves with diarrhoea appears to be of secondary importance.

Although some efficacious vaccines have been developed against the various enteropathogenic agents, the multiplicity of agents causing neonatal diarrhoea leads one to suspect that improved management methods for parturient beef cows and their calves would contribute to a major increase in beef calf survival.

INTRODUCTION

The calving period remains the most critical period for beef cattle production. The mortality rates during calving and the first 2 months after calving may average 5% for cows and up to 25% for calves. Usually a cow produces one calf each year. When the calf dies at or soon after calving, the production of beef for the given year is lost.

Beef calf mortality averaged 8.9% during calving and the first day post partum (Laster and Gregory, 1973) in Nebraska; and diarrhoea caused the death of 6.5% of Montana beef calves (Meyers, 1976). Michigan beef producers (Oxender et al.,1973) lost 13.2% of their calves within 60 days of calving. Thus calving losses and early neonatal deaths appear to cause the major losses in beef cattle production. Good management during the calving period is very critical in preventing these losses in beef production. This paper has been prepared as a review of Canadian and United States research concerning the major causes of calving period mortality in beef cattle. We have selected a few representative papers to illustrate major factors in each of the following areas:

1. Dystocia.
2. Neonatal diseases.
3. Colostrum.
4. Clinical treatment.
5. Preventive measures.

1. DYSTOCIA

The derm 'dystocia' in this paper is used to describe a parturition which is abnormally prolonged or requires assistance to complete the delivery of the calf. The risk of traumatic injury to the cow and calf is significantly increased when dystocia occurs. Laster and Gregory (1973) have shown that four times as many calves die within 24 hours of birth in cows with dystocia (Table 1).

TABLE 1

PERINATAL[+] CALF MORTALITY FOR BEEF CALVES.(Laster and Gregory, 1973)

Type of parturition	Calves dead %
Normal (n = 3877)	5.0%
Dystocia (n = 1187)	20.4%
Average (n = 5064)	8.6%

[+] Dead at birth or died within 24 hours of birth.

Normally about 5.0% of the calves die at birth or are stillborn. Moreover calf mortality appears to be similar in beef and dairy cattle (Oxender et al., 1973). The 5% mortality at birth for calves appears to be a stable figure and it is doubtful if this incidence of mortality can be significantly reduced. However, for mortality rates greater than 5%, such as the 20.4% mortality rate for dystocia cases, reduction is feasible if dystocia can be minimised.

Disproportion in size between the foetus and dam appears to be the major cause of dystocia. Thus there is a critical need for adequate growth in heifers to allow an increase in pelvic area. To illustrate: when Angus heifers were maintained on a growth ration, the pelvic area increased in size from 147 cm^3 at breeding to 232 cm^2 at the time of parturition (Rice and Wiltbank, 1970). These same authors reported dystocia rates of 28% for heifers with pelvic areas greater than 200 cm^2 and 68.7% for heifers with less than 200 cm^2 of pelvic area.

Male calves also increase the number of dystocias and thus the mortality rate (Table 2). The greater birth weight for males appears to be the major cause of the increase in mortality for male calves.

TABLE 2

THE INFLUENCE OF THE SEX OF THE CALF ON DYSTOCIA AND PERINATAL[+] MORTALITY
(Laster et al., 1973)

Sex	Dystocia %	Mortality %
Male	28	10.4
Female	17	6.6

[+] Dead within 24 hours after birth.

Formulae have been developed by several researchers, for
the effect of increased birth weight on the dystocia rate.
The breed of the sire is a significant factor contributing to
birth weight and dystocia. Charolais, Simmental, Limousin and
South Devon bulls sire calves having more dystocia than calves
sired by Hereford, Angus and Jersey bulls. Thus genetics and
nutrition contribute to beef production through a direct
influence on dystocia.

2. NEONATAL DISEASES OF BEEF CALVES

Infectious diseases in newborn calves cause significant
losses in beef production. What are the major disease problems?
In a survey of 35 Canadian beef herds (Acres, 1976), enterotoxigenic *E. coli* was found in calves from 11 herds, Reo-like
virus in 13 herds and both agents in 5 herds (Table 3). Only
6 of 35 herds were negative with respect to these two pathogens.
Both of these disease agents produce diarrhoea in newborn
calves.

In addition to the two infective agents mentioned above,
another (Coronavirus) has also been identified as a cause of
neonatal diarrhoea in calves (Mebus, 1976).

One cannot question that these enteric viruses cause
severe diarrhoea in calves, particularly when the pathological
lesions in the small intestine are studied with electro microscopy. The villi are shortened; some villi are fused with

adjacent villi and many villar tips are denuded (Mebus et al., 1975). Upon inspection of the pathological lesions in the gut of the calf caused by these Coronaviruses, one is surprised that any calves survive the effects of these enteric viruses.

TABLE 3

THE HERD PREVALENCE RATES OF ENTEROTOXIC *E. coli* AND REO-LIKE VIRUS IN CANADIAN BEEF CATTLE (Acres, 1976).

	Infected herds	
	Number	%
E. coli (enterotoxic)	11	31.4
Reo-like virus	13	37.2
Both	5	14.3
Neither	6	17.1

In order to limit this paper to major beef calf problems we have only mentioned the above three major diarrhoeal agents. Why the focus on diarrhoeal agents? Radostits and Acres (1974) reported that 51.8% of the newborn calf problems occur in the digestive system (Table 4).

TABLE 4

FREQUENCY OF DISORDERS IN VARIOUS BODY SYSTEMS DIAGNOSED AS THE CAUSE OF ILLNESS IN CALVES FROM BIRTH TO TWO MONTHS OF AGE (Radostits and Acres, 1974)

System	Total no.	% of total	Age of calf (weeks)	
			0-2	2-8
Digestive	855	51.8%	34.2%	17.6%
Cardiovascular	212	12.9	9.0	3.8
Respiratory	183	11.1	4.9	6.2
Musculo-skeletal	150	9.1	5.6	3.5
Nervous	111	6.7	4.5	2.2
All other	139	8.4	3.5	5.0
Total	1650	100.0	61.7	38.3

This predominance of problems in the digestive system is twice as great in calves under 2 weeks of age as in older calves. Although other systems accounted for many newborn calf disease problems - cardiovascular 12.9%, respiratory 11.1% and musculo-skeletal 9.1% - the major problem in newborn calves is diarrhoeal disease. We also feel that the majority of diarrhoeal epizootics will be associated with the presence of multiple enteropathogens.

3. COLOSTRUM FOR NEWBORN CALVES

We have outlined some evidence indicating that diarrhoeal diseases cause major problems in newborn calves. One of these, enterotoxigenic *E. coli*, can cause severe diarrhoea and rapid death. Will colostrum protect calves from colisepticaemia?

A study (Johnston et al, 1977) using 20 colostrum-deprived calves was designed to test the effect of a colostrum diet on prevention of colisepticaemia. Four diets were tested: 1) pooled colostrum, 2) commercial milk replacer, 3) PVP (polyvinyl-pyrrolidine, a large inert macromolecule) and 4) saline solution. Blood samples were collected every 12 hours and the average serum immunoglobulin G concentrations are shown in Table 5. As expected, serum immunoglobulins increased significantly only in calves on the colostrum diet.

TABLE 5

SERUM IMMUNOGLOBULIN G VALUES AT 0, 12, 24 AND 36 HOURS AFTER BIRTH IN CALVES ORALLY EXPOSED TO *E. coli* AT 27 HOURS (Johnston et al., 1977).

Diet	Age of calves (hours)			
	0[+]	12 (mg/ml)	24	36
Colostrum	2.6	17.3	26.5	34.8
Milk replacer	0.8	0.8	1.1	1.7
Polyvinylpyrrolidine	3.6	4.1	1.3	2.4
Saline solution	0.0	0.0	0.0	0.0

[+] Blood sample collected just prior to first feeding within 4 hours of birth.

These calves were orally exposed to a septicaemia-producing
E. coli (serotype O26:K60-NM) at 27 hours of age and within 8
hours all calves had diarrhoea. Colisepticaemia (*E. coli* cultured
from liver, spleen or cardiac blood) was present in 1 of 5 calves
fed colostrum in 5 of 5 calves fed milk replacer, in 5 of 5
calves fed PVP and in 4 of 5 calves fed saline solution (Table
6). At necropsy of the calves, the same organism was isolated
from small intestines of 19 of the 20 calves.

TABLE 6

BACTERIOLOGIC CULTURAL EXAMINATIONS OF NEONATAL CALVES ORALLY EXPOSED TO
E. coli AT 27 HOURS AFTER BIRTH (Johnston et al., 1977).

Diet	Cultural isolation of *E. coli*			
	Liver	Spleen	Cardiac blood	Small intestine
Colostrum	1/5	1/5	1/5	1/5
Milk replacer	5/5	5/5	5/5	5/5
Polyvinylpyrrolidine	5/5	5/5	4/5	5/5
Saline solution	4^+/5	4^+/5	4^+/5	4^{++}/5

$^+$ 1 calf had *Klebsiella* and *Proteus* spp.

$^{++}$ Another calf had *Proteus, Streptococcus,* and *Citrobacter*.

Thus, serum immunoglobulin G concentrations increased in
calves fed colostrum in sharp contrast to the agammaglobulinaemia
occurring in calves fed milk replacer, PVP and saline diets.
These results indicate that colostrum fed to the calf soon after
birth provides protection from colisepticaemia, but does not
prevent the diarrhoea of colibacillosis. Many other studies
show the necessity for ensuring that newborn calves ingest
adequate colostrum. In concluding this discussion on the benefits
of colostrum it seems advisable to recommend that each calf
receive 2 to 3 kg of colostrum within 2 hours after birth. The
colostrum can be given via stomach tube if the calf is unable
to nurse the cow. Since colostrum provides the newborn calf with
antibodies for preventing diseases, it is extremely important
for colostrum ingestion to precede the challenges from the
environment.

4. CLINICAL TREATMENT OF CALVES

In the preceding sections we have described diarrhoeal disease as the major clinical problem in newborn calves. What happens to the calf with diarrhoea? Tennant et al. (1972) have determined the blood chemical changes in calves with diarrhoeal diseases (Table 7). The normal ranges for the various parameters are also presented for comparison.

TABLE 7

BLOOD CHEMICAL ANALYSIS OF NEONATAL CALVES WITH ACUTE ENTERIC INFECTIONS (Tennant et al., 1972)

Analysis	No. of observations	Mean ± SD	Normal range
Packed cell volume	28	45.3 ± 7.0	25-40
Blood pH	11	7.1 ± 0.1	7.3-7.4
Bicarbonate (mEq/L)	11	13.7 ± 4.2	23-28
Potassium (mEq/L)	28	7.4 ± 1.6	4.5-5.5
Phosphorus (mEq/L)	25	9.2 ± 3.6	2.9-5.8
Blood urea nitrogen < (mg/100 ml)	28	50.1 ± 30.5	10-20
Sodium, chloride, calcium, magnesium - Normal range			

These blood values were obtained from young calves (1 to 10 days old) with diarrhoea and indicate that acidosis and dehydration are the primary clinical problems.

Assuming that preventive medicine would be successful in preventing some but not all neonatal calf diarrhoea what is the preferred approach to clinical treatment?

Certainly the rapid fluid loss and metabolic acidosis associated with neonatal diarrhoea can force newborn calves into a clinically critical condition within hours. Few people question the need for rapid replacement of body fluids and correction of the acidosis. However, other questions remain. Should the milk diet be replaced? Should antibiotics be used? Which antibiotics?

The results of various clinical treatment regimes are shown in
Table 8.

TABLE 8

THE EFFECTS OF VARIOUS CLINICAL TREATMENT METHODS[+] ON SURVIVAL RATES FOR
CALVES WITH DIARRHOEA (Radostits et al., 1975).

Antibacterial agent	Feeding regime	
	Milk starved for 24 h and given oral fluids	Whole milk
Ampicillin (oral)	45[++] (60%)[+++]	38 (50%)
Nifuraldezone (oral) and Chloramphenicol (iv)	44 (53%)	45 (90%)
No antibacterial	48 (61%)	34 (60%)

[+] All calves received fluids and electrolytes intravenously.

[++] No. of calves treated.

[+++] Percentages represent survival rates.

Three groups of calves were milk starved for 24 hours and
given oral electrolyte solutions while three other groups
received a milk diet; there were no significant differences in
survival between the groups without milk and milk-fed calves.
Perhaps even more surprising to many people was the fact that
survival rates were the same for calves receiving no antibacterial
treatment as for calves on antibacterial treatments.

The results from these clinical studies indicate that
replacing body fluids and correcting the metabolic acidosis are
of primary importance when treating calves with diarrhoeal
disease. Whether antibacterial therapy will be beneficial or
not depends on the infective agents. Many strains of *E. coli* are
resistant to antibiotics; thus, cases caused by resistant
bacterial or viral infections fail to respond to antibiotic
therapy. Therefore it appears that the most beneficial treatment
of calves with diarrhoeal disease is an intensive effort to
replace fluids and correct acidosis.

5. REDUCTION OF CALF MORTALITY BY PREVENTION

Prevention of calf mortality appears much more rewarding in terms of beef production and respective costs involved than developing improved clinical treatments for diseased calves. Providing adequate nutrition for the pregnant cow and experienced supervision during parturition are fundamental steps in reducing mortality in beef herds. Consideration of ease of calving when making genetic selection can also reduce mortality. In addition calf survival and beef production are enhanced by adequately controlling diseases such as brucellosis, leptospirosis, infectious bovine rhinotracheitis and bovine viral diarrhoea. Methods are presently known for controlling the above mentioned factors which decrease beef production.

What methods are available for preventing infections by other enteropathogens? Several methods of increasing the *E. coli* resistance in newborn calves have been investigated recently. Olson and Waxler (1976) used a bacterin to immunise bovine foetuses and neonates against an *E. coil* (O26:K60:NM). Both foetuses vaccinated in utero and calves vaccinated as neonates had increased resistance to oral challenge organisms. The results of quantitative and qualitative radioimmunoassay indicated that the increased immune response to the O26 antigen was mainly the IgM, although there were also demonstrable changes in IgGl and G2. Moreover these actively acquired immune responses were serotype specific.

Myers (1976) reported a reduction in calf mortality for Montana beef herds due to the use of a mixed strain *E. coli* bacterin. The mixed bacterin was prepared from six enterotoxigeni *E. coli* strains prevalent in Montana. The pregnant cows received two injections of the bacterin or a placebo prior to calving. Calving mortality due to diarrhoeal disease was 2% lower in calves from vaccinated dams versus calves from cows given the placebo. These two experiments serve to represent the present methods being studied to increase the newborn calf's resistance to pathogenic *E. coli*.

Methods of preventing viral infections in newborn calves by the Reo-like virus and Coronavirus have received considerable study. Mebus (1976) has reported on several attempts to use modified viral vaccines to prevent infections by these two enteric viruses. Efficacy appears to be better for the Reo-like virus vaccine than for the Coronavirus vaccine. It appears possible to stimulate humoral antibody production in the case of the Coronavirus yet this offers little protection against the viral invasion of the gut epithelium.

In another attempt to determine the efficacy of prevention through vaccination, Acres and Radostits (1976) used both an *E. coli* bacterin and a modified live Reo-like virus vaccine. They concluded that diarrhoeal disease in the calves was not decreased by vaccination when compared with animals receiving placebo. A Coronavirus was found in three of the herds under study and may have contributed to the failure of the vaccination procedure to prevent disease.

We conclude that numerous enteropathogens contribute to diarrhoeal diseases. In addition the effects of these pathogens are modified by environmental factors; thus we are faced with a complex problem requiring other methods in addition to selected vaccines to prevent diarrhoea.

6. CONCLUSIONS

Calf mortality in beef herds can be reduced and beef production increased by applying these principles:

1. Providing adequate nutrition for pregnant females.

2. Separate animals due to calve from the rest of the herd.

3. Use calving areas separated from constantly used areas to minimise contamination at the time of birth.

4. Provide constant surveillance and experienced assistance at calving to ensure that stress and traumatic injuries are minimal during parturition.

5. Avoid overcrowding in the calving area.

6. Ensure that each newborn calf obtains 2 to 4 kg of colostrum immediately after birth.

7. Remove sick calves immediately to an isolated hospital area.

8. Minimise environmental stress factors.

At current market prices beef calf mortality represents a loss of about $ 25.00 per cow each year in the United States. Reducing calf mortality by 50% would increase the net returns by about $ 15.00 per cow yearly in addition to significantly increasing beef production.

REFERENCES

Acres, S.D. 1976. The epidemiology of the calf scours complex in western Canada. Proceedings of Minisymposium on Neonatal Diarrhoea in Calves and Pigs, University of Saskatchewan, p. 2.

Acres, S.D. and Radostits, O.M. 1976. The efficacy of a modified live Reo-like virus vaccine and an *E. coli* bacterin for prevention of acute undifferentiated neonatal diarrhoea of beef calves. The Canadian Veterinary Journal 17, 197.

Johnston, N.E., Estrella, R.A. and Oxender, W.D. 1977. The resistance of neonatal calves on colostrum diet to oral challenge with a septicaemia-producing *Escherichia coli*. American Journal of Veterinary Research. (in press).

Laster, D.B. and Gregory, K.E. 1973. Factors influencing peri- and early postnatal calf mortality. Journal of Animal Science 37, 1092.

Laster, D.B., Glimp., H.A., Cundiff, L.V. and Gregory, K.E. 1973. Factors affecting dystocia and the effects of dystocia on subsequent reproduction in beef cattle. Journal of Animal Science 36, 695.

Mebus, C.A. 1976. Calf diarrhoea induced by Coronavirus and Reovirus-like agent. Proceedings of Minisymposium on Neonatal diarrhoea in Calves and Pigs, University of Saskatchewan, p. 13.

Mebus, C.A., Newman, L.A. and Stair, E.L. 1975. Scanning electron, light, and immunofluorescent microscopy of intestine of gnotobiotic calf infected with calf diarrhoeal Coronavirus. American Journal of Veterinary Research 36, 1719.

Myers, L.L. 1976. Vaccination of cows with an *Escherichia coli* bacterin for the prevention of naturally occurring diarrhoeal disease in their calves. American Journal of Veterinary Research 37, 831.

Olson, D.P. and Waxler, G.L. 1976. Immune responses of the bovine foetus and neonate to *Escherichia coli:* quantitation and qualitation of the humoral immune response. American Journal of Veterinary Research 37, 639.

Oxender, W.D., Newman, L.E. and Morrow, D.A. 1973. Factors influencing dairy calf mortality in Michigan. Journal of the American Veterinary Medical Association 162, 458.

Radostits, O.M. and Acres, S.D. 1974. Disease of calves admitted to a large animal clinic in Saskatchewan. The Canadian Veterinary Journal 15, 82.

Radostits, O.M., Rhodes, C.S., Mitchell, M.E., Spotswood, T.P. and
    Wehkoff, M.S. 1975. A clinical evaluation of antimicrobial agents and
    temporary starvation in the treatment of acute undifferentiated
    diarrhoea in newborn calves. The Canadian Veterinary Journal 16, 219.
Rice, L.E. and Wiltbank, J.N. 1970. Dystocia in beef heifers. Journal of
    Animal Science 30, 1043.
Tennant, B., Harold, D. and Reina-Guerra, M. 1972. Physiologic and metabolic
    factors in the pathogenesis of neonatal enteric infections in calves.
    Journal of the American Medical Association 161, 993.

DISCUSSION

R.D. Politiek *(Netherlands)*

I would like to ask Dr. Brolund Larsen or Dr. Sejrsen whether their average heifer of 450 kg applies to the period just after calving or one week later. What is the standard deviation? If it is about 7%, I expect that there are also some heifers calving below 400 kg body weight. Do you have a procedure to avoid these extremes or at what weight of the heifer are you inseminating? Is it 350 kg? Perhaps this is the way to reach the target of 450 kg calving weight.

K. Sejrsen *(Denmark)*

In the case of our groups fed on barley straw, insemination is performed when the animals have passed a body weight of 300 kg.

R.W.J. Plenderleith *(UK)*

Do you have any figures on cow wastage in relation to the calving age of heifers? If you calve heifers earlier, are you more likely to get less longevity than when you calve later?

K. Sejrsen

No, we haven't seen any difference in longevity.

E. Grunert *(West Germany)*

One question to Dr. Oxender. You recommended that 2 - 3 kg of colostrum should be given to the calf within the first hours of parturition. How much do you recommend to be given within the first 24 hours?

W. Oxender *(USA)*

In order to get maximum immunoglobulin levels we had to feed between 2 and 3 litres of colostrum during the first day. Our recommendation is to start with 2 or 3 kg colostrum.

P. Larvor *(France)*

You recommended separating mother and calf rather early. But isn't it well known that calves staying with their mother have a higher immunoglobulin titre?

W. Oxender

Yes, I am aware of the quoted mother effect. But I think there is a misunderstanding. I did not mean to recommend separating the mother from the calf; our recommendations are, with both dairy and beef cows, that the cow which is about to calve should be separated from the rest of the herd so that she can be observed and assisted if necessary and then the calf also has the opportunity to get colostrum. We find this a significant factor in dairy as well as in beef operations. We also recommended to the dairy men to leave the cow and calf together for 24 hours at least.

D.K. Hammer (West Germany)

Do you really think there is a need for circulating antibody after intake of colostrum? Perhaps a certain amount of antibody in the intestinal tract would be sufficient to resist E. coli infections. Colostrum contains only a very small amount of IGE, which is known to resist enzymatic degradation.

W. Oxender

I cannot answer that question since we have no data. We have speculated that we might be able to prevent colibacillosis by feeding colostrum to calves. But I agree that there might be a beneficial local effect of the antibody. It was clearly demonstrated that systemic or circulating antibodies in the case of the coronavirus infections have no protective ability on the intestinal epithelium which is destroyed by the virus. To follow that I have to refer to Mebus (1976) who quantified the local or gut level versus the systemic level and came to the conclusion that the systemic or circulating level of immunoglobulin may be less important in the case of some agents than the local antibacterial or anti-infective ability of the colostrum.

J. Brolund Larsen (Denmark)

I would like to ask Dr. Hoffmann about something we have recognised in the early calving heifers. Apparently in these heifers labour isn't expressed so well; it seems to be weaker than in older animals and apparently affects the viability of the calf. Couldn't that be an effect of decreased hormone production?

B. Hoffmann *(West Germany)*

I would rather put that question to Dr. Taverne because I
don't think that we have obtained any information from our hormone data to answer it. Dr. Taverne also says no!

I was quite impressed by Dr. Lowman's statement that 90% of
the cases of dystocia observed occurred in animals with prolonged gestations. If I understood this correctly, this to me
would be the first indication that dystocia has something to do
with genetics. I hope I am stepping on somebody's toes and will
thus provoke a discussion, but to me all the other things seem
to be more environmental or management in nature such as nutrition, early breeding and selection of the right sire.

H.O. Gravert *(West Germany)*

Well, I guess we are already playing ping-pong between the
geneticists and the non-geneticists. We know that the heritability for the maternal effect of gestation length is rather low.
I feel that there is a real random term in the gestation length,
something like a random chance for parturition, and this is why
I think that we might be a little more optimistic in managing
the gestation length by our techniques.

F. Pirchner *(West Germany)*

I don't think heritability for gestation length is low; on
the contrary it is quite high. I think there are two components.
I agree with Dr. Gravert that heritability for the first component, the maternal effect - if it exists at all - is quite low,
but for the direct effect it is quite high. So take them both
together and you get quite a sizeable genetic determination.

H.J. Langholz *(West Germany)*

From nine publications we calculated a heritability of 0.3.

B. Hoffmann

Does anybody know how gestation length is affected in cases
of egg transplantation studies - transplanting a purebred embryo
to the dam of another breed? Does the foetus emerge according to

the genotype of the foetus, or does the mother determine the end of gestation?

M.A.N. Taverne (The Neverlands)

There are the famous results from the experiments crossing donkeys with horses and in the first place the genotype of the foetus was the factor which determined the gestation length.

R. Bar Anan (Israel)

We also found a heritability of 0.3 for gestation length (direct effect). We looked at it very carefully for years, but then the variation was so small that you couldn't rely on it. You could get bulls with very short gestations of their offspring and quite a lot of calving trouble, and also the opposite. So it is very dangerous to rely on the length of pregnancy for reducing difficult calving. On the other hand, breeding for easy calving reduced the length of pregnancy by about two days in roughly 10 years, but this was not intentional.

H.J. Langholz

I want to come back to the suggestion from Dr. Oxender to put 2 - 3 kg of colostrum into the calf within the first hours of life. We tried to do this in our suckler herd and I advised the herdsmen to try to make the calf suck if it is not up within the first hours. However, it took us up to three hours and more to succeed. Would you recommend putting the colostrum into the calf by force if it does not suck voluntarily?

W. Oxender

Commonly a stomach tube is used to put the first 2 or 3 kg directly into the calf if it is one that is not nursing straight away. I think one of the most important factors we need to consider with colostrum is that it is a prevention and not really a treatment. I don't think it helps much unless we get the colostrum into the calf before the environmental challenge arrives. For that reason if the calf is not nursing vigorously right away one should not hesitate to use a stomach tube to introduce the colostrum.

NUTRITION AND MANAGEMENT OF THE DAM IN RELATION TO CALVING PROBLEMS, PART 2

Chairman: R. Smidt

EFFECT OF PLANE OF NUTRITION DURING LATE PREGNANCY ON THE INCIDENCE OF CALVING PROBLEMS IN BEEF COWS AND HEIFERS

M.J. Drennan

The Agricultural Institute, Grange, Dunsany, Co. Meath, Ireland.

ABSTRACT

In two experiments, beef heifers (approximately 370 kg initially) were fed on three planes of nutrition during the last three months of pregnancy. Daily weight increases pre-partum were 0.55, 0.14 and -0.30 kg for high, moderate and low plane heifers respectively in Experiment 1. The corresponding figures for Experiment 2 were 0.57, 0.24 and -0.17 kg. In both experiments, birth weights of calves from high-plane heifers were significantly greater than their moderate- and low-plane counterparts. Three calves out of 16 were dead at birth from the high-plane treatment in Experiment 1 and there were no further losses in this experiment. In Experiment 2, the incidence of calving problems (including calves dead at birth) was higher in the low-plane heifers (10 in 19) with the moderate plane intermediate (5 in 20) and the high plane least (1 in 20).

In three experiments (experiment 3, 4 and 5) mature beef cows were fed on different planes of nutrition during the last 3 - 4 months of pregnancy. Despite variations in pre-partum weight increase from 0.61 to -0.16 kg per day, there were no major effects of nutrition on either calf birth weights or calving problems. These results are discussed in relation to other findings in the literature.

1. INTRODUCTION

The importance of beef production and dairying in Ireland is illustrated by the fact that over 80% of the land is used for these enterprises. There are 1.9 million cows in the country and a total cattle population of 6.0 million (Central Statistics Office publication, February 1977). There is no movement of cattle for feeding into the country but a small proportion of calves and unfinished animals are exported.

Seventy-two per cent of the cows are in dairy herds which are predominantly Friesian. The remaining 28% of cows are in beef herds which are of mixed breeding with a high proportion of Hereford crosses. Sires used are mainly Friesian and Hereford with only an estimated 5% of the total calf crop in 1976 being Charolais, Simmental and Limousin (Cunningham, 1976). However, there is increasing interest in using these larger breeds particularly in beef herds with the result that the incidence of calving difficulties is likely to increase.

In single suckling, economic returns do not generally justify high feeding levels for the cows in winter. In practice, substantial body weight losses by cows are permitted in winter and the cows make up this weight loss during the subsequent season at pasture. As feed requirements are lower during pregnancy than during lactation the cows are often calved in spring and are lactating during the grazing season. In these circumstances weaning generally occurs at housing and thus winter feed requirements for cows are minimised.

2. EXPERIMENTS WITH HEIFERS

The objective of the two experiments with beef heifers was to obtain informaton on the optimum feeding levels during late pregnancy for heifers in a once-bred heifer system (McCarrick and Crowley, 1967, 1968). The Hereford-cross heifers were in calf to Hereford bulls and calved in spring at approximately 2 years of age. In each experiment the heifers were group fed

on three different planes of nutrition from housing until
calving.

Experiment 1

The three feeding levels were:

A.   High   - leafy grass silage (high quality) fed to
appetite.

B.   Moderate - stemmy grass silage fed to appetite.

C.   Low   - stemmy grass silage restricted.

Average daily dry-matter intakes per animal for high,
moderate and low plane heifers were 6.1, 4.8 and 3.4 kg
respectively. A total of 49 heifers calved and the duration
of the period of differential feeding was approximately 105
days (Table 1). The average live weight of the heifers was
380 kg initially.

TABLE 1

EFFECT OF PLANE OF NUTRITION DURING LATE PREGNANCY ON HEIFER AND CALF
WEIGHTS IN EXPERIMENT 1

	Plane of nutrition		
	High	Moderate	Low
No. of heifers	16	17	16
Expl. period (days)	101.3 ± 3.8^1	107.9 ± 3.7	106.6 ± 3.8
Initial wt (kg)	375 ± 8.0	378 ± 7.8	395 ± 8.0
Wt change pre-calving (kg/day)	$0.55^{a2} \pm 0.035$	$0.14^b \pm 0.034$	$-0.30^c \pm 0.035$
Wt post-calving (kg)	$373^a \pm 7.9$	$341^b \pm 7.7$	$324^b \pm 7.9$
Birth weights (kg)	29.1^a	27.2^b	26.1^b
Calves dead at birth	3	-	-
Calf mortality (1st 6 weeks)	-	3	2

In this and subsequent tables: 1 - $\pm$ refers to SE for treatment means

2 - Values on the same line with different
superscripts are significantly different (P< 0.05)

Their weight losses to immediately post-calving were -2,
-36 and -71 kg for high, moderate and low plane heifers

respectively. The corresponding figures for weight changes per
day (kg) during late pregnancy were 0.55, 0.14 and -0.30.
These treatment differences were highly significant. Calf
birth weights (adjusted for sex) were 29.1, 27.2 and 26.1 kg
from heifers fed the high, moderate and low planes of
nutrition respectively. Calves from heifers fed on a high
plane of nutrition during late pregnancy were significantly
heavier at birth than calves from the other two treatment groups.
The only serious incidence of calving problems was in the high
plane group where 3 calves out of 16 were dead at birth.

Experiment 2

The three feeding levels were:

A. High   - medium quality grass silage fed to appetite.
              plus 1.83 kg barley per animal daily.

B. Moderate   - medium quality grass silage fed to appetite.

C. Low   - medium quality grass silage restricted.

Average daily dry-matter intakes of silage for high,
moderate and low plane groups were 4.9, 5.5 and 3.9 kg per
animal respectively. The total number of heifers calving was
59 and the duration of the differential feeding period was 87
days (Table 2). The heifers were approximately 360 kg live
weight initially and weight losses to post-calving were -3, -32
and -47 kg for high, moderate and low plane animals respectively.
The corresponding figures for weight changes per day (kg) pre-
calving were 0.57, 0.24 and -0.17. These differences in heifer
weight changes were again significant. Calf birth weights
(adjusted for sex) were 30.7, 27.3 and 26.8 kg from heifers
fed high, moderate and low planes of nutrition respectively.
Calves from high plane heifers were significantly heavier at
birth than those from the other two treatments. The incidence
of calving problems and calves dead at birth was far higher in
the low plane heifers (10 in 19) with the moderate plane intermediate (5 in 20) and the high plane least (1 in 20). There
were no calf losses in the first 6 to 8 weeks of life for the
high plane treatment group in either experiment but losses
occurred in the other two groups in both years (Tables 1 and 2).

TABLE 2

EFFECT OF PLANE OF NUTRITION FOR HEIFERS DURING LATE PREGNANCY ON HEIFER
AND CALF WEIGHTS IN EXPERIMENT 2

	Plane of nutrition		
	High	Moderate	Low
No. of heifers	20	20	19
Expl. period (days)	87.4 ±3.4	86.5 ± 3.4	87.2 ±3.5
Initial wt (kg)	364 ±7.0	367 ± 7.0	346 ±7.1
Wt change pre-calving (kg/day)	0.57^a ±0.035	0.24^b ± 0.035	-0.17^c ±0.036
Wt post-calving	361^a ±7.2	335^b ± 7.2	298^c ±7.4
Birth weights (kg)	30.7^a	27.3^b	26.8^b
Calving problems	1	3	9
Calves dead at birth	-	2	2
Calf mortality (1st 8 weeks)	-	2	3

Despite similar effects of nutrition on heifer weight
changes and calf birth weights in the two experiments the
incidence of calving problems varied. The higher calf losses
at birth with high-plane heifers in Experiment 1 compared with
Experiment 2 could be due to heavier initial heifer weights and
the longer period the heifers remained on the higher feeding
level in Experiment 1 compared with Experiment 2. This longer
feeding period on high-quality silage could result in the
heifers being in better body condition at calving, McCarrick
(1966) has shown that silage diets result in greater fat
deposition than hay diets and Wiltbank (1971), in reference to
earlier work by Wiltbank, Bond and Warwick (1965), has indicated
the deleterious effects of excess fatness in heifers on calf
losses at or near calving. The extremely high incidence of
calving problems in low plane heifers in Experiment 2 is
probably as much a reflection of actual body weight as of
weight change. These heifers weighed only 298 kg post-calving.

The report by Wiltbank (1971) stated that most of the high-
plane heifers appeared to have large amounts of fat in the pelvic

region which could have decreased the size of the pelvic opening. Pelvic area measurements by Rice and Wiltbank (1970) have shown negative correlations between dystocia in heifers and size of the pelvic opening. Similar results were obtained by Bellows et al. (1971) and Corah, et al. (1975). Bond and Wiltbank (1970) and Arnett et al. (1971), also reported that calving difficulties and calf mortality were greatest in heifers fed on high planes of nutrition. However, the latter workers reported that over-feeding mature cows during late pregnancy did not have similar detrimental effects. Other studies with heifers by Christenson et al. (1967) report similar findings.

In a study by Absher and Hobbs (1968) the figures for the percentage of heifers (initial weight 350 kg) requiring assistance at calving were 65.0, 50.0 and 38.1 for those which gained 0.90, 0.68 and 0.45 kg/day respectively in late pregnancy. These differences in calving difficulty were, however, not significant. The mean birth weights for Angus and Hereford calves requiring assistance were 2.35 kg (P< 0.01) and 2.76 kg (P <0.01) greater respectively than calves born unassisted. It was also noted that the proportion of animals requiring assistance at calving was greater for younger animals: 64% of 22 heifers under two years old at calving required assistance compared with 41% of 41 heifers over two years.

Laster et al. (1973) in a study of factors affecting dystocia again found that birth weight had a significant effect. They also found that the incidence of dystocia in two-year-old heifers was 36% higher than in three-year-old animals and 45% higher than in four - and five-year-olds. It was also noted that cows with longer gestation lengths experienced calving difficulty and this was associated with breed of sire. Rice and Wiltbank (1970), Bellows et al. (1971) and Nelson and Huber (1971) also found that increased birth weights resulted in increased calving difficulty.

In all of the above studies referring to the effects of feeding levels on the incidence of dystocia, the heifers continued to gain weight during late pregnancy but at varying rates. In such circumstances, it was excessive feeding that resulted in calving difficulties. However, where light heifers are allowed to lose weight in late pregnancy, as in Experiment. 2, this can also have undesirable effects. In work carried out by Corah et al. (1975) two groups of heifers (415 kg initially) were used which either gained 36 kg or lost 6 kg live weight during the last 100 days of pregnancy. Calves born to heifers on restricted feeding were 2 kg lighter (P< 0.05) at birth but there was no effect of nutritional level on the percentage of assisted births. However, 10% of the calves born to restricted heifers died at birth compared with 3% of the calves from heifers fed the high energy ration.

3. EXPERIMENTS WITH MATURE COWS

In two experiments (Drennan and Bath, 1976), Hereford-cross cows were individually fed the following rations during late pregnancy:

A. Grass silage to appetite
B. Grass silage restricted to 27.3 kg per animal daily

The quality of the silage used and its dry matter content were higher in Experiment 3 than in Experiment 4. Consequently, the two planes of nutrition will be referred to as high and moderate in Experiment 3 and moderate and low in Experiment 4. In Experiment 3, cows were mated with two Hereford bulls whereas two Friesian bulls were used in Experiment 4.

Experiment 3
There were 13 cows (mean initial live weight 508 kg) in each of the two treatment groups (Table 3). The average duration of the experimental period was approximately 100 days. Daily weight changes for high and moderate plane animals pre-calving were + 0.58 and - 0.12 kg respectively. These differences were highly significant (P< 0.001). High plane cows were 1 kg heaver post-calving than at the start of the

study whereas the moderate plane cows lost 57 kg over the same period. Plane of nutrition during late pregnancy had no effect on either calf birth weights or the incidence of calving difficulties.

TABLE 3

EFFECT OF PLANE OF NUTRITION FOR COWS ON COW AND CALF WEIGHTS IN EXPERIMENT 3

| | Plane of nutrition | | |
	High	Moderate	F-test
No. of cows	13	13	
Expl. period (days)	103.3 + 4.9	97.5 + 4.7	NS
Initial wt (kg)	506 + 15.0	509 + 14.4	NS
Wt change pre-calving (kg/day)	0.58+ 0.06	-0.12+ 0.06	***
Post-calving wt (kg)	507 + 12.6	452 + 12.1	**
Calf birth weights (kg)	35.4 + 1.6	34.1 + 1.5	NS
No. of difficult calvings	2	2	
Calves dead at birth	1	-	

Experiment 4

There were 13 and 16 cows (mean initial live weight 500 kg) in the medium and low plane treatments respectively. The duration from the start of the experiment to calving was 78 days (Table 4). Daily weight changes pre-calving for medium and low plane cows were + 0.15 and - 0.16 kg respectively.

These differences in weight gain were significant (P< 0.001) The actual live-weight losses from the start to post-calving were 48 and 84 kg for medium and low plane cows respectively. Despite the considerable weight loss by low plane cows there was no effect of treatment on calf birth weight. Although there were five and four assisted calvings on the moderate and low planes of nutrition respectively, only two of those could be described as difficult both of which were with cows from the moderate plane of nutrition.

TABLE 4

EFFECT OF PLANE OF NUTRITION FOR COWS ON COW AND CALF WEIGHTS IN EXPERIMENT 4

| | Plane of nutrition | | |
	Moderate	Low	F-test
No. of animals	13	16	
Expl. period (days)	79.2 ± 3.8	77.3 ± 3.4	NS
Initial wt (kg)	500 ±17.8	502 ±16.0	NS
Wt change pre-calving (kg/day)	0.15 ± 0.04	-0.16 ± 0.04	***
Post-calving wt (kg)	452 ±14.5	418 ±13.1	NS
Calf birth weight (kg)	37.4 ± 0.9	37.3 ± 0.8	NS
Difficult calvings	2	-	

Experiment 5

In this experiment 42 third-calving cows were used, all of which were in calf to one Charolais bull. The cows were mainly Hereford x Friesian. All cows were fed barley straw to appetite plus the following supplements (containing 3% minerals/vitamins) per cow daily:

A.   1.8 kg barley
B.   1.8 kg barley/soyabean meal (15% crude protein)
C.   2.7 kg barley/soyabean meal (15% crude protein)

The initial weight of the cows, which were group fed, was approximately 450 kg (Table 5). The duration of the differential feeding period was 115 days. All treatment groups gained weight pre-calving and weight changes for groups A, B and C to immediately post-calving were -54, -37 and + 11 kg respectively. There were no serious calving problems and only four cows were assisted at calving, three of which were from treatment A and one from treatment C.

The results from the three experiments would suggest that with mature cows in relatively good body condition initially,

TABLE 5

EFFECT OF PLANE OF NUTRITION FOR COWS ON COW AND CALF WEIGHTS IN EXPERIMENT 5

	Treatment		
	A	B	C
No. of cows	14	14	14
Expl. period (days)	114.4	115.2	118.1
Initial wt (kg)	454	446	443
Wt change pre-calving (kg/day)	0.17	0.26	0.61
Post-calving wt (kg)	400	409	454
Calf birth wt (kg) Males	43.9	41.2	41.1
Females	38.7	36.9	41.1
Assisted calvings	3	-	1

ranges in weight change to post-calving from a gain of 11 kg to a loss of 84 kg did not have any influence on the incidence of calving problems. The results of Jordan et al. (1968) and Powell and Matravers (1975) support these findings. However, more severe nutritional restriction during late pregnancy was shown in other studies to reduce calf birth weights. Tudor (1972) fed cows (378 kg initially) on high and low planes of nutrition during the last 100 days of pregnancy which resulted in weight changes pre-calving of 0.62 and -0.38 kg per day respectively. In this study, calf birth weights were significantly ($P< 0.001$) reduced by low plane feeding but there was no effect of treatment on calf mortality or on the incidence of dystocia. Wiltbank et al. (1962) fed 500 kg cows on moderate and low planes of nutrition during the last 130 days of pregnancy. Weight changes for moderate- and low-plane cows pre-calving were 0.27 and -0.44 kg respectively. Although calf birth weights were significantly ($P< 0.01$) reduced by the low feeding level, there was no effect of treatment on the incidence of calving difficulties. The data of Pinney et al. (1962) suggest that calving difficulties may be increased by feeding cows on excessively high planes of nutrition during late pregnancy.

In three studies by Hight (1966, 1968a and 1968b) Aberdeen-Angus cows (400 - 430 kg initially) were fed on different planes of nutrition for periods ranging from the last 83 to 107 days of pregnancy. Daily weight losses pre-calving for low plane cows were 0.66, 0.50 and 0.15 kg for the three studies. In each instance calf birth weights were significantly decreased (P< 0.001) by lowering the plane of nutrition in late pregnancy. In these latter studies, the number of calves present at weaning as a percentage of the number of cows at the start of the experiment was less for the low pre-calving treatments. The calves were born outdoors and the lighter calves from the low pre-calving treatments may have been unable to withstand the prevailing environmental stress in early post-natal life. Work with sheep by Alexander (1962) and Alexander et al. (1956) has shown the importance of lamb birth weight on fat reserves which is a major energy source in early postnatal life. Corah et al. (1975) found that weight losses of 10 and 65 kg in two groups of cows (450 kg initially) during the last 100 days of pregnancy resulted in significantly (P< 0.05) lower calf birth weights from the severely restricted group. Although the percentage of assisted births was the same for both groups, 10% of the calves from severely restricted cows were dead at birth and a further 19% died because of diarrhoea. There were no calf losses in the group fed at the higher nutritional level.

4. CONCLUSIONS

Work with heifers has clearly shown that high plane of nutrition during late pregnancy increases the incidence of calving difficulties. This increase in calving difficulties has been attributed to various reasons which include excess fat deposition in the pelvic region which may reduce the size of the pelvic opening. Pelvic size measurements have shown that as pelvic area decreases the incidence of calving problems increase. Increased gestation length with a resultant increase in calf birth weight was also shown to increase the incidence of calving problems. With heifers calving at approximately 2 years, the incidence of calving problems was shown to be greater in heifers under 24 months compared with older heifers. At the other extreme, calf losses at birth and later have increased where heifers are allowed to lose weight during late pregnancy. Overall the studies would suggest that the optimal nutritional plane for heifers is that which provides a daily gain of approximately 0.5 kg in late pregnancy but initial heifer weight must also be taken into consideration.

Mature cows in general are far less susceptible to the effect of nutritional influences in late pregnancy on calving difficulties and calf losses than are heifers. However, it has been demonstrated that excessively high planes of nutrition resulting in very fat cows at calving can cause increased calving problems. At the other extreme, low planes of nutrition for light cows, resulting in a significant reduction in calf birth weights, does increase calf deaths. However, with cows initially in good body condition (weighing 500 kg) maintenance of live weight, or losses up to 0.1 to 0.2 kg daily in late pregnancy, can be tolerated and do not reduce overall calving performance.

REFERENCES

Absher, C.W. and Hobbs, C.S. 1968. Pre-calving level of energy in first
    calf heifers. Journal of Animal Science 27: 1130-1131 (abstr).
Alexander, G. 1962. Energy metabolism in the starved new-born lamb.
    Australian Journal of Agricultural Research 13: 144.
Alexander, G., McCance, I. and Watson, R.H. 1956. The relation of maternal
    nutrition to neonatal mortality in Merino lambs. Proceedings 3rd
    International Congress on Animal Reproduction, Section 1: 5-7.
Arnett, D.W., Holland, C.L. and Tolusek, R. 1971. Some effects of obesity
    in beef females. Journal of Animal Science 33: 1129-1136.
Bellows, R.A., Short, R.E., Anderson, D.C., Knapp, B.W. and Pahnish, O.F.
    1971. Cause and effect relationships associated with calving
    difficulty and calf birth weight. Journal of Animal Science 33, 407-415.
Bond, J. and Wiltbank, J.N. 1970. Effect of energy and protein on oestrus,
    conception rate, growth and milk production of beef females. Journal
    of Animal Science 30: 438-444.
Christenson, R.K., Zimmerman, D.R., Clanton, D.C., Jones, L.E., Tribble,
    R.L. and Sotomayor, R. 1967. Effect of pre-calving energy levels on
    performance of beef heifers. Journal of Animal Science 26: 916 (abstr).
Corah, L.R., Dunn, T.G. and Kaltenbach, C.C. 1975. Influence of pre-partum
    nutrition on the reproductive performance of beef females and the
    performance of their progeny. Journal of Animal Science 41: 819-824.
Cunningham, P.C. 1976. Breed composition of the 1976 calf crop. Farm and
    Food Research 7: 16-17.
Drennan, M.J. and Bath, I.H. 1976a. Single suckled beef production. 3.
    Effect of plane of nutrition during late pregnancy on cow
    performance. Irish Journal of Agricultural Research 15: 157-167.
Drennan, M.J. and Bath, I.H. 1976b. Single suckled beef production. 4.
    Effect of plane of nutrition during late pregnancy on subsequent calf
    performance. Irish Journal of Agricultural Research 15: 169-176.
Hight, G.K. 1966. The effects of under-nutrition in late pregnancy on
    beef cattle production. New Zealand Journal of Agricultural Research
    9: 479-490.
Hight, G.K. 1968a. Plane of nutrition effects in late pregnancy and during
    lactation on beef cows and their calves to weaning. New Zealand
    Journal of Agricultural Research 11: 71-84.

Hight, G.K. 1968b. A comparison of the effects of three nutritional
        levels in late pregnancy on beef cows and their calves. New
        Zealand Journal of Agricultural Research 11 : 477-486.

Jordan, W.A., Lister, E.E. and Rowlands, G.J. 1968a, b. I. Effect of
        planes of nutrition on wintering pregnant beef cows. Canadian
        Journal of Animal Science 48: 145-154. II. Effect of varying planes
        of winter nutrition of beef cows on calf performance to weaning.
        Canadian Journal of Animal Science 48: 155-161.

Laster, D.B., Hudson, A.G., Glimp, A., Cundiff, L.V. and Gregory, K.E. 1973.
        Factors affecting dystocia and the effects of dystocia on subsequent
        reproduction in beef cattle. Journal of Animal Science 36: 695-705.

McCarrick, R.B. 1966. Effect of method of grass conservation and herbage
        maturity on performance and body composition of beef cattle.
        Proceedings 10th International Grassland Congress pp 575-580.

McCarrick, R.B. and Crowley, J.P. 1967. Nutrition of beef heifers during
        pregnancy and a 6-week lactation in a cow-heifer beef production
        system. Research Report, Animal Production Division, An Foras
        Taluntais, Dublin, pp. 16-17.

McCarrick, R.B. and Crowley, J.P. 1968. Nutrition of beef heifers during
        pregnancy and an 8-week lactation in a cow-heifer beef production
        system. Research Report, Animal Production Division, An Foras
        Taluntais, Dublin, pp 15-17.

Nelson, L.A. and Huber, D.A. 1971. Factors influencing dystocia in Hereford
        dams. Journal of Animal Science 33, 1137-1138 (abstr).

Pinney, D., Pope, L.S., Van Cotthem, C. and Urban, K. 1962. Effect of winter
        plane of nutrition on the performance of three and four year old beef
        cows. Miscellaneous Publication Oklahoma Agricultural Experimental
        Station, No. 67: 50-57.

Powell, T.L. and Matravers, C. 1975. Feeding levels in late pregnancy and
        early lactation for spring calving single suckler cows. Experimental
        Husbandry 29: 29-37.

Rice, L.E. and Wiltbank, J.N. 1970. Dystocia in beef heifers. Journal of
        Animal Science 30 : 1043 (abstr).

Tudor, G.D. 1972. The effect of pre- and post-natal nutrition on the
        growth of beef cattle. 1. The effect of nutrition and parity of the
        dam on calf birth-weight. Australian Journal of Agricultural
        Research, 23 : 389-395.

Wiltbank, J.N. 1971. Relationship of energy, cow size and sire to calving difficulty. Proceedings 5th Congress on Artificial Insemination of Beef Cattle. pp 18-24.

Wiltbank, J.N., Bond, J. and Warwick, E.J. 1965. Influence of total feed and protein on reproductive performance of the beef female through second calving. USDA Technical Bulletin, 1314. Cited by Wiltbank, J.N. 1971.

Wiltbank, J.N., Rowden, W.W., Ingalls, J.E., Gregory, K.E. and Koch, R.M. 1962. Effect of energy levels on reproductive phenomena of mature Hereford cows. Journal of Animal Science 21 : 219-225.

THE INFLUENCE OF PRE—PARTAL FEEDING ON ENERGY METABOLISM IN EARLY LACTATION

E. Farries

Institute of Animal Husbandry and Animal Behaviour, Mariensee,
Agricultural Research Centre, Braunschweig-Völkenrode,.
Federal Republic of Germany.

ABSTRACT

It is generally known that dairy cows with a high milk yield at the beginning of lactation often suffer from serious disturbances in energy metabolism, eg bovine ketosis, probably originating in late pregnancy.

The nutrient intake of dairy cows during a dry period of 6 - 8 weeks normally exceeds the requirements for the foetus, the uterus and the mammary glands. A surplus of energy is mainly stored in the form of body fat.

At the beginning of lactation this body fat is mobilised by hormonal regulation, but it can only be utilised for milk production if there is sufficient oxaloacetate for a combination with acetyl Co A from the break-down of body fat. Under this condition citrate will be formed.

Oxaloacetate is mainly used for the synthesis of glucose and later for lactose, if the output of lactose is high at the beginning of lactation, since the lactose level in milk is nearly constant.

If acetyl Co A cannot be converted to citrate, it is condensated or decarboxylated to ketone bodies and results in bovine ketosis. Only a small amount of body fat can be utilised directly for the milk fat synthesis.

Therefore it cannot be recommended to raise the energy supply during pregnancy more than necessary for the development of the embryo.

INTRODUCTION

The standard of performance in dairy cows has been continuously increased by selection and improved nutrient supply.
Milk yields of more than 7 000 kg FCM in 305 days are no rarity.
We should not only aim at an increased milk production with a
high concentration of milk fat, but the animal should also be
in production for many lactations with regular calvings, ie
with an optimum of fertility and a minimum of stress on the
organism. The higher the standard of performance, the more
disturbances in metabolism can be observed, mainly at the
beginning of lactation.

One of these disorders in energy metabolism is the so-
called bovine ketosis or acetonaemia. It appears mainly in the
first 6 weeks of lactation in high yielding cows (Jazbec, 1967).

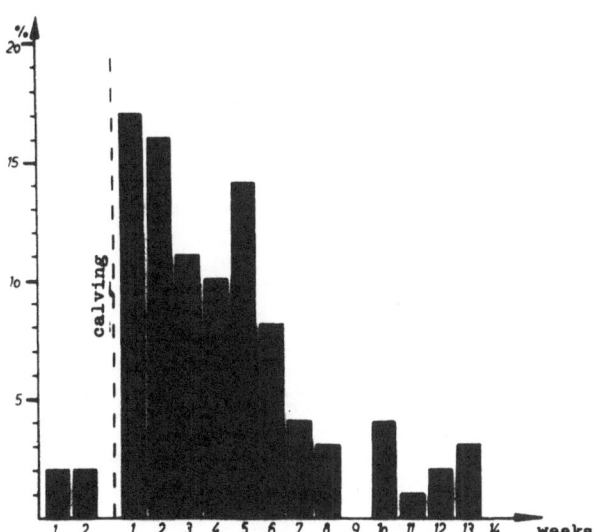

Fig. 1. Frequency of ketosis in relation to stage of lactation.

The origin of this disease is often described as a lack of available energy at the time of rapidly increasing milk production, since at the beginning of lactation a dairy cow with a high milk yield is hardly ever able to take up sufficient nutrients to meet the requirements for milk production.

If the animals get ketosis, feed consumption is decreased immediately and milk production drops conspicuously.

At this point it is not possible to compensate the energy deficiency by feeding methods, but only through application of energy-rich substances by a veterinarian.

Under practical farm conditions it can often be observed that the animals which show disorders in energy metabolism, are those which have been well prepared for the next lactation during pre-partal feeding, ie which have stored sufficient energy in the form of body fat. It is a very common belief that this is necessary for high producing dairy cows in order to compensate for the deficiency energy intake at the beginning of lactation.

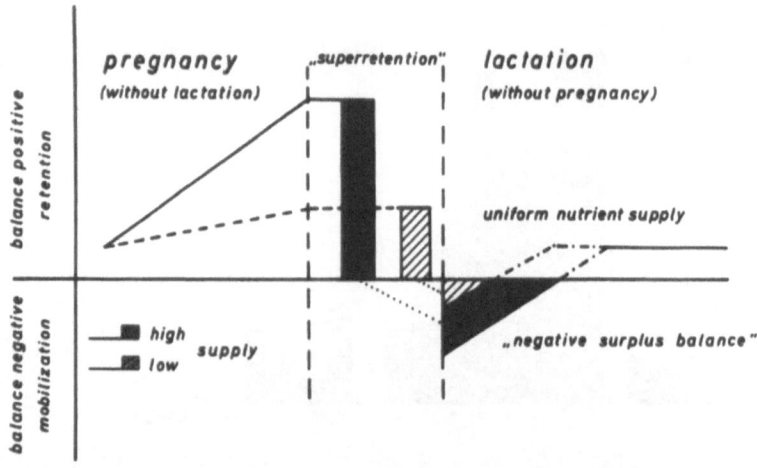

Fig. 2. Retention and mobilisation during pregnancy and lactation.

The more the energy supply exceeds the requirements for the foetus, the uterus and the preparation of the mammary glands, the more energy is stored, mainly in the form of body fat and with only small amounts of glycogen in liver and muscles. Immediately after calving the mobilisation of body fat is initiated by the increasing production of prolactin. In relation to the energy which is stored during the last part of pregnancy, a more or less long and intensive period of mobilisation can be observed, which means a period with a negative nutrient balance. This could be confirmed in many experiments with dairy cows, sows and goats (Piatkowski, 1962, 1964; Oslage and Farries, 1966, 1970; Lenkeit et al., 1955, 1956; Kalaissakis, 1958, 1959).

Simultaneously milk production increases and consequently the availability of these mobilised body substances for milk synthesis seems to be obvious.

Biochemical pathways show, however, that this conversion cannot be effective if there is not sufficient oxaloacetate available in combination with acetyl Co A (Coenzyme A) in order to form citrate.

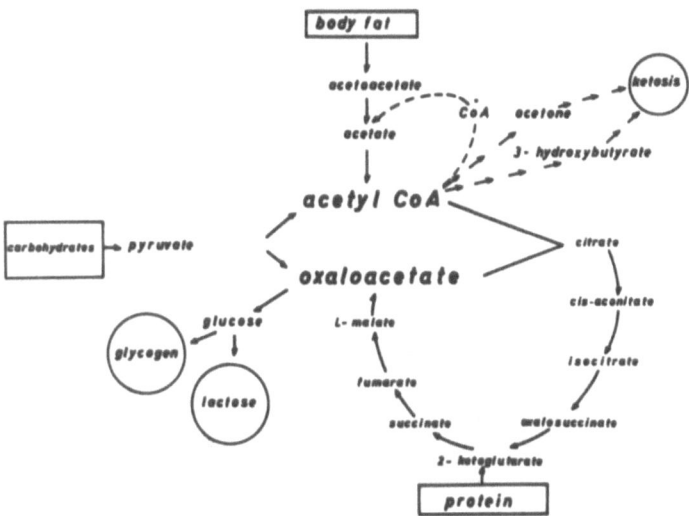

Fig. 3. Biochemical pathways.

The precursor for oxaloacetate is mainly pyruvate from the carbohydrates and a small amount of the glucogenic amino acids. But oxaloacetate is the most important precursor for glucose and glucose is necessary for the synthesis of lactose in the mammary glands. The concentration of lactose in milk is nearly constant, which means that the output of lactose is directly related to the milk yield.

In highly lactating dairy cows therefore the oxaloacetate-glucose pathway is strongly stressed, especially if there is a reduction in feed intake.

In this situation no oxaloacetate remains for a combination with acetyl Co A from the β-oxydation of body fat. Acetyl Co A is then condensated or decarboxylated to the so-called ketone bodies which are enriched in the blood and excreted with urine and milk. The higher the concentration of ketone bodies, the more the feed intake decreases and the more evident becomes the difference between oxaloacetate necessary for the formation of lactose and the utilisation of body fat. Thus oxaloacetate holds a very important key position in the energy metabolism of the ruminant (Kronfeld, 1976; Drepper, 1976).

EXPERIMENTS WITH DAIRY COWS

In order to confirm the statement that cows fed a high energy level during pregnancy often suffer from disturbances in energy metabolism at the beginning of lactation, 60 dairy cows of the Institute's herd were fed at very different energy levels during the last 8 weeks of pregnancy and during the first 10 weeks of lactation.

The experimental design in Table 1 shows that the animals were enabled to store quite different amounts of energy. The energy mobilisation at the beginning of lactation was stimulated in one group of the amply fed cows by an extreme underfeeding for observation of the utilisation of body fat for the milk production and its influence on the energy metabolism. Another group, fed on a low level during the dry period, was better fed during lactation to enable a better conversion of nutrients for milk production or for weight gain. The cows of the other groups were fed according to milk yield.

TABLE 1

EXPERIMENTAL DESIGN (KETOSIS EXPERIMENT 72/73)

Exp. group	n	Energy supply during last 10 weeks ante-partum	Exp. group	n	Energy supply first 10 weeks post-partum
I	30	maintenance + 16 kg FCM	I a	15	norm
			I b	15	norm - 30%
II	30	maintenance + 2 kg FCM	II a	15	norm
			II b	15	norm + 30%

Depending on the level of energy supply during pregnancy, a very different concentration of glucose in blood could be observed during lactation.

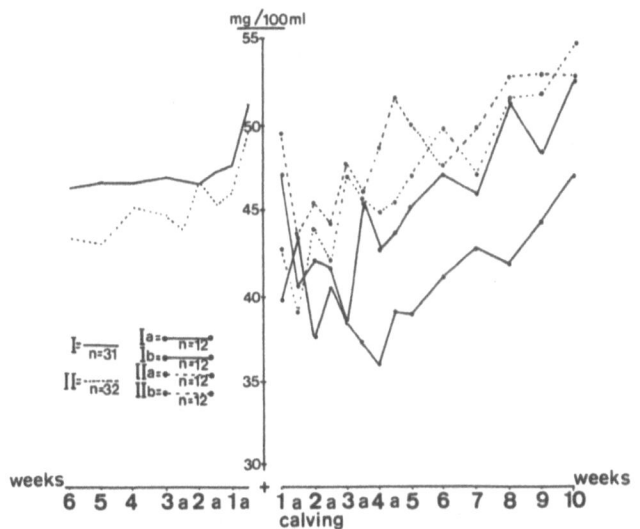

Fig. 4.   Ketosis experiment 1972/1973; glucose in blood.

There was no difference in blood glucose during pregnancy; the concentration varied in the normal range between 40 and 60 mg/100 ml. At the beginning of lactation a more or less intensive decline in glucose concentration could be observed, especially during the first 4 weeks of lactation.

This decline was most obvious in animals fed amply during pregnancy and low rations during lactation, subsequently glucose concentration went down below 40 mg/100 ml. On the contrary, in cows fed low rations during pregnancy and ample rations during lactation the glucose concentration in the blood was higher. It is evident, that the energy metabolism is especially stressed during the first 4 weeks of lactation, which means during the period of intensive mobilisation of body substances combined with increasing milk yield.

Both processes need oxaloacetate as mentioned before. Oxaloacetate can only be provided by a sufficient supply with energy rich components of the ration. If energy supply is reduced at the beginning of lactation, oxaloacetate cannot be formed sufficiently to meet the requirements for a compensation of acetyl Co A as well as for the synthesis of lactose. Consequently the concentration of ketone bodies will rise.

The content of all ketone bodies is inversely related to glucose and is highest in animals with an intensified mobilisation of body fat, whereas the other animals show only a slightly increased concentration compared to prepartal data.

The highest level is reached about the 4th - 5th week of lactation in all cows regardless of feed supply, ie at the peak of lactation. Most cases of ketosis generally appear during this period.

During the period of ketosis feed intake is retarded, which means that the difference between energy needed for compensation of acetyl Co A from body fat and synthesis of lactose and energy available for these processes becomes more and more evident.

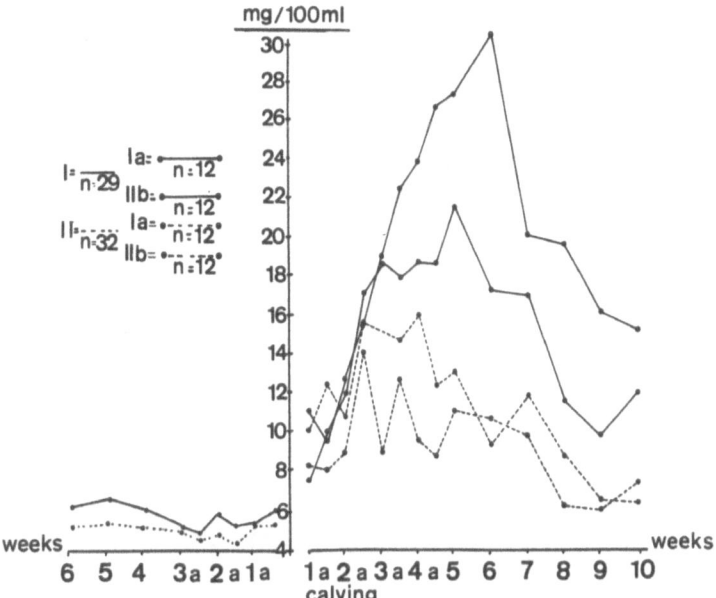

Fig. 5.   Ketosis experiment 1972/73; total ketone bodies in blood.

Consequently, the milk production will decrease and the mobilisation of body mass will be intensified, but in spite of this there is only little chance to balance energy metabolism.

Therefore, from the physiological point of view it cannot be recommended to produce energy reserves during pregnancy in the form of body fat.

However, the results of these experiments have shown that the milk fat content is higher in animals with body fat reserved from pregnancy, since direct conversion from the fatty acids of body fat to milk fat seems to be possible.

Consequently, the milk production in terms of kg FCM (fat corrected milk) is also higher whereas the difference in total milk yield is not as evident.

This higher milk fat content is combined with an increased production of ketone bodies; therefore most of the cows fed well during pregnancy suffer from a subclinical ketosis at the

452

beginning of lactation.

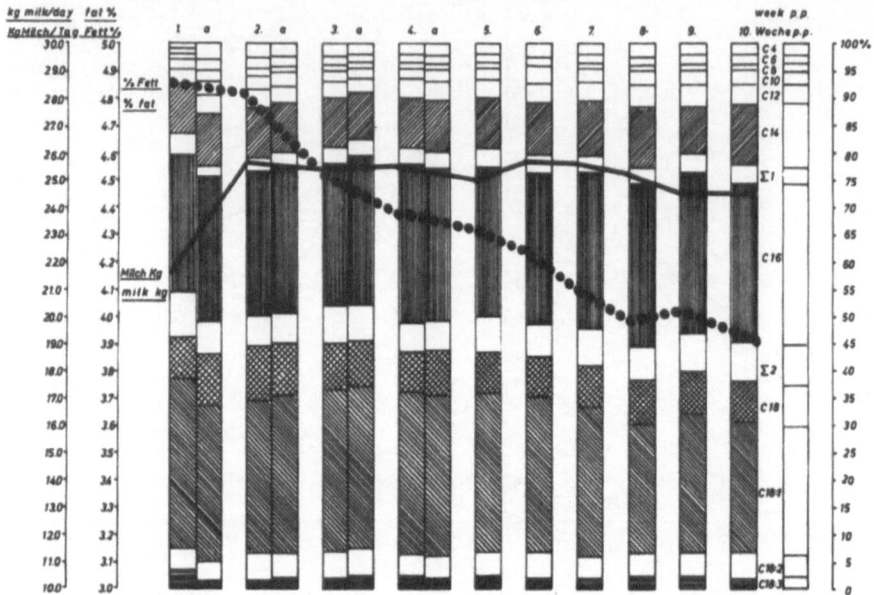

Fig. 6. Ketosis experiment 1972/73; Distribution of fatty acids in milkfat. Group 1b.

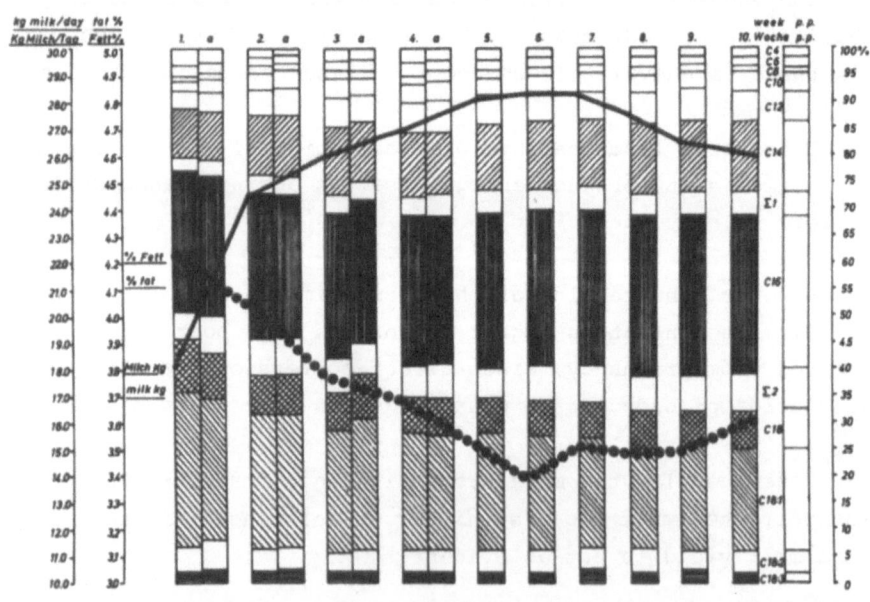

Fig. 7. Ketosis experiment 1972/73; Distribution of fatty acids in milkfat. Group IIb.

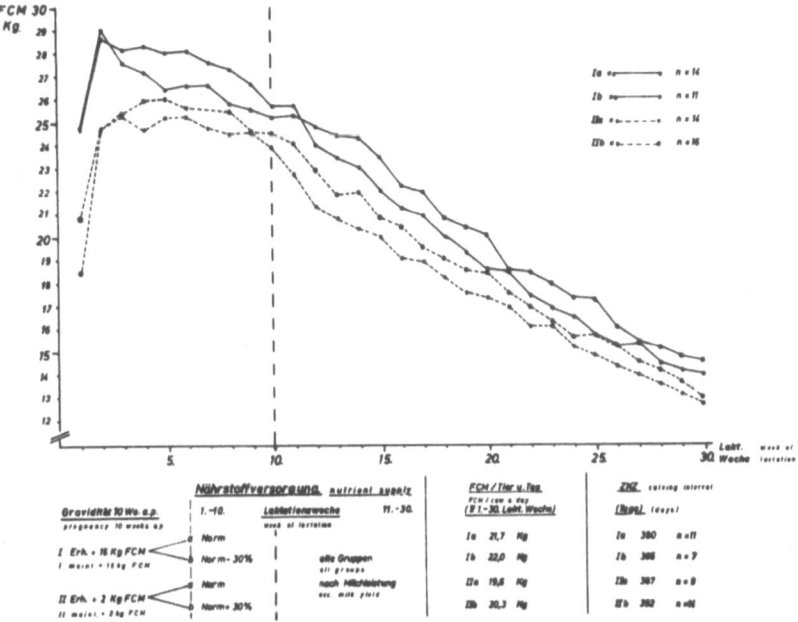

Fig. 8. Ketosis experiment 1972/73; Course of lactation.

Summarising a pre-partal surplus of energy stored in the form of body fat, may be the origin of disturbances in the energy metabolism at the beginning of lactation.

REFERENCES

Drepper, K. 1976. Bovine Ketosis bei Hochleistungskühen. (Versuch einer
Literaturauswertung). Beiheft: Zeitschrift für Tierphysiologie,
Tierernährung, Futtermittelkunde 7: 27-42.

Jazbec, J. 1967. Häufigkeit, Äthiologie und Therapie einer subklinischen
Azetonämie in einem Rinderbestand. Deutsche Tierärztliche Wochenschrift
74: 597-603.

Kalaissakis, P. 1958. Langfristige Untersuchungen zum äusseren und inneren
Stoffwechsel von graviden und laktierenden Ziegen. 1. Mitteilung.
Zeitschrift für Tierphysiologie, Tierernährung, Futtermittelkunde 13:
355-366.

Kalaissakis, P. 1959. Langfristige Untersuchungen zum äusseren und inneren
Stoffwechsel von graviden und laktierenden Ziegen. 2. Mitteilung.
Zeitschrift für Tierphysiologie, Tierernährung, Futtermittelkunde.
14: 204-214.

Kronfeld, D.S. 1976. The potential importance of the proportions of glucogenic, lipogenic and aminogenic nutrients in regard to the health and
productivity of dairy cows. Beiheft: Zeitschrift für Tierphysiologie,
Tierernährung, Futtermittelkunde 7: 5-26.

Lenkeit, W., Gütte, J.O. and Streutter-Petermöller, A. 1955. Langfristige
Untersuchungen zum äusseren und inneren Stoffwechsel des graviden und
laktierenden Schweines. 2. Mitteilung: Der Ablauf der N-Bilanz vom
Ende der Gravidität bis zum Ende der Laktation bei gleichbleibender
Ernährung. Zeitschrift für Tierphysiologie, Tierernährung, Futtermittel-
kunde 10: 228-237.

Lenkeit, W., Gütte, J.O. and Kirchhoff, W. 1956. Weitere Untersuchungen zur
Abhängigkeit des N-Umsatzes während der Laktation von der Nährstoff-
versorgung während der Gravidität. 4. Mitteilung: Nährstoffspeicherung
und Leistung. Zeitschrift für Tierphysiologie, Tierernährung, Futtermittelkunde 11: 337-352.

Oslage, H.J. and Farries, F.E. 1966. Beiträge zum Stoffwechsel von Kühen im
Ablauf von Trächtigkeit und Laktation. 2. Mitteilung: Untersuchungen
zum N-Stoffwechsel. Landbauforschung Völkenrode 16: 53-64.

Oslage, H.J. and Farries, F.E. 1970. Beiträge zum Stoffwechsel von Kühen
im Ablauf von Trächtigkeit und Laktation. 3. Mitteilung: Untersuchungen zum Mineralstoffwechsel. Landbauforschung Völkenrode 20: 13-24.

Piatkowski, B. 1962. Untersuchungen über den Stoffwechsel an hochleistenden
kühen während der Trockenperiode und Laktation. 1. Mitteilung:
Proteinstoffwechsel. Archiv für Tierernährung 12: 75-92.

Piatkowski, B. 1964. Untersuchungen über den Proteinstoffwechsel an
hochträchtigen Zwillingsfärsen. Archiv für Tierarnährung 14: 47-54.

THE INFLUENCE OF PRE-CALVING FEEDING AND MANAGEMENT OF THE COW ON EASE OF CALVING AND CALF VIABILITY

K. Sejrsen and A. Neimann-Sørensen

National Institute of Animal Science, 25 Rolighedsvej, DK 1958
Copenhagen V. Denmark

ABSTRACT

The effect of pre-calving feeding and management on ease of calving and calving difficulties are investigated on the basis of the literature and Danish results.

Calving performance and calf mortality of dairy cows do not seem to be influenced to any great extent by pre-calving feeding level. Birth weight which is an important factor in determining calving performance, is only affected by severe undernutrition leading to lighter calving. Results indicate that cows in fat condition at calving have more difficult calvings than cows in average condition. Results concerning thin or lean animals at calving are conflicting.

Calving difficulties as well as calf mortality are higher in the winter than in the summer. The difference may at least partly be due to the increased exercise the cows get during the grazing season. There does not seem to be any striking effect of type of housing on calving performance and calf mortality, but the calf mortality is found to increase with increasing herd size. A small experiment indicated a positive effect on calving performance by letting the cows calve in a special calving pen rather than in the tie stall.

1. INTRODUCTION

Viable offspring and high milk production are the ultimate
demands from the modern dairy cow, and if the cow is going to
be able to fulfil these demands it is important that feeding
and management are optimal.

In the last part of pregnancy the size of the foetus is
increasing rapidly. At the same time extensive mammary growth
and regeneration are taking place in heifers and cows
respectively. At parturition the cows are submitted to severe
stress by delivery of the calf and by initiation of lactation.

Against this background the purpose of pre-calving feeding
is to supply the cow with sufficient nutrients for foetal and
mammary development and to build up body reserves to meet the
high demand for energy at the initiation of lactation. Furthermore an adequate supply of specific nutrients has to be given
in order to prevent metabolic disorders such as milk fever and
ketosis, which often occur in connection with calving. In the
light of the drastic changes in the cow's metabolism in the last
part of pregnancy and around calving it is important that cows
are not submitted to unnecessary stress by bad management.

The changes are especially drastic for the animals calving
for the first time, and the greatest problems associated with
the calving therefore, not unexpectedly, occur at first calving.
This is underlined in the calving statistics showing 3 - 4 times
as high calf mortality at first calving as at second or later
calvings (Elleby and Rasmussen, 1971). Age at first calving
is well known to have a marked influence on ease of calving
and calf mortality. As this topic is covered elsewhere in the
programme of this symposium it will not be treated here.

2. PRE-PARTUM LEVEL OF ENERGY

The results in the literature concerning the effect of
feeding in late pregnancy on ease of calving and viability of

the offspring are very limited. In the extensive literature
reviewed by Broster (1971) no effect of heavy pre-partum feeding
was reported. Lodge (1972) states in a review that the results
concerning effect of pre-calving feeding on calf survival are
conflicting. Gardner (1969), offering two levels of energy
pre-partum to Holstein cows, found no difference in calving
difficulties and calf mortality. Corah et al. (1975), in
contrast, found that more calves from Hereford heifers and cows
fed restricted level died at or near birth. This happened in
spite of the fact that no differences in ease of calving were
observed.

This, however, is in disagreement with results from a Danish
crossbreeding experiment (Stenbaek, 1977). Two levels of feeding
during the winter period (5 - 6 v 7 - 8 Scandinavian feed units
per day) were compared and calf mortality as well as calving
difficulties were highest at the highest level of feeding
(Table 1). Cows of F_1 crosses between Danish dual-purpose breeds
as dam breeds and eight different European beef breeds as sire
breeds were inseminated with semen from the sire breed. All
heifers, irrespective of sire breed, were bred with Aberdeen-
Angus semen. The duration of parturition measured as time from
showing of placenta to delivery of the calf, was recorded on 25
calvings in each group. The variation within groups was large.
However, on average the duration was lowest in the group fed the
low level of energy; the time spent in calving in the group on
the high level of energy is probably decreased by the greater
assistance offered this group.

Birth weight of the offspring is one of the important
factors associated with ease of calving (Freeman, 1976;
Philipsson, 1976a), and the effect of pre-calving feeding on
birth weight is reported in several experiments. The results,
however, are conflicting. Lodge et al. (1975) found no
significant difference in birth weight between two levels of
pre-calving feeding to Holstein cows (maintenance and 1.8 times
maintenance). Gardner (1969) and Ekern (1972) with dairy
breeds and Ferrell et al. (1976) and Wallace and Raleigh (1964)

with beef breeds comparing two levels of feeding above
maintenance found no differences in birth weight either. Hight
(1966), Wiltbank et al. (1962) and Corah et al. (1975) on the
other hand reported lower birth weight of the calves from the
cows on a low level of nutrition prior to calving.

TABLE 1

EFFECT OF LEVEL OF FEEDING DURING THE WINTER ON CALF MORTALITY, CALVING
DIFFICULTIES AND DURATION OF CALVING (STENBAEK, 1977).

	Low level		High level	
	1st calvers	2nd calvers	1st calvers	2nd calvers
Calf mortality:				
No. of calvings	39	11	40	11
No. of dead calves	0	0	2	2
Ease of calving:				
No. of calvings	39	11	40	11
1 - no help	33	9	22	6
2 - minimal assistance	4	1	5	1
3 - chains used	2	1	9	1
4 - veterinary assistance	-	-	4	-
5 - Caesarean section	-	-	-	1
Average score	1.2	1.3	1.9	1.5
Duration of calvings:				
No. of calvings	25	-	25	-
Average time, min	68	-	73	-
Range	18-155	-	13-217	-

Joubert and Bonsma (1957) concluded from a study with four
breeds that birth weight, when the dam had reached 70 - 80% of
mature weight, is affected by level of nutrition in the pregnancy
only if the ration falls below the maintenance level of the
dam. Hight (1966), Morris (1970) and Tudor (1972) also found
that calf birth weight only decreased after severe undernutrition.

3. CONDITION OF THE COW

It is often believed that heavy feeding resulting in fat in
the birth canal leads to more difficult calving. This, however,
has not been reported in any of the experiments concerning pre-
partum feeding level even with very high pre-partum feeding
intensity. Philipsson (1976c) on the other hand found that
calving difficulties were affected by the condition of the cow
at calving (Table 2). The results show that thin as well as
fat cows had more difficult calvings than cows in medium
condition at calving. Arnett et al. (1971) found in agreement
with this, that fat beef heifers required assistance six times
as often as heifers in moderate lean condition. Nelson and
Huber (1971) reported for beef heifers that 19% of heifers in
average condition required assistance, while 28% of the fat
heifers required assistance. In contrast to Philipsson (1976c),
Nelson and Huber (1971) found that lean heifers needed less
assistance than heifers in average condition.

TABLE 2

CALVING DIFFICULTY IN RELATION TO CONDITION SCORE (PHILIPSSON, 1976c)

Condition score	No. of observations	Calving difficulty %
1 (thin)	18	23.8
2	157	18.9
3	555	16.0
4	290	19.8
5 (fat)	42	29.5

4. SEASON

In almost all studies where seasonal differences have
been investigated, calving difficulty is associated with season
of calving (Freeman, 1976). There is also a large seasonal
variation in calf mortality. In Denmark a higher number of
stillbirths are found in the winter than in summer (Table 3)

TABLE 3

SEASONAL VARIATION IN CALF MORTALITY (% STILLBIRTHS) IN DENMARK (ELLEBY AND MYGIND-RASMUSSEN, 1971)

| Month | Breed | | | | | |
| | RDM | | SDM | | Jersey | |
	1st calvn.	2nd and later	1st calvn.	2nd and later	1st calvn.	2nd and later
January	11.4	4.0	9.5	3.4	5.1	4.0
February	10.6	3.4	7.9	2.8	5.5	3.4
March	9.6	2.9	7.6	2.6	6.2	3.5
April	9.5	3.2	6.8	3.1	6.2	4.0
May	9.1	3.6	6.8	3.7	6.0	3.9
June	9.9	4.0	7.9	3.8	5.1	3.1
July	9.9	3.9	8.6	3.6	4.7	3.4
August	8.0	3.5	6.5	3.3	4.1	3.2
September	7.5	3.3	6.2	3.2	3.7	3.1
October	9.2	3.6	6.7	3.2	4.9	3.2
November	11.4	3.6	8.3	3.4	5.2	3.3
December	11.8	3.9	8.7	3.4	5.5	3.8

(Elleby and Mygind-Rasmussen, 1971). From the Table it also appears that there are considerably more stillborn calves at the first calving than at later calvings. Furthermore it can be seen that the frequency of stillbirths in heifers is lower in the Jersey breed compared with the Danish dual-purpose breeds. Bar-Anan et al. (1976) in Israel and Philipsson (1976b) in Sweden found a similar seasonal pattern in calf mortality and calving performance. Philipsson suggested that the increased daylight and increased amount of exercise during summer are important factors leading to easy calvings. That exercise is of importance agrees with results from the Danish pilot farms (Table 4). The results show that the number of abnormal calvings are 6.4% when the cows are on pasture, against 10.9% when the cows are stabled all the year (Landøkonomisk Driftsbureau, 1976). In a Danish experiment three management systems were compared: grazing, housing all the year, and

TABLE 4

CALVING PERFORMANCE AND CALF MORTALITY IN RELATION TO SUMMER GRAZING AND
HOUSING ALL THE YEAR (LANDØKONOMISK DRIFTSBUREAU, 1976)

	Summer grazing	Housing all year
No. of herds	18	18
Cows per herd	42	51
Abnormal calvings[1]	6.4	10.9

1) Cows requiring veterinarian assistance (%).

housing all the year plus daily exercise. No differences in
calf mortality between these systems were observed (Table 5)
(Klausen, 1977). The reason why no differences were observed
in this experiment, may be the good management and more
thorough observation of the cows on the experimental farm.

5. TYPE OF HOUSING AND HERD SIZE

The viability of the calves and the ease of calving do
not seem to be influenced to any great extent by the housing
system. In current Danish experiments ordinary tie stalls
are compared with two different loose housing systems, namely
insulated barn with cubicles and uninsulated barn with deep
bedding (Konggaard, 1977). The calving results are shown in
Table 6. From the results it can be seen that there does not
seem to be any significant difference between housing systems
neither in calving performance nor in calf viability.

Data from the Danish pilot farm have been arranged
according to the following types of housing - traditional tie
stall, tie stall with dung grit, loose housing with solid floor
and loose housing with slatted floor, (Table 7). (Landøkonomisk
Driftsbureau, 1976). Although it is difficult to draw any
strong conclusions from these results, it will be seen that
more abnormal calvings have been observed in the traditional
tie stalls. Bar-Anan et al. (1976) observed that calving
difficulties as well as calf mortality increased with herd size.
The same was observed in Sweden (SHS, 1974) (Table 8).

TABLE 5

CALF MORTALITY IN RELATION TO SUMMER GRAZING (KLAUSEN, 1977)

	1st calving		2nd and later	
	No. of calvings	No. of dead calves	No. of calvings	No. of dead calves
Summer grazing	19	1	49	2
Indoors all year	18	1	41	0
Indoors all year + exercise	16	1	37	2

TABLE 6

EASE OF CALVING AND CALF MORTALITY IN RELATION TO TYPE OF HOUSING
(KONGGAARD, 1977)

	Tie stall	Cubicles	Deep bedding
Calving difficulty 1973-1976			
No. of calvings	67	85	94
Average difficulty[1]	1.7	1.7	1.5
Calf mortality 1970-1976			
No. of calvings	133	158	168
No. of dead calves[2]	13	17	18
Dead calves - %	9.8	10.8	10.7

1) 1 - no help, 2 - minimal assistance, 3 - moderate assistance

    4 - heavy assistance, 5 - veterinary assistance.

2) Stillborn calves + calves dead during the first 3 days of life.

TABLE 7

CALVING PERFORMANCE AND CALF MORTALITY IN RELATION TO TYPE OF HOUSING
(LANDØKONOMISK DRIFTSBUREAU, 1976)

	Tie stalls		Loose housing	
	Traditional	Dung grit	Solid floor	Slatted floor
No. of herds	8	8	8	8
Cows per herd	43	45	58	61
Abnormal calvings in %[1]	8.1	5.2	4.8	5.5

1) Cows requiring veterinarian assistance.

TABLE 8

FREQUENCY OF STILLBORN CALVES IN RELATION TO HERD SIZE (SHS, 1974)

Herd size	Per cent stillborn	
cows	Heifers	Bull calves
<10	2.78	3.63
10 - 15	2.78	3.82
15 - 20	2.92	3.81
20 - 30	3.00	4.17
30 - 40	3.22	4.41
40 - 50	3.79	4.74
>50	4.12	5.84

6. MANAGEMENT AROUND CALVING

In some tie stall barns the cows calve in the tie stall
while the cows in other herds are brought to a calving pen
sometime before calving. These two systems have been compared
in a small experiment with Black Pied heifers calving for the
first time (Konggaard et al., 1967). The results (Table 9)
show that bringing the animals to a calving pen one week
before calving decreased the number of cases with calving
difficulties. In the case of heifers calving in the pen only
13% required more than minimal assistance, while this was
required for 50% of the heifers calving in the tie stall.

TABLE 9

EASE OF CALVING OF HEIFERS IN RELATION TO PARTURITION IN THE TIE STALL OR
IN CALVING PEN (KONGGAARD ET AL., 1967)

Calving performance	No. of those calving in the tie stall	No. of those calving in calving pen
1 - without assistance	5	7
2 - minimal assistance	2	6
3 - medium assistance	3	1
4 - heavy assistance	4	1
5 - veterinary assistance	0	0
Av. difficulty	2.4	1.7
Heifers requiring more than minimal assistance (%)	50	13

466

REFERENCES

Arnett, D.W., Holland, G.L. and Totusek, R. 1971. Some effects of obesity in beef females. Journal of Animal Science 33: 1129-1136.

Broster, W.H. 1971. The effect on milk yield of the cow of the level of feeding before calving. Dairy Science Abstracts 33: 253-270.

Bar-Anan, R., Soller, M. and Bowman, J.C. 1976. Genetic and environmental factors affecting the incidence of difficult calving and perinatal calf mortality in Israeli-Friesian dairy herds. Animal Production 22: 299-310.

Corah, L.R., Dunn, T.G. and Kaltenbach, C.C. 1975. Influence of prepartum nutrition on the reproductive performance of beef females and the performance of their progeny. Journal of Animal Science 41: 819-824.

Ekern, A. 1972. Feeding of high yielding dairy cows. I. The effect of different levels of feeding before and after calving on milk yield and composition. Agricultural University of Norway, Technical bulletin No. 172. 79 pp.

Elleby, F. and Rasmussen, V. Mygind. 1971. Kaelvningsstatistik. 4. medd. fra LPH-udvalget, Arhus. 107 pp.

Ferrell, C.L., Garrett, W.N. and Hinman, N. 1976. Growth, development and composition of the udder and gravid uterus of beef heifers during pregnancy. Journal of Animal Science 42: 1477-1489.

Freeman, A.E. 1976. Management traits in dairy cattle. Dystocia, udder characteristics related to production, and a review of other traits. Livestock Production Science 3: 13-26.

Gardner, R.W. 1969. Interaction of energy levels offered to Holstein cows prepartum and postpartum. II. Reproductive performance. Journal of Dairy Science 52: 1985-1987.

Hight, G.H. 1966. The effect of undernutrition in late pregnancy on beef cattle production. New Zealand Journal of Agricultural Research 9: 479-490.

Joubert, D.M. and Bonsma, F.N. 1957. The effect of nutrition on the birth weight of calves. Science bulletin No. 371, Department of Agriculture, Union of South Africa.

Klausen, S. 1977. Unpublished results.

Konggaard, S.P., Larsen, J. Brolund, Klausen, S. and Kirsgaard, E. 1967. Kunsttørrede graesmarksafgrøder. 1. Kløvergraespiller kontra kraftfoder. Yearbook of National Institute of Animal Science. p.482-489.

Konggaard, S.P. 1977. Preliminary results.

Landøkonomisk Driftsbureau. 1976. Nogle kvaegsygdommes økonomiske betydning. Undersøgelse nr. 31 fra Det Landøkonomisk Driftsbureau.

Lodge, G.A. 1972. Energy and nutrient requirements for pregnancy. In: Handbuch der Tierernährung. 2. Band. W. Lenkeit and K. Breirem (editors) Paul Parey, Hamburg and Berlin, pp. 157-189.

Lodge, G.A., Fisher, L.J. and Lessard, J.R. 1975. The influence of pre-partum feed intake on performance of cows fed ad libitum during lactation. Journal of Dairy Science 58: 696-702.

Morris, J.G. 1970. The survival feeding of pregnant and lactating beef cows on all sorghum grain rations: the effect of early weaning of the calves. Journal of Agricultural Science 75: 479.

Nelson, L.A. and Huber, D.A. 1971. Factors influencing dystocia in Hereford dams. Journal of Animal Science 33: 1137.

Philipsson, J. 1976a. Studies on calving difficulty, stillbirth and associated factors in Swedish cattle breeds. 1. General introduction and breed averages. Acta Agriculturae Scandinavica 26: 151-164.

Philipsson, J. 1976b. Studies on calving difficulty, stillbirth and associated factors in Swedish cattle breeds. II. Effects of nongenetic factors. Acta Agriculturae Scandinavica 26: 165-174.

Philipsson, J. 1976c. Studies on calving difficulty, stillbirth and associated factors in Swedish cattle breeds. III. Relationships between calving performance, precalving body measurements and size of pelvic opening in Friesian heifers. Acta Agriculturae Scandinavica 26: 221-229.

SHS. 1974. Kontroll- och seminverksamhet 1972-1973. Meddelande nr. 64. Svensk Husdjursskötsel, Hällsta.

Stenbaek, B. 1977. Preliminary results.

Tudor, G.D. 1972. The effect of pre- and postnatal nutrition on the growth of beef cattle. I. The effect of nutrition and parity of the dam on calf birth weight. Australian Journal of Agricultural Research 23: 389.

Wallace, J.D. and Raleigh, R.J. 1964. Calf production from Hereford cows wintered at different nutrition levels. Journal of Animal Science. 23: 605.

Wiltbank, J.N., Rowden, W.W., Ingalls, J.E., Gregory, K.E. and Koch, R.M. 1962. Effect of energy level on reproductive phenomena of mature Hereford cows. Journal of Animal Science 21: 219.

CLINICAL ASPECTS OF THE NUTRITIONAL STATUS OF THE DAM AND PARTURITION

E. Grunert

Clinic of Obstetrics and Gynaecology of Cattle, School of
Veterinary Medicine, 3000 Hannover, Germany.

ABSTRACT

The nutritional status of the dam has only limited influence on parturition itself. However, in extremely underfed heifers the pelvic diameter will be less than in well fed animals; the calf birth weight will also be less and vice versa. Therefore some authors were not able to show that nutrition has an influence on the occurrence of dystocia. When an influence was obvious it was always related to the breed, eg Angus, Holstein and Devon. In extremely underfed dams poor relaxation of the connective tissues in the pelvic region led to constriction of the birth canal negatively influencing parturition. In extremely overfed heifers calf mortality was found to be high. This was caused by protracted parturition due to excessive deposition of fat in the pelvic region. Keeping in mind that parturition is influenced by many factors, and due to the small number of experiments related to nutrition and calving, a definite conclusion considering all breeds cannot be drawn.

1. EXPERIMENTAL RESULTS

Dam age and calf birth weight are two of the most important
factors affecting dystocia in cattle. Especially in young heifers
a small diameter of the pelvic opening can influence parturition
negatively. Calf birth weight is primarily influenced by genotype (sire breed and dam breed) (Kunert et al., 1971; Laster,
1974; Meyer 1964; Pinney et al., 1972), sex of calf (Bellows et
al., 1971; Bond and Wiltbank, 1970; Donald et al., 1962; Gianola
and Tyler, 1974; Laster et al., 1973; Meyer, 1964; Schwark et al.,
1972; Tudor 1972), gestation length (Bellows et al., 1971;
Gianola and Tyler, 1974) and age of dam (Laster, 1974; Meyer,
1964; Pinney et al., 1972). Obstetricians confirm the incomplete
correlation between the nutritional status of dam and difficulties
at parturition.

Only a limited number of experiments dealing with the
influence of the nutritional status on parturition have been
published. Some authors fed different diets eg 6.3 or 9.0
Mcal net energy per head per day (Hafez et al., 1968), low
quality hay at 10 lb or good quality hay at 16 lb plus 6 lb
sorghum grain per head per day (Young, 1970); 3.5 kg Rhodes
grass hay or 3.5 kg Rhodes grass hay and barley per head per
day (Tudor, 1972); 4.9, 6.2 or 7.7 kg TDN per head per day
(Laster, 1974). They observed a significant influence of the
diet on calf birth weight, dam weight and diameter of the
pelvic opening at the time of delivery.

TABLE 1

BIRTH WEIGHT IN RELATION TO LEVEL OF NUTRITION OF HEIFERS AND SEX OF CALF
(Young, 1970)

Sex of Calf	No.	Low plane Birth wt (lb $\pm$ SE)	No.	High plane Birth wt (lb $\pm$ SE)
Male	23	53.8 $\pm$ 1.31	25	59.3 $\pm$ 1.25
Female	18	48.8 $\pm$ 1.48	14	54.1 $\pm$ 1.68

Low v. High plane males : P <0.01
Low v. High Plane females : P< 0.02

TABLE 2

WEIGHT CHANGES AND PELVIC AREA OF HEIFERS IN RELATION TO LOW AND HIGH
NUTRITIONAL PLANES (YOUNG, 1970)

	Low plane	High plane
Number of heifers	46	44
Weight at parturition (lb $\pm$ SD)	540. 4 $\pm$ 63.36	748.8 $\pm$ 67.51
Pelvic area at parturition (cm^2 $\pm$ SD)	227.4 $\pm$ 17.89	236.7 $\pm$ 14.14

Weight at parturition   : P <0.01

Pelvic area at parturition   : P <0.01

The above mentioned authors did not find a significant
influence of feeding and nutritional status of the dam on
parturition and calf mortality in their experiments using
Holstein heifers (Reid et al., 1957), Hereford heifers (Tudor,
1972) and Hereford and Angus heifers (Hafez et al., 1968;
Schultz, 1969 and Young, 1970). According to Young (1970) out
of the eleven cases of dystocia six were in the high-plane group
and five in the low-plane group. Ten of the eleven cases were
male calves and one was a female calf.

An influence of undernutrition on the rate of dystocia in
cattle was found by Young (1968) for Devon heifers, by Schultz
(1969) for Holsteins and by Bellows et al. (1971) for Angus
heifers. Young (1968) observed differences in the occurrence
of dystocia between heifers fed a diet of silage (a low quality
silage, which proved to be a bare maintenance ration) and
lucerne-fed heifers (12 lb of good quality lucerne hay per head
per day).

According to Bellows et al. (1971) pre-calving body weight
of the Angus dam exerted a highly significant negative effect
on calving difficulty score. However, this effect was not
significant in Hereford dams. In addition, significant negative
correlations were found between body weight at certain periods

TABLE 3

DYSTOCIA IN DEVON HEIFERS IN RELATION TO THE DEATH OF THE FOETUS AND SEX OF THE FOETUS (YOUNG, 1968)

Group	Total Assisted	% of Total	Calf Death	% of total	Sex of calves % of dystocia
Silage fed	7	24.1	2	7.1	42.9 %♂ 57.1 %♀
Lucerne fed	5	14.3	3	8.6	80.0 %♂ 20.0 %♀

(end of breeding season, mid-gestation, pre-calving) and calving difficulty score. This suggests that pre-calving body size of the Angus dams may have been below optimum and resulted in more calving difficulty. The authors found also a highly significant negative correlation between calving difficulty score and pelvic width in Angus dams but this correlation was not significant in Herefords. This may suggest that lower body weights were reflected in narrower pelvic openings which become a limiting factor resulting in more calving difficulty in the Angus.

According to Schultz (1969) underfeeding results in calves only slightly smaller than normal at first calving but decreases the body weight of the dam and therefore increases calving difficulties. These experimental data correspond with field observations.

2. CLINICAL EFFECTS OF UNDERFEEDING

Based on clinical findings the following nutrition-related factors appear to influence parturition negatively.

A. Small size of the dam with a small pelvic opening which, besides other factors, may be caused by breeding a poorly grown, under-fed female that may be old enough to breed but whose body has been greatly retarded due to poor nutrition, parasitism or disease. Excessive undernutrition of the dam will decrease the birth weight of the calf (Hight, 1966; Joubert, 1954;

Lees et al., 1948; Ryley, 1961; Ryley and Garner, 1962 and
Wiltbank et al., 1962). But, on the other hand, it affects the
body weight of the dam more than that of the calf leading to
dystocia (Reid, 1953). In heifers with a small pelvic opening
the necessity for Caesarean section or foetotomy is frequent.
Applying excessive traction (eg with five men or more, or using
a calf puller) the following complications can be caused:

a)   Paralysis due to injuries to the obturator or perineal
     nerve.

b)   Compression of spinal cord.

c)   Fractures of the pelvis.

d)   Dislocation of the sacro-iliac articulation.

B.   Stenosis or constriction of the vulva and vestibule is
usually observed in heifers with genital hypoplasia due to
improper growth of the dam because of chronic disease or poor
nutrition.

A poor relaxation of the pelvic tissues during parturition
in silage-fed 24 - 27 - month-old heifers was observed by
Young (1968), when starvation was carried to the point of
inanition.

In these cases the forced extraction of the foetus is
often necessary. The following complications are observed:

a)   Haemorrhage or haematomas in the birth canal.

b)   Rupture of the pelvic blood vessels.

c)   Lacerations of the genital tract with secondary
     infections (necrotic vestibulitis and vaginitis).

d)   Rupture of the perinaeum or recto-vaginal perforation.

C.   Uterine inertia as a result of poor nutritional status of
the dam, frequently associated with disease. It is observed more
often in older dairy cows.

3. EFFECTS OF OVERFEEDING

On the other hand overfeeding or excessive fattening can also lead to dystocia. Calving difficulties and high calf mortality in heifers with excessive body fatness (high energy level) was found by Holland (1961), Arnett (1963), Williams (1968), Bond and Wiltbank (1970), Kunert et al. (1971) and Roberts (1971). Bond and Wiltbank (1970) observed a heavy death loss of first calves from heifers receiving high energy diets. Many of these calves were in posterior presentation or in some other abnormal position. Some calves were born alive but breathed only once or twice after birth. No abnormalities were found at autopsy and findings indicate the deaths were caused by slow or difficult delivery. The protein level of the diet had no significant effect on the survival or birth weight of calves, although the low protein level resulted in the poorest survival and lowest birth weight. These experimental data also correspond with field observations. In excessively fat heifers the following causes appear to lead to dystocia.

A. Excessive deposition of fat in the pelvic region due to high feeding levels (Gianola and Tyler, 1974; Laster et al., 1973) reduces its lumen and makes parturition more difficult (Franzos, 1970; Williams, 1968), leading to fatigue of the musculature of the uterus and retarded involution of the organ. Over-feeding seems also to predispose to stillbirth and puerperal metritis (Franzos, 1970).

Other complications are:

a) Vaginal rupture with prolapse of the perivaginal fat and, later on, in some cases formation of tumours.

b) Infections of the perivaginal region and sometimes development of pelvic phlegmons.

B. Relative foetal oversize. Overfeeding increases the weight of the foetus (Hafez et al., 1968; Laster, 1974) but also influences to a certain extent the weight of the dam leading in young heifers to enlarged pelvic diameter.

The clinical aspects can be summarised as follows:

The nutritional status of the dam has generally only
limited influence on parturition itself. In extremely
underfed heifers the pelvic diameter will be less than in
well fed animals, however, the calf birth weight will also
be less and vice versa. The experimental data showed that
when an influence was obvious it was related to the breed,
eg Angus, Holstein and Devon. In extremely underfed dams
poor relaxation of the connective tissues in the pelvic
region lead to constriction of the birth canal negatively
influencing parturition. In these cases the withdrawal
of the foetus by excessive traction should be avoided as
severe complications may occur. In underfed heifers with
dystocia Caesarean section is indicated. To prevent
calving difficulties induction of premature parturition
should be taken into consideration. Especially in
extremely overfed heifers calf mortality and injuries to
the birth canal were found to be high. This was caused by
retarded parturition or forced extraction due to excessive
deposition of fat in the pelvic region. In order to deliver
a viable foetus and to prevent injury to the extremely
fat dam with dystocia, Caesarean section should be
performed.

REFERENCES

Arnett, D.W. 1963. Studies with twin beef females. I. Some effects of level
of nutrition. PD. Thesis, Oklahoma State University, Stillwater,
Oklahoma.

Bellows, R.A., Short, R.E., Anderson, D.C.,Knapp, B.W. and Pahnish, O.F. 1971.
Cause and effect relationships associated with calving difficulty
and calf birth weight. Journal of Animal Science 33, 407-415.

Bond, J. and Wiltbank, J.N. 1970. Effect of energy and protein on estrus,
conception rate, growth and milk production of beef females. Journal
of Animal Science 30: 438-444.

Donald, H.P., Russell, W.S. and Taylor, St. C.S. 1962. Birth weights of
reciprocally cross-bred calves. Journal of Agricultural Science
58: 405-412.

Franzos, G. 1970. Observations on the relationship between overfeeding and
the incidence of metritis in cows after normal parturition. Refuah
Veterinariah 27: 148-155.

Gianola, I. and Tyler, W.J. 1974. Influences on birth weight and gestation
period of Holstein-Friesian cattle. Journal of Dairy Science 57: 235-240.

Hafez, E.S.E., Dyer, J.A. and Jainudeen, M.R. 1968. Effect of maternal
caloric intake on foetal development in beef cattle. American Journal
of Veterinary Research 29: 2281-2285.

Hight, G.K. 1966. The effects of undernutrition in late pregnancy on beef
cattle production. New Zealand Journal of Agricultural Research
9: 479-490.

Holland, G.L., 1961. The influence of excessive body fatness on the
performance of beef females. PD. Thesis, Oklahoma State University,
Stillwater, Oklahoma.

Joubert, D.M. 1954. The influence of winter nutritional depressions on
the growth, reproduction and production of cattle. Journal of
Agricultural Science 44: 5-66.

Kunert, G., Schwark, H.J. and Leuschner, W. 1971. Die Vorverlegung der
Erstkonzeption in ihrem Einfluss auf Befruchtungsergebnis und Geburts-
verlauf. Wissenschaftliche Zeitschrift Universität Jena, Mathematische
Naturwissenschaftliche Reihe 20: 436-442.

Laster, D.B. 1974. Factors affecting pelvic size and dystocia in beef
cattle. Journal of Animal Science 38, 496-503.

Laster, D.B., Glimp, H.A., Cundiff, L.V. and Gregory, K.E. 1973. Factors affecting dystocia and the effects of dystocia on subsequent reproduction in beef cattle. Journal of Animal Science 36: 695-705.

Lees, F.T., McMeekan, C.P. and Wallace, L.R. 1948. Proceedings of the 8th Conference of the New Zealand Society of Animal Production, p. 60.

Meyer, H. 1964. Das Geburtsgewicht beim Kalb und die Ursachen seiner Variation. Züchtungskunde 36: 303-316.

Pinney, D.O., Stephens, D.F. and Pope, L.S. 1972. Lifetime effects of winter supplemental feed level and age at first parturition on range beef cows. Journal Animal Science 34: 1067-1074.

Reid, J.T. 1953. Cited by Meyer, H. 1964.

Reid, J.T., Loosli, J.K., Turk, K.L., Trimberger, G.W., Asdell, S.A. and Smith, S.E. 1957. Effect of nutrition during early life upon the performance of dairy cows. Proceedings of the Cornell Nutrition Conference pp. 65-71.

Roberts, S.J. 1971. Veterinary obstetrics and genital diseases. Edwards Brothers Inc., Ann Arbor, Michigan.

Ryley, J.W. 1961. Drought feeding studies with cattle. 6. Sorghum silage, with and without urea, as a drought fodder for cattle in late pregnancy and early lactation. Queensland Journal of Agricultural Science 18: 409-424.

Ryley, J.W. and Gartner, R.J.W. 1962. Queensland Journal of Agricultural Science 19: 309.

Sagebiel, J.A., Krause, G.F., Sibbit, B., Langford, L., Comfort, J.E., Dyer, A.J. and Lasley, J.F. 1969. Dystocia in reciprocally crossed Angus, Hereford and Charolais cattle. Journal of Animal Science 29: 245-250.

Schultz, L.H. 1969. Relationship of rearing rate of dairy heifers to mature performance. Journal Dairy Science 52: 1321-1329.

Schwark, H.J., Oehler, H. and Kunert, G. 1972. Die Körpermasse des Rindes bei der Geburt - Bedingtheit und Folgewirkung. Fortschrittsberichte für die Landwirtschaft und Nahrungsgüter-wirtschaft 10: 1-86.

Tudor, G.D. 1972. The effect of pre- and post-natal nutrition on the growth of beef cattle. I. The effect of nutrition and parity of the dam on calf birth weight. Australian Journal of Agricultural Research 23: 389-395.

Williams, K.R. 1968. A study of dystocia in the heifer (Primipara).
    Veterinary Record 83: 87-92.

Wiltbank, J.N., Rowden, W.W., Ingalls, J.E., Gregory, K.E. and Koch, R.M.
    1962. Effect of energy level on reproductive phenomena of mature
    Hereford cows. Journal of Animal Science 21: 219-225.

Young, J.S. 1968. Breeding patterns in commercial beef herds. 3.
    Observations on dystocia in a Devon herd. Australian Veterinary
    Journal 44: 550-556.

Young, J.S. 1970. Studies on dystocia and birth weight in Angus heifers
    calving at two years of age. Australian Veterinary Journal 46: 1-7.

DISCUSSION

B. Hoffmann *(West Germany)*

Dr. Sejrsen observed a seasonal effect on calving difficulties which was more obvious in heifers than in cows. Couldn't
this also be indirectly related to the onset of puberty, which
is influenced by season, and hence to early breeding of heifers?

K. Sejrsen *(Denmark)*

I agree that the effect in those data was most pronounced
on heifers, but I think the tendency was quite clear in later
calvings also. And in the latest data there is no doubt about
the seasonal effect.

R.D. Politiek *(Netherlands)*

Dr. Farries mentioned the differences in fat-corrected milk.
Did you also look at the protein content in your experiments
with underfed animals?

E. Farries *(West Germany)*

No, not in this experiment. But it seems most likely from
the experiments of Kaufmann in Kiel that the bacterial protein
production in the rumen of underfed animals is reduced and this
may reduce the protein concentration in milk.

J. Brolund Larsen *(Denmark)*

Dr. Farries' low fat measurements in the milk were quite
surprising to me because our latest data (and I think some results from Norway also), feeding different levels before and after
calving, showed that there were really no differences in yield.
Could you tell us a little more about the composition of the
feed you used? What amount of fat have you been feeding to the
cows in the daily ration?

A second question: what were the pH values in the rumen?
Was the pH-value decreased and if so, why? We know that there
is a very close correlation between the pH-value in the rumen
and the butterfat concentration in the milk. Have you tried to

control the pH in the rumen at the same time as you got this decrease in butterfat concentration?

E. Farries

To the first question: the composition of the ration was quite normal, meeting practical requirements. All the animals received 20 kg of corn silage and 4 kg of hay and then concentrates according to milk yield.

The pH in the rumen was not estimated in this experiment because with 60 unfistulated cows it seemed rather difficult to get the exact pH values. The difference in milk fat concentrations could be explained by the hypothesis that there is a direct switch-over from utilising fatty acids for body fat to milk fat. So the difference in body and milk fat is not directly influenced by the ration but by the mobilisation of body fat.

J. Brolund Larsen

How much fat are you feeding per cow per day?

E. Farries

These data are not available at the moment.

A. Osinga (Netherlands)

I would like to return to the question Dr. Hoffmann put to Dr. Sejrsen. I think he is right that the seasonal influences are only on the heifer calvings and not on the mature calvings, an observation confirmed by the figures Professor Politiek has shown. But I think the reason for this is that stillbirth and dystocia are much more correlated in the heifers than in the mature cows where lots of other reasons for stillbirth besides dystocia exist. The seasonal influences are on dystocia and not on stillbirth and they are predominantly found in heifer calvings and not so much in mature calvings.

K. Sejrsen

I am not a specialist in this field. But I refer to the work of Dr. Philipsson and what he has referred to and also

Dr. Bar-Anan has shown some figures, and from all of this - at least to me - the seasonal effect on calving difficulties is well established in dairy cattle.

H.O. Gravert *(West Germany)*

We have to be careful because in some regions there is also a certain correlation of season and age. There are more 2-year-old heifers calving in autumn and there are more heifers of about 2½-years-old calving in spring.

H. Kräusslich *(West Germany)*

We have analysed more than 2 million calvings in Bavaria and we do not have the calving pattern Dr. Gravert has mentioned. We found a strong influence of the season or time of year on both heifer and cow calvings, for example in relation to veterinary aid or calf mortality. The relationship is more pronounced for veterinary assistance but it is also significant for mortality. Higher losses occur during winter calvings, considerably lower losses during summer, as was also reported from Denmark.

R.D. Politiek

Dr. Sejrsen suggested that the animals get more exercise in summer. During the winter months of December and January the highest rate of stillbirths is observed; but the figure is much lower in March and April and I don't think that the Danish cows are already outside during these months. I have data on the birth weight of calves in relation to season from about 700 registered births of heifers calving at two years. Spring calves are 0.7 kg lighter, which may have an effect on the occurrence of dystocia. So we have to consider weight of calf and season, but of course there may also be an effect of the dam.

H.J. Langholz *(West Germany)*

Despite the fact that there is a constant tendency for problems to occur with underfed or overfed heifers, the results are conflicting as was pointed out especially by Dr. Sejrsen. I wonder if this can be due to the different breeds investigated and their different reactions to feeding, depending on their growth potential. Maybe we should define the feeding level in relation to the mature weight of the breed.

K. Sejrsen

I did not say that all of the seasonal effect was due to exercise but only part of it. From the literature it is well known that there are other seasonal influences like the length of the day. Also prolactin exhibits a seasonal effect but about its physiological action I don't know anything. I agree with Dr. Langholz that a lot of the differences in the results might be explained if it were possible to define the feeding level in relation to the breed of the cow, but this is not possible from the data in the literature.

I.L. Mason

What also came out from some of the figures about the seasonal effect is that the pregnancies are longer in winter; this would be another effect on calving difficulties via birth size.

R.T. Berg (Canada)

I would like to follow up the points made about low level feeding. Dr. Drennan was worried about lowering the birth weights and I think this was associated with a higher calf mortality, probably after birth. Dr. Sejrsen was talking about feeding the pregnant cow and how the mammary development is dependent on the kind of feeding. I think the question is: what is a high, what is a medium and what is a low ration; and, what is a fat and what is a medium and what is a thin animal? In beef cattle suckler herds in Canada - and I think generally in North America - the first calving heifer is often badly fed. We have quite a bit of information on losses that would be similar to the kind that Dr. Drennan observed in his experiments. If a heifer is losing weight and is in poor condition at calving, quite often one of the results is lack of colostrum and lack of milk. This is where Dr. Oxender comes in with his calf losses later on, because the calf may not have had either colostrum or milk. Maybe this is a little hard to understand for people who are feeding dual-

purpose cattle for milk production but some of these beef
heifers come in with no milk at all.

R.D. Politiek

I want to come back to the point Mr. Mason made concerning
the season. As already mentioned, calves born between November
and February were significantly heavier by about 0.7 kg than
those born from March to May. There were no differences in
gestation length but dystocia was less frequent between March
and May. But when we keep weight constant we find almost no
difference between autumn and spring calving, and only a very
small effect due to sex of the calf. Thus, in my opinion, most
of the differences can be explained by the difference in birth
weight.

B. Hoffmann

Talking about the season to me means that we are talking
about a biological clock which is regulated by external factors
within the animal. Here I think 'season' in many cases means
management due to the season. This should be pointed out very
clearly; we have to talk about specific factors and not just
about season.

E. Farries

I was asked to give a short comment about the fertility of
the animals used in our experiments. These data have been collected by Dr. Lotthammer and Professor Grunert. As shown in
the table, one of the parameters was contractility of the uterus
and it was better in cows on a low plane nutrition. The involution of the uterus was completed within 4 weeks post partum in
82% of the low-fed cows and in only 46% of the high-fed cows. In
general, cows fed on the low energy level during the last 8 weeks
of pregnancy showed the better conditions. (See Table 1).

H. Kräusslich

I have a question for Dr. Farries. Did you control feed
intake after parturition in the two groups, and specially intake
of roughage?

TABLE 1

PUERPERIUM, HEALTH- AND FERTILITY-DISORDERS IN COWS AFTER DIFFERENT NUTRIENT
SUPPLIES BEFORE CALVING (n = 30 cows/group)

	Maintenance ration		Test
	+16 kg FCM	+2 kg FCM	
Uterine contractility (graded 0 - 5)	2.0	3.8	$F = 28.3***$
Involution of uterus completed at 4 weeks post partum (%)	46.4	82.8	$Chi^2 = 8.3**$
Frequency of puerperal endometritis (%)	70.8	26.9	$Chi^2 = 9.6**$
Av. number of treatments required	2.9	1.6	$F = 5.64*$
Av. days of disease	32.7	27.7	$F = 4.16*$
Frequency of genital tract infections post partum (%)	55.1	22.5	$Chi^2 = 6.7**$
Frequency of follicular cysts (%)	44.8	18.7	$Chi^2 = 4.8*$
Frequency of paralysis uteri post partum (%)	25.6	6.3	$Chi^2 = 4.5*$

Source: Lotthammer, 1975

E. Farries

Yes, feed intake was controlled exactly because all the
animals were in individual feeding systems. I have shown that
it was possible to increase the feed intake in those cows which
were on a low plane before calving more than 15% above the level
necessary for their milk production. The intake of roughage was
the same in all cows because we only fed 4 kg of hay and 20 kg
of corn silage in order to have no residues left. The intake of
this ration was nearly 100%. This also answers Dr. Brolund
Larsen's question: between the groups there was almost no difference in fat intake, because the basic ration was the same and
concentrate according to milk yield.

J. Brolund Larsen

As far as I understood, you were feeding the two groups according to different milk yields. That means that there must have been a difference in the feeding. This question always arises when you feed according to yield, especially in the first 6 weeks after calving. You are feeding 'plus' because, when feeding to yield, you always have to be ahead. So I still wonder about the results of the milk-fat measurements. If there isn't an adequate buffer effect, a decrease in the pH-value in the rumen may be observed which is then causing this drop in the milk fat.

E. Farries

Possibly, but we haven't determined the pH in the rumen.

STATUS, NUTRITION AND MANAGEMENT OF THE NEWBORN CALF, PART 1

Chairman: H. Thornberry

CONCLUSIONS FROM THE EEC SEMINAR ON   PERINATAL ILL-HEALTH IN
CALVES*

H. Thornberry
Veterinary Research Laboratory, Abbotstown, Castleknock,
Co. Dublin, Ireland.

This seminar was held at the Institute for Research on
Animal Diseases, Compton, England, from September 22nd to 24th,
1975. The term perinatal was applied to the period extending
from late pregnancy to approximately three months of age. In
all 32 papers were presented, each followed by a full
discussion. They covered the wide fields of infections enteritis,
immunity and immune prophylaxis against infection in the calf,
the physiopathology and therapeutics of diarrhoea, non-infectious
factors and economic aspects. The EEC has produced a 193 page
booklet including the lectures and discussions. As it is not
possible to deal with any but a small fraction of the lectures
and discussions, I will generalise on some points of importance
presented at the seminar without mentioning the names of the
contributors. I must, however, refer to the Editor, Dr. Rutter,
with part of whose excellent summary of the seminar I shall
commence my brief resumé.

Calf mortality is an important problem in all European
countries, accounting for 9 - 13% of calves born. Deaths are
divided into four periods:

a)   Deaths in utero,

b)   Perinatal deaths, including stillbirths and deaths in
the first 24 h of life,

c)   Neonatal deaths from one day to three weeks after
birth when the calf is dependent to a variable degree
on passive maternal protection,

d)   Deaths after three weeks.

Results in the UK suggest that calf mortality in the beef
herd is about half of that in the dairy herd (5.5% compared with

*See Reference

11.8%). This may reflect the effects of extensive husbandry
systems, lower planes of nutrition with consequent lower birth
weight of calves, and better opportunities for colostral intake
in the beef herd.

Perinatal deaths are very important, accounting for
between 50% and 62% of deaths of calves born to heifers in some
European countries. Dystocia associated with large calves
account for a significant number of perinatal deaths. In
addition, prolonged calvings may reduce the viability of calves
which survive the perinatal period.

There was considerable discussion as to whether the
available figures of up to 60% of mortality in calves being due
to *E. coli* are valid. It seems probable that rotavirus and
possible other virus infections may be the primary pathogens in
some cases. Certainly the view was not unanimously supported
that dehydration is significant as the cause of death in *E. coli*
disease.

Salmonellosis remains a problem in some countries, but
valuable lessons may be learnt from Denmark and the Netherlands
where attempts to eradicate this disease have been remarkably
successful.

The production of specific antibodies and their role in
protection against infectious diseases requires further
elucidation, both in mammary secretions and in the calf.
Consideration of these points raises the question as to how
much disease would occur if colostrum intake, nutrition and
management were satisfactory. If present knowledge were
effectively applied there is little doubt that a significant
improvement in calf viability would be achieved. Regarding
salmonellosis, the results of surveys and experiments suggest
that very young calves are commonly infected in the immediate
post partum period with material (eg on the teats) contaminated
by the faeces or vaginal discharge of the dam or some other adult
animal.

Occasionally direct transfer of organisms via the milk may occur. Once individual animals are infected and become excretors, spread to calves of approximately the same age is likely to occur by the oral route if animals are kept in groups. However, infection could occur by other routes although this would be unlikely to produce severe disease immediately.

The epidemiology of viral enteritis is not completely understood. Once the disease is established on a farm, repeat outbreaks usually occur annually. The rotavirus is extremely resistant to environmental factors and may survive for several years in buildings, etc. in dried dung. Usually an outbreak of viral diarrhoea in calves is preceded by diarrhoea in some of the adults of the herd. It would seem that a reo-like virus can also be associated with the disease.

Immunoprophylaxis in the newborn can be achieved in several ways, some of which are not very practical:

1) Vaccination of the dam.

2) Vaccination of the calf in utero.

3) Vaccination of the newborn calf.

4) Infection of the calf with immune sera.

5) The maximal transfer of immune globulins to the calf by natural means.

It involves ensuring the maximum amount of sucking in the first 6 h of life. The passive immunity which calves receive from colostrum is of overriding importance in their survival, especially in protection from septicaemic infections. On the other hand the main nutritional factor which affects the incidence of diarrhoea in the first three weeks of life is the feeding of milk substitute diets which contain 'severely' preheated spray-dried skim-milk powder. However, the recent introduction of non-milk proteins into such diets, as a result of the high cost of skim-milk powder, has similar and often greater effects on the digestive processes than that of the 'severely' heat-treated milk powder.

Under intensive experimental conditions it is possible to eliminate perinatal diarrhoea and mortality in bucket-reared calves removed from their dams at birth, by feeding:

1) 7 kg colostrum from the first two milkings after calving, with the first feed given as soon after birth as possible.

2) Milk substitute containing 200 mg fat/kg dry matter after the colostrum feeding period.

Thus, continued mortality in dairy calves under practical conditions is due to the failure of calves to ingest enough immunoglobulin sufficiently soon after birth and to the fact that the quality of the milk substitute is outside the control of the farmer.

The failure of calves to ingest sufficient immunoglobulin is primarily due to the lack of realisation by the stockman of the changes which occur in the composition of colostrum in the time after calving and of the changes which occur in the ability of the calf to absorb immunoglobulin during time after birth and to a lack of supervision of calves after birth. The importance of the ingestion of colostrum at the earliest age possible so as to allow complete protection has been highlighted by the suggestion that closure to absorption by the calf of IgM and IgA occurs earlier than to that of IgG.

The errors in management that can occur during this critical period of the calf's life include:

1) Removal of the calf at birth and insufficient colostrum of the first milking given by bottle or bucket early enough after birth.

2) Removal of the calf at birth and the calf being allowed to suck only twice daily, while the dam is milked out after suckling.

3) The calf does not suck because of weakness or inability

to stand, or there is lack of 'mothering' by the dam, or there may be difficulty for the calf in reaching the teats on a pendulous udder.

4) Where the colostrum from various milkings of cows calving at about the same time are bulked together, and this bulked colostrum of low Ig content is fed to newborn calves.

5) Where the colostrum is diluted with water, so that the concentration of Ig is reduced, or is heated sufficiently (> 50°C) to denature the Ig.

6) Where the dam has been milked before parturition so that the post partum secretion is only of similar composition to that of milk.

It now seems probable that in addition to the accepted short-term importance of colostral antibody, it also has a longer term value. It is well recognised that even on farms where the calves are single suckled and allowed to remain with their dams for at least 24 h after birth there are great differences in the amount of immunoglobulin received by the different calves. One factor, for example, which could be in part responsible for this difference, and which it is known has a high heritability, is the milking rate. This could well affect the ease and speed with which a calf receives colostrum. Data are available which point to the existence of genetic factors which affect the survival of the calf for the first few hours of life and later. This is obviously a field demanding further study.

The study of the metabolism of milk-fed calves shows that the potential growth of an animal is positively correlated with its globulin level at eight days of age.

The results obtained from metabolic profiles enable classification of herds into three groups:

1) The herds in which pregnant cows are deficient in

minerals. The profile is characterised by a low plasma
globulin level (< 2.3 g/100 ml) and also by anaemia. The
calves are weak after birth, and the mortality rate is
high. Therapeutic treatments are generally ineffective
in this group.

2) Normal herds, in which postnatal mortality is below 5%.

3) Herds in which there are no detectable nutritional
deficiencies, but in which there are many infected animals
(average globulin levels are <5.1 g/100 ml in the dams).
The calves born to these dams respond more satisfactorily
to treatment than those in group (1) despite a high
disease incidence.

Lastly we come to the economic aspects of perinatal ill-
health in calves. Here I think that some actual figures give
the clearest picture. In Britain the total number of calves
available annually for beef production is 3.5 million. The
total calf loss due to mortality is 9.4%. This means a total
loss in potential beef value of £66 m. An improvement of even
1% in the mortality rate would amount to £7 m. I should stress
that these figures refer to the years before 1975. Since then
inflation and other factors will have considerably increased
the financial figures.

In addition to the various causes of perinatal illness and
mortality already mentioned is the breed of the sire. Heavier
birth weights, though commercially desirable, are a major
factor in the complex interrelationships which result in dystocia
and calf mortality. However, in beef herds calf mortality
levels associated with the breed of the sire are generally
lower than in dairy herds. This is probably a reflection of the
cow type involved (mainly crossbred cows) and the fact that
feeding levels before calving are appreciably lower than in
the dairy herds.

It will be noted that I have not referred to the various
causes and effects of abortions and stillbirths, because it is

such a wide subject which could not be dealt with in a resumé
of this type. However, we all accept that it is a most
important factor in calf losses.

In conclusion I might mention that, very conscious of the
importance of colostrum, the committee of the EEC Beef Production
Pathology Group is planning a seminar on the subject.

REFERENCE

Perinatal Ill-Health in Calves, Ed. J.M. Rutter. Commission of the European
Communities, D6 VI-E-4: Co-ordination of Agricultural research, 200
rue de la Loi, B-1040-Brussels.

IMMUNE MECHANISMS IN THE NEWBORN CALF

D.K. Hammer

Max-Planck-Institüt für Immunbiologie,
78 Freiburg i. Br., West Germany.

ABSTRACT

In the bovine spcies, there is a selective transmission of maternal immunoglobulins to the offspring which takes place entirely postnatally through the colostrum and thus via the intestinal epithelium. It should be recognised, however, that humoral antibodies acquired by the newborn are only effective in combating infections by organisms that replicate outside cells, and in neutralising bacterial toxins. Consequently cell-mediated immunity is most likely the major immune defence mechanism against infection by facultative intracellular parasites. As shown in lower animals, there are four principal types of cellular interaction playing important roles in the regulation of immune responses:

a) between macrophages and thymus-dependent (T) lymphocytes playing a critical role in the antigen-stimulated activation of T cells;

b) the helper effect exerted by specific T lymphocytes in the activation of thymus-independent (B) lymphocytes and their differentiation into antibody-secreting cells;

c) negative regulation mediated by suppressor cells, a subpopulation of T lymphocytes, and

d) collaboration of distinct types of T cells in the proliferation of specific cytotoxic effector cells.

Recent experiments have shown that lymphoid cell interactions are apparently controlled by genes coded within the major histocompatibility complex (MHC).

The inability of newborn mammals to mount an efficient immune response has been attributed to lack of functional macrophages and T cells as well as to the existence of suppressor T cells. These observations suggest that one mechanism for immune deficiency of newborn calves is associated with populations of immunocompetent cells which are functionally still immature at the early stage of development.

Specific resistance to infection is afforded by two types of defence mechanism, humoral antibody production and cell-mediated immunity. It is increasingly clear that humoral antibodies are particularly effective in combating infections by microorganisms that replicate outside cells, and in neutralising bacterial toxins. Cell-mediated immunity is now recognised to be also the major mechanism for destroying bacteria and viruses that replicate intracellularly and thus are protected from humoral antibodies (Campbell, 1976). In calves there is practically no transfer of immunoglobulin across the placenta and since none is produced by the foetus the young are born essentially agammaglobulinaemic. They acquire maternal antibodies for passive protection through the colostrum and thus via the intestinal epithelium (Miller, 1966). It is well documented that at least three IgG class immunoglobulins, characterised by both individual specificity and different net electrical charge, are present in bovine serum; these represent real subclasses denoted by IgG_{2a}, IgG_{2b} and IgG_1 (Kickhöfen, et al., 1968).

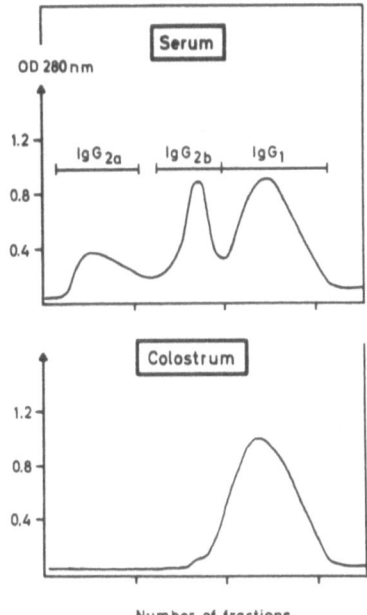

Fig. 1, Elution pattern of the IgG fraction isolated from bovine serum and colostrum from DEAE-Sephadex.

496

Evidence has been provided that IgG passes unchanged from the maternal plasma to the mammary gland and there is a selective transport of the more negatively charged IgG_1 fraction. The recognition part of this transport system may be associated with structural moieties or conformations of the IgG_1 molecule that adapt them for recognition by surface receptors of the acinar epithelium (AE) of the colostrum-forming mammary gland (Kemmler et al., 1975). Evidence that we are in fact dealing with a receptor involved in the transport process comes from the observation that AE has a high affinity for the IgG_1 subclass but lacks any binding capacity for IgG_{2a}, IgG_{2b}, IgM and IgA.

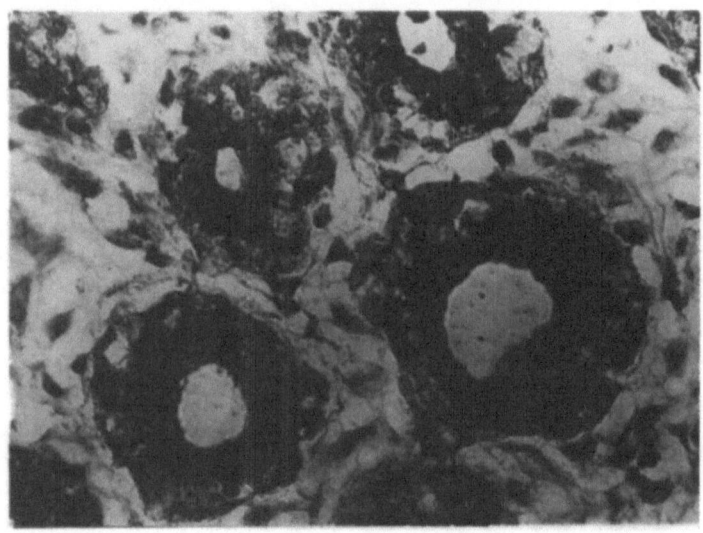

Fig. 2a. Localisation of IgG anti-horseradish peroxidase antibody.

It should be remembered, however, that the susceptibility of newborn calves to infections can be attributed among other reasons to the selective production of antibodies by the dam which cannot be transferred to the offspring via colostrum. Support for this hypothesis came from the observation that there is an inverse relationship between the charge of antigen and that of the antibodies preferentially provoked by it (Mossmann et al., 1973). The significant finding to emerge from experimental studies is that immunisation of cattle with a

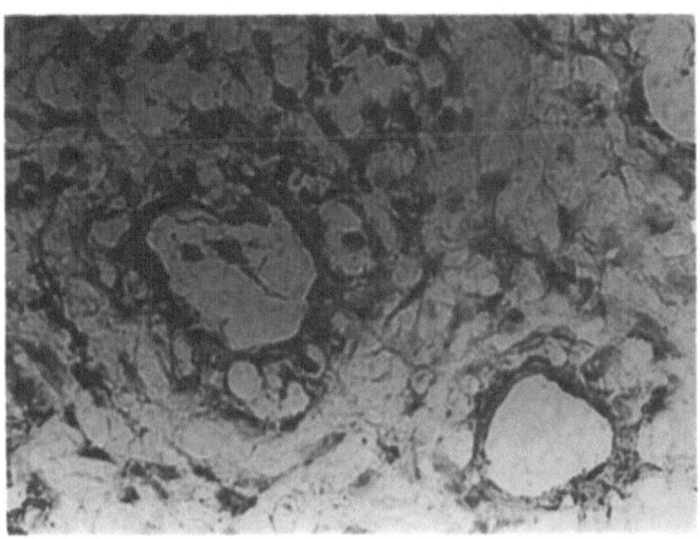

Fig. 2b. Tissue section from colostrum-forming bovine mammary gland reacted with a) IgG$_1$ and b) IgG$_{2a}$.

negatively charged antigen (azobenzenearsonate conjugate of maleylated edestin or ABA-Mal-edestin) provokes the production of more positively charged IgG$_{2a}$ antibodies which cannot be transferred to the colostrum.

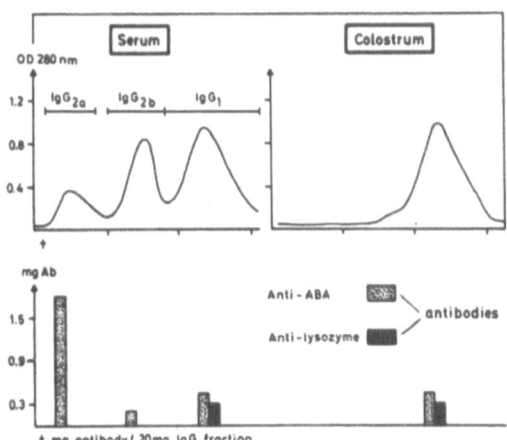

Fig. 3. Distribution of precipitating antibodies against differently charged antigens in IgG subclasses.

Studies on the cellular mechanisms of the immune response in cattle are still in their infancy and little is known about the sequence of maturation of cell types involved in responses to antigenic stimulation.

In some species, particularly the mouse, it is now recognised that differentiation and activation of immunocompetent cells is regulated by a complex series of events involving interactions between distinct cell types. These mechanisms are presumed to exist also in the bovine species by some inferential observations.

The differentiation of immunocompetent cells can be conveniently divided into two stages:

1) The differentiation of stem cells into immunocompetent B lymphocytes without the requirement of exogenous antigenic stimulation but with the cooperation of a central lymphoid organ (Bursa equivalent).

B cells can be identified as lymphocytes with surface membrane-bound immunoglobulins in high density, being apparently receptors through which antigens are recognised. Following interactions of antigen with their respective receptors on the surface of B cells, second stage events are triggered which include B cell proliferation and terminal differentiation into antibody-secreting plasma cells.

2) The differentiation of stem cells into competent T lymphocytes affected by the thymus.

Evidence was obtained that the thymus produces a family of polypeptides (thymosin) that play an important role in the maturation, differentiation and function of T cells (Goldstein et al., 1975). In response to thymus-dependent antigens T cells proliferate and differentiate into effector cells, attacking the antigen directly or by

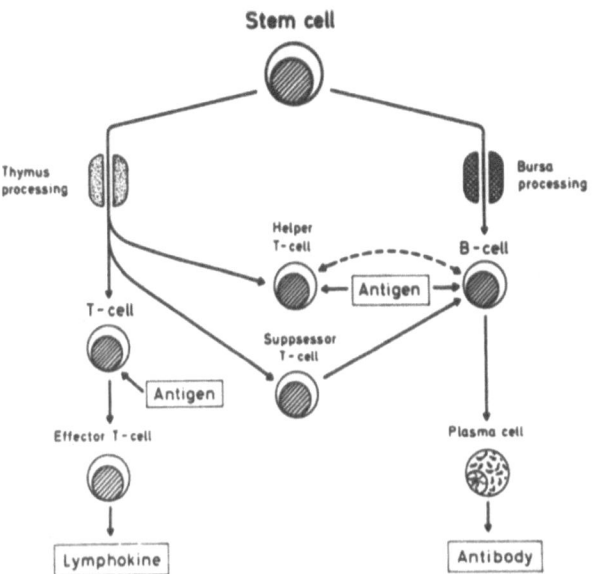

Fig. 4. Model of events of cellular maturation, cellular interaction and cellular biosynthesis required for normal immune response.

secretion of soluble products, ie lymphokines which can activate macrophages to function in an unspecific way. Effector T cells are representative for cell-mediated immune reactions playing a major role including tumour immunity, delayed-type hypersensitivity, graft rejection and resistance to intracellularly replicating organisms.

On the other hand T lymphocytes have been shown to exert a powerful regulatory effect on B-cell differentiation. Cells expressing a positive regulatory effect serve as helper T cells required for the optimal production of antibody to certain but not necessarily to all antigens. More recently it has been recognised that a subpopulation of T cells may act as negative regulators (suppressor T cells) of B-cell maturation by inhibiting this process (Paul and Benacerraf, 1977). As found recently effector-helper and suppressor T cells belong to distinct populations of T lymphocytes, which may be discriminated from each other by the expression of distinctive differentiation antigens, termed Ly antigens (Cantor and Boyse, 1975).

An important feature of immune responses to thymus-dependent

antigens is that they are strictly controlled in an antigen-
specific manner by the operation of a group of genes found in
the major histocompatibility complex (MHC) of higher vertebrates
(Munro and Bright, 1976). The MHC is divided into different
regions mainly on the basis of genetic recombinants. The
probable arrangement of these regions in the mouse are shown
in Figure 5.

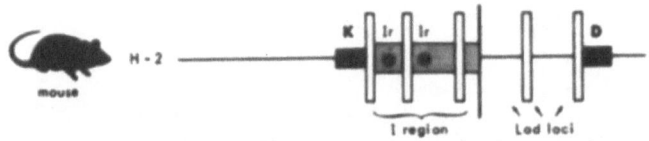

Fig. 5. Diagrammatic representation of the MHC of the mouse (H-2).

Through the study of structurally simple antigens it was
demonstrated that the immune response is controlled by immune
response (Ir) genes which are associated with the I region.
Ir genes have been found associated with the MHC in all
species in which they have been looked for but so far in the
bovine species the evidence for them is more circumscribed. It
is increasingly apparent that the action of Ir genes is not
limited to positive regulation but genes of this type regulate
specific suppressive responses as well (Katz and Armerding,
1976).

The first line of defence associated with humoral antibody
formation may be trapping and/or processing of microorganisms
by macrophages. Antigen bound to macrophage surfaces or
internalised by the cell may be more immunogenic than antigen

that has not been processed in this way (Unanue, 1972). It
is not clear, however, whether all antigens must undergo
processing and presentation by macrophages.

The second step in humoral antibody formation is mediated
by most but not necessarily all antigens which are thymus-
dependent, ie cannot normally trigger B cells to synthesise
antibodies in the absence of T cells. Thymus-independent
antigens on the other hand can stimulate B cells when T cells
are severely depleted or absent. Most evidence suggests that
thymus-dependent antigens trigger T lymphocytes via antigen-
specific receptors. In interactions between antigen, macrophages
and T cells it seems that I region products may play a part
in recognition (Munro and Bright, 1976). At the same time,
factors with antigen specificity have been obtained from primed
lymphocytes (Taussig and Munro, 1976). These factors are able
to mediate some of the specific regulatory activities that T
cells exert on the immune response. There are antigen-specific
as well as non-specific factors induced by antigens, which
either enhance or suppress the antibody response. These
suppressor or helper factors are also good candidates for
involvement in the antigen recognition system of T lymphocytes.
Several lines of evidence now suggest that the T cell factor
can trigger B cells in the presence of antigen.

The triggering of B cells that produce the humoral antibody
involves the binding of antigen Ig receptors in the B cell
surface followed by the binding of the T cell factor to the
antigen and to the B cell acceptor site.

It becomes increasingly clear that cell-mediated immune
mechanisms play a major role in destroying microorganisms that
replicate intracellularly and thus are protected from humoral
antibodies. Although these responses were originally considered
to be due only to direct cell interactions it is now recognised
that humoral factors termed lymphokines may also be involved.

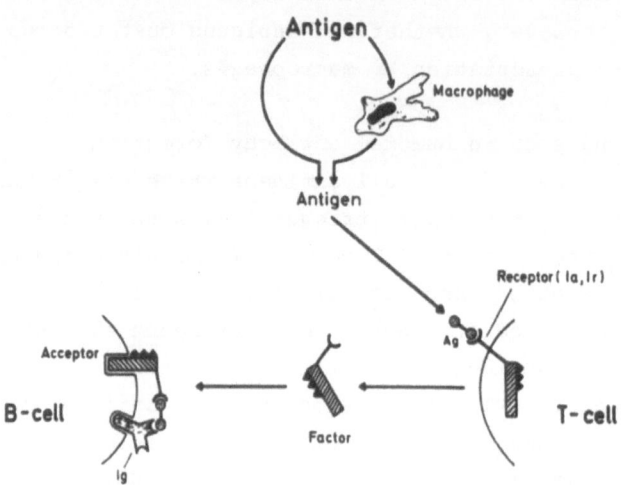

Fig. 6 Cell interaction in the humoral immune response. A model showing
the possible mechanism of B cell activation by T cell factor and
antigen processed and/or presented by macrophages and the role of
Ir gene products.

The first event, antigen processing by macrophages and/or
polymorphnuclear leucocytes, has not been well studied for
cell-mediated immunity. But it is reasonable to assume that
triggering of T cells by antigens for cell-mediated immunity is
quite likely not different from the mechanisms in the stimulation
of T cells for antibody production.

As a consequence of T cell stimulation they may attack the
antigen directly. These reactions are antigen-specific,
require direct contact between cells and microorganisms and
have not been shown to involve any humoral factors.

The alternative mechanism by which T cells bring about
cell-mediated reactions is through secretion of lymphokines.
These secretion products of T cells act on non-immunocompetent
cell populations and enhance the ability of macrophages to
resist bacterial infection (Campbell, 1976).

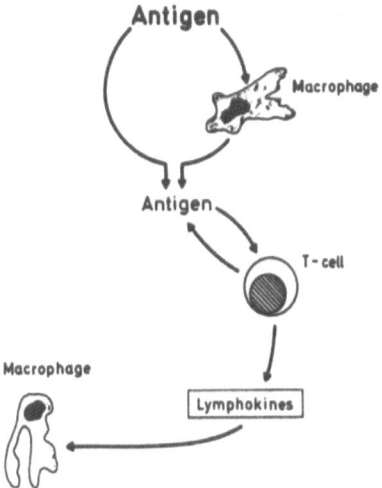

Fig. 7. Cell-mediated immune responses. A model showing a) stimulation of T-cells by macrophage processed antigen and b) activation of macrophages by lymphokines released from stimulated T-cells.

The issue now to be resolved is the inefficiency of new-born mammals, including calves, to cope with invading organisms, although passively transferred maternal immunoglobulins are present in a sufficient amount. This immune deficiency may be attributed to an immature state in the events of cellular differentiation, cellular interaction and cellular biosynthesis.

In this regard it is of interest that newborn mice lack functional macrophages, ie they are immature in their activity to antigenic signals (Hardy et al., 1973). Further studies have led to the conclusion that specific suppressor T cells are present in excess during early development (Mosier and Johnson, 1975). Thus functional activity of neonatal T lymphocytes seems to be dominated by a suppressive influence exerted both on other T cells and on B cells.

Finally it is known that certain individuals within a species fail to respond to a specific antigen and this deficiency maps within the I region. So far, in different mouse strains, defects have been found in the response to more than 40 antigens. In every case the absence of response is recessive - that is response is dominant (McDevitt and Landy, 1972).

In summary, it should be recognised that the regulation of the immune response by a complex series of events was demonstrated in a variety of animal species particularly in the mouse system and it might be presumed to exist in the bovine species by some inferential observations. However, there is still much to be learnt about the immune status of the newborn calf and it is at the present stage premature to attempt a rational synthesis of all the facts.

The understanding of T and B lymphocyte specificity involves the consideration of different requirements for distinct specific defence mechanisms. The B lymphocyte system is characterised by an enormous variety of immunoglobulins designed to deal with unpredictable and unforeseen microbial and toxic agents.

Humoral protection is afforded to the offspring by passive transfer of maternal immunoglobulins. In cases in which an increase of passive immunity is needed, care should be taken that antibodies are selected by vaccination of the dam; they can then be transferred to the newborn via the colostrum. By contrast the T lymphocyte system appears to have arisen in order to recognise antigens on the cell surface and in combating infections by microorganisms that replicate intracellularly. This is valid for virus and most of the bacterial infections. Knowledge about the principal types of cellular interactions and functions, which play important roles in the regulation of the immune response, has great significance and practical implications in manipulation of the immune system.

One theme threading through the new immunology concerns genetic control. This control is multifactorial and operates at several different levels. We are just beginning to decipher its complexity, but it can be assumed that the principles, when they emerge, will lead into new approaches for immunological manipulation for the prevention of infectious disease.

REFERENCES

Campbell, P.A. 1976. Immunocompetent cells in resistance to bacterial
  infections. Bacteriological Reviews 40: 284.

Cantor, H. and Boyse, E.A. 1975. Functional subclasses of T lymphocytes
  bearing different Ly antigens. I. The generation of functionally
  distinct T-cell subclasses is a differentiative process independent
  of antigen. Journal of Experimental Medicine 141: 1376.

Goldstein, A.L., Thurman, G.B., Cohen, G.H. and Hooper, J.A. 1975. In
  'Biological activity of thymic hormones', ed. van Bekkum, D.W.
  (Kooyken Scientific Publications, Rotterdam), pp. 173-197.

Hardy, G., Globerson, A. and Danon, D. 1973. Ontogenic development of
  the reactivity of macrophages to antigenic stimulation. Cellular
  Immunology 9: 282.

Katz, D.H. and Armerding, D. 1976. The role of histocompatibility gene
  products in lymphocyte triggering and differentiation. Federation
  Proceedings 35: 2053.

Kemler, R., Mossmann, H., Strohmaier, U., Kickhöfen, B. and Hammer, D.K.
  1975. In vitro studies on the selective binding of IgG from
  different species to tissue sections of the bovine mammary gland.
  European Journal of Immunology 5: 603.

Kickhöfen, B., Hammer, D.K. and Scheel, D. 1968. Isolation and characterisation
  of γG type immunoglobulins from bovine serum and colostrum. Hoppe-
  Seyler's Zeitschrift für Physiologische Chemie 349: 1755.

McDevitt, H.D. and Landy, M. 1972. In 'Genetic control of immune
  responsiveness'. Academic Press, New York.

Miller, J.F.A.P. 1966. Immunity in the foetus and the new born. British
  Medical Bulletin 22: 21.

Mosier, D.E. and Johnson, B.M. 1975. Ontogeny of mouse lymphocyte function.
  II. Development of the ability to produce antibody is modulated by
  T lymphocytes. Journal of Experimental Medicine 141: 216.

Mossmann, H., Bartsch, K., Rüde, E., Kickhöfen, B. and Hammer, D.K. 1973.
  The influence of the net charge of antigen on the distribution of
  bovine IgG class antibodies. European Journal of Immunology 3: 293.

Munro, A. and Bright, S. 1976. Products of the major histocompatibility
  complex and their relationship to the immune response. Nature 264:145.

Paul, W.E. and Benacerraf, B. 1977. Functional specificity of thymus-dependent
    lymphocytes. Science 195: 1293.
Taussig, M.J. and Munro, A.J. 1976. Antigen-specific T-cell factor in cell
    cooperation and genetic control of the immune response. Federation
    Proceedings 35: 2061.
Unanue, E.R. 1972. The regulatory role of macrophages in antigen stimulation.
    Advances in Immunology 103; 71.

THE EFFECT OF DIFFERENT METHODS OF FEEDING COLOSTRUM ON CALF BLOOD SERUM IMMUNOGLOBULIN LEVELS

R.J. Fallon

The Agricultural Institute, Grange, Dunsany, Co. Meath, Ireland.

ABSTRACT

The blood serum immunoglobulin (Ig) levels of 1100 calves purchased at public auction showed that 45% had Ig levels of less than 16 when determined by the Zinc Sulphate Turbidity Test (ZST). Ig levels of less than 16 ZST units indicate inadequate blood antibody levels. The effect of different methods of feeding colostrum on Ig levels in the newborn calves was investigated in three experiments in which the colostrum feeding was followed by whole milk feeding from the bulked supply.

Experiment 1

    Treatments:

1) *one nipple feed 4 - 6 h after birth;*

2) *one pail feed 4 - 6 h after birth;*

3) *two nipple feeds, first 4 - 6 h after birth, second 4 - 6 h later;*

4) *two pail feeds, first 4 - 6 h after birth, second 4 - 6 h later.*

The mean Ig levels, at 96 h after birth, for these treatments were 10.0 ± 0.91, 10.6 ± 0.97, 20.1 ± 0.45 and 22.6 ± 0.76 ZST units respectively. The corresponding colostrum intakes, as a percentage of body weight were 4.3 ± 0.41, 5.3 ± 0.32, 10.7 ± 0.51 and 10.9 ± 0.39.

Experiment 2

    Treatments:

1) *two pail feeds in the presence of the dam;*

2) *two pail feeds in the absence of the dam.*

The respective 96 h Ig levels were 26.1 ± 0.90 and 20.5 ± 1.10 ZST units and the colostrum intake $10.7\% \pm 0.34$ and $9.9\% \pm 0.28$.

Experiment 3

   Treatments:

1) two pail feeds at body temperature (38°C);

2) two pail feeds at ambient temperature (14°C).

Ig levels were 23.9 ± 1.10 and 21.91 ± 1.73 ZST units and the colostrum intakes 10.1% ± 0.45 and 9.6% ± 0.44 respectively. Different methods of increasing blood serum Ig levels in the calf are discussed.

1. INTRODUCTION

The importance of colostrum as a source of immune globulin
protection for the newborn calf is well established (Ingram et
al., 1956; Fey and Margadant, 1961; Fisher, 1971; Logan and
Penhale, 1971; Fisher et al., 1975; Fisher et al., 1976). Many
surveys have shown a positive relationship between absorbed
colostrum immune globulins and calf survival (Gay et al., 1965;
Smith et al., 1967; Selman et al., 1971c; Fisher, 1971; Oxender
et al., 1973; Thomas and Swann, 1973; Irwin, 1974; Williams
et al., 1975). A survey conducted in Ireland over the last
four years on 1 250 calves purchased direct from farms and
from calf auction markets shows this positive relationship
(Figure 1 and Table 1). The blood serum immune globulin (Ig)
level of the calves was determined on arrival by the Zinc
Sulphate Turbidity (ZST) Test (McEwan et al., 1970a). Forty per
cent of the calves had readings of less than 15 ZST units
indicating that these calves had absorbed insufficient colostral
immune globulins (Table 1).

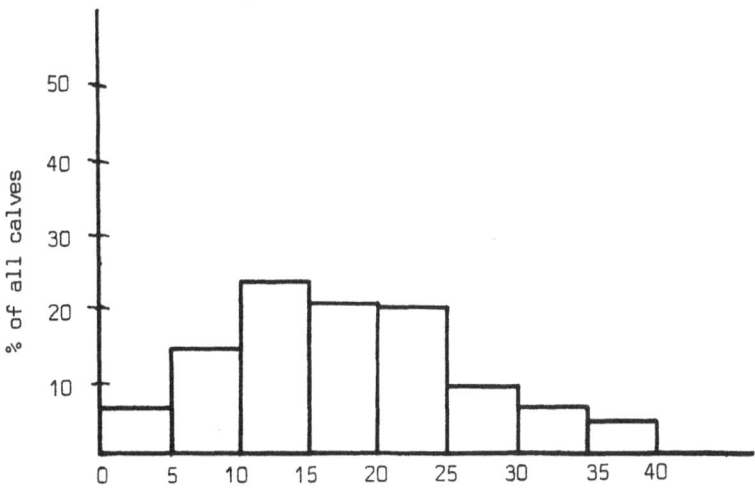

Blood serum Ig-level (ZST units)

Fig. 1. Immunoglobulin (Ig) status of 1 250 purchased calves

510

Low Ig levels may be due to:

a) failure of the calf to suck colostrum,

b) calf separated from dam prior to suckling,

c) calf having difficulty in obtaining colostrum from a large teat,

d) cow milked out immediately after calving with no colostrum being available for the calf,

e) variation in immunoglobulin concentration of colostrum offered to the calf. (Table 2).

TABLE 1

EFFECT OF IMMUNOGLOBULIN STATUS ON CALF MORTALITY

ZST units	0-4	5-9	10-14	15-19	20-24	25-29	30-34
Mortality %	18	8	6.5	5	3	3	3

TABLE 2

IMMUNOGLOBULIN CONCENTRATIONS (g/100 g) OF FRIESIAN COW COLOSTRUM

	Colostrum sample		
	1st milking	2nd milking	3rd milking
No. of cows sampled	56	56	56
Mean (g/100 g)	11.5	6.0	2.5
Range (g/100 g)			
0-4.9	7*	45	91
5-9.9	29	44	9
10-14.9	38	11	-
15-19.9	22	-	-
20 Plus	4	-	-

* = %

The importance of providing the calf with an opportunity to ingest colostrum within 6 h of birth has been well established (Kruse, 1970b; Selman 1973; Penhale et al., 1973).

While agreeing with the statement: '.......arrange the maximal transfer of immune globulins to the calf by natural means. This is the most practical means of achieving high levels of immune globulins in the calf and involves management of the cow and calf to ensure the maximum amount of suckling in the first 6 h of life' (Fisher and Martinez, 1975), it appears that farmers may not necessarily adopt and apply the simple routine of assisting the calf in suckling. As a result, it was decided to develop a method of artificially feeding colostrum to the newborn calf which would have the desired effect of providing a high level of immune globulins in the calf. Three experiments were designed to examine a number of factors which might affect immune globulin levels in the calf:

a) feeding method,

b) feeding frequency,

c) presence of dam and

d) feeding temperature.

The calves were separated from their dams at birth and fed colostrum from a pooled store. The pooled store was made up by combining first milking of 6 to 8 cows in a container. First milked colostrum which was heated to 38°C by immersing a pail in a water bath was fed. At each feed, the calves were offered colostrum ad libitum. At the end of the colostrum feeding period, the calves were fed whole milk twice daily for the following 7 days. All calves were blood sampled 96 h after birth and the blood serum immune globulin concentration determined.

2. EXPERIMENTAL

2.1. Experiment 1. Method and Frequency of feeding

One hundred and six Friesian calves were allocated to the following treatments:

a) One feed via nipple, 4 - 6 h after birth (22 calves)

b) One feed via pail, " " " 22 calves)

c) Two feeds via nipple, first 4 - 6 h after birth, second feed 4 - 6 h later (34 calves)

d) Two feeds via pail, first 4 - 6 h after birth, second feed 4 - 6 h later (28 calves).

The respective intakes of first-milked colostrum (expressed as a percentage of calf body weight) for one feed via nipple and one feed via pail were 4.3 and 5.3% with serum Ig levels of 10.1 and 10.6 ZST units (Table 3).

TABLE 3

EFFECT OF METHOD AND FREQUENCY OF FEEDING COLOSTRUM ON BLOOD SERUM Ig LEVELS AT 96 H AFTER BIRTH

| | Treatment | | | |
	One nipple feed	One pail feed	Two nipple feeds	Two pail feeds
No. of calves	22	22	34	28
Intake (% birth weight)	4.3 ± 0.3	5.3 ± 0.4	10.8 ± 0.5	11.1 ± 0.5
Ig level	10.1 ± 1.0	10.6 ± 1.0	19.9 ± 0.8	22.6 ± 0.9

Both of these Ig levels are generally considered inadequate for normal immune protection (Gay et al., 1965; Fey, 1971; Fisher et al., 1975; McEwan et al., 1970b). Two feeds with intakes of 10.8 and 11.1% gave serum Ig values of 19.9 and 22.6 ZST units for nipple- and pail-fed calves respectively. Two feeds result in a significant improvement (P < 0.05) in the Ig levels. Ig levels obtained from two feeds would be regarded as satisfactory. Pail and nipple feeding were not significantly different in relation to colostrum intake of serum Ig levels.

2.2. Experiment 2: Feeding in the presence or absence of the dam

Studies conducted by Selman et al. (1971a) showed a significant effect of artificially feeding colostrum in the presence of the dam. They obtained Ig levels of 17.7 ZST units for presence of dam compared with 10.3 for absence of dam.

Forty Friesian bull calves were allocated to the following treatments:

a) Two feeds by pail in presence of dam (calves muzzled immediately after birth, muzzles removed only at feeding times)

b) Two feeds by pail in absence of dam (calves separated from dam at birth).

The intake of colostrum for two feeds in presence and absence of dam were not significantly different: 10.6 and 9.3% respectively (Table 4).

TABLE 4

EFFECT OF PRESENCE OF DAM ON BLOOD SERUM Ig LEVELS 96 h AFTER BIRTH

| | Treatment | |
	Absence of dam	Presence of dam
No. of calves	20	20
Intake (% of body weight)	9.3 ± 0.5	10.6 ± 0.4
Ig level	20.6 ± 1.2	26.1 ± 1.0

The Ig levels of 26.1 and 20.6 ZST units for presence and absence of the dam were significantly different (P < 0.05), thus confirming the earlier findings of Selman et al. (1971a).

2.3. Experiment 3: Feeding temperature

Thirty-eight calves were allocated to the following treatments:

a) Two feeds by pail at body temperature, $38^{\circ}C$ (19 calves)

b) Two feeds by pail at room temperature, $14^{\circ}C$ (19 calves)

The intakes of colostrum for feeding at body temperature and room temperature were not significantly different: 9.7 and 9.2% respectively (Table 5). The Ig levels of 23.4 and 21.9 ZST units for body temperature and room temperature were not significantly different.

TABLE 5

EFFECT OF FEEDING TEMPERATURE ON BLOOD SERUM Ig LEVELS 96 h AFTER BIRTH

| | Treatment | |
	Body temperature	Room temperature
No. of calves	19	19
Intake (% of body weight)	9.7 ± 0.6	9.2 ± 0.6
Ig level	23.4 ± 1.2	21.9 ± 1.8

3. DISCUSSION

These results suggest that two feeds of first-milked colostrum fed to appetite (the first feed 4 - 6 h after birth and second feed 4 - 6 h later) provide the calf with adequate immune globulin protection. It has been suggested that certain calves prematurely lose their ability to absorb the immune globulin fraction of colostrum (Fey, 1971; Gay, 1965; Gay et al., 1965). In this study there was no evidence to support this view and the findings would agree with results obtained by Kruse (1970a) and Fisher (1971). The Ig levels obtained were much greater than the 4.3 ZST units obtained by de la Fuente (1970).

The intake of 10% of body weight obtained, ie 40 kg calves ingesting 4 kg of colostrum in two feeds, is similar to the recommendation of Kruse (1970b) who suggested feeding colostrum ad libitum (at least 2 kg) and feeding as soon as possible after birth (within 5 h).

The effective improvement which resulted in feeding first milked colostrum in two feeds by pail over the base obtained from the 1 250 purchased calves (Figure 2) suggests that a practical method of artificially feeding colostrum can be implemented at farm level.

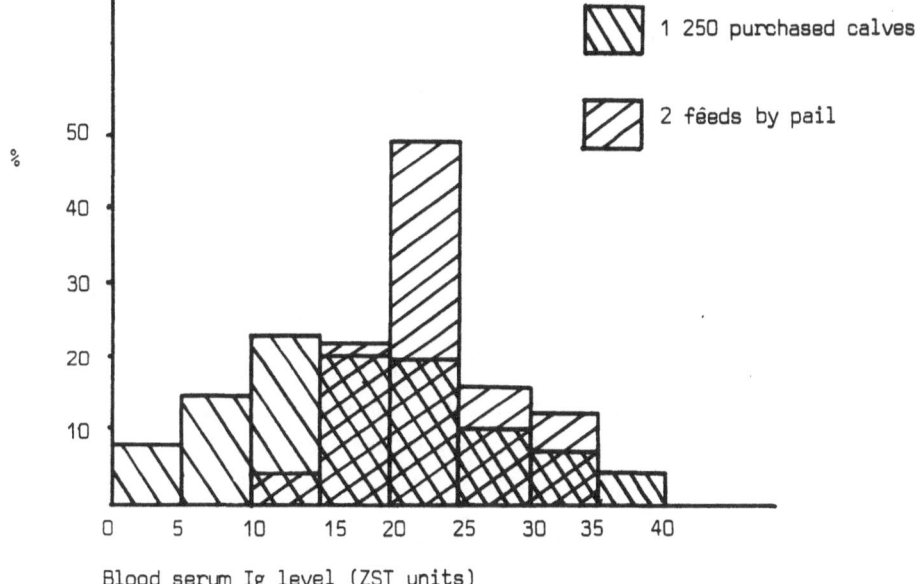

Fig. 2. Effect of two feeds by pail on improving the blood serum Ig status of calves.

An Ig level of 20 ZST units or more compares quite favourably with the minimum quantity of immune globulin that must be absorbed, namely 1.43 g/kg body weight (McEwan et al., 1970b) and with the value of 1.47 g/kg body weight obtained for healthy calves (Fisher et al., 1975).

REFERENCES

de la Fuente, cited by Fisher, E.W. and Martinez, A.A. 1975.

Fey, H. 1971. Immunology of the newborn calf: its relationship to colisept-
icemia. Annals of New York Academy of Science 176: 49-63.

Fey, H. and Margadant, A. 1961. Hypogammaglobulinamie bei der Colispesis
des Kalbes. Pathologia et Microbiologia 24: 970-976.

Fisher, E.W. 1971. Hydrogen ion and electrolyte disturbances in neonatal
calf diarrhoea. Annals of New York Academy of Science 176: 223-230.

Fisher, E.W. and Martinez, A.A. 1975. Immunity and neonatal calf diarrhoea.
Proceedings of the CEC seminar on 'Pathology', Compton pp. 83-89.

Fisher, E.W., Martinez, A.A., Trainin, Z. and Meirom, R. 1975. Studies of
neonatal calf diarrhoea. II: Serum and faecal immune globulins in
enteric colibacillosis. British Veterinary Journal 131: 402-415.

Fisher, E.W., Martinez, A.A., Trainin, Z. and Meirom, R. 1976. Studies of
neonatal calf diarrhoea. IV: Serum and faecal immune globulins in
neonatal salmonellosis. British Veterinary Journal 132: 39-48.

Fisher, E.W. and Martinez, A.A. 1976. Immune globulins and enterotoxic
colibacillosis. Veterinary Record 98: 31.

Gay, C.C., Anderson, N., Fisher, E.W. and McEwan, A.D. 1965. Gammaglobulin
levels and neonatal mortality in market calves. Veterinary Record
77: 148-149.

Gay, C.C. 1965. Escherichia coli and neonatal disease of calves.
Bacteriological Reviews 29: 75-101.

Ingram, P.L., Lovell, R. and Wood, P.C. and Aschaffenburg, R., Bartlett, S.,
Kom, S.K., Palmer, June., Roy, J.H.B. and Shillam, K.W.G. 1956.
Bacterium coli antibodies in colostrum and their relation to calf
survival. Journal of Pathology and Bacteriology 72: 561-568.

Irwin, V.C.R. 1974. Incidence of disease in colostrum deprived calves.
Veterinary Record 94: 105.

Kruse, V. 1970a. Absorption of immune globulins from colostrum in newborn
calves. Animal Production 12: 627-638.

Kruse, V. 1970b. A note on the estimation by stimulation technique of the
optimal colostrum dose and feeding time at first feeding after calf's
birth. Animal Production 12: 661-664.

Logan, E.F. and Penhale, W.J. 1971. Studies on the immunity of the calf
    to colibacillosis. Veterinary Record 88: 222-228.

McEwan, A.D., Fisher, E.W., Selman, I.E. and Penhale, W.J. 1970a. A turbidity
    test for the estimation of immune globulin levels in neonatal calf
    serum. Clinica Chimica Acta 27: 155-163.

McEwan, A.D., Fisher, E.W. and Selman, E. 1970b. Observations on the immune
    globulin levels of neonatal calves and their relationship to disease.
    Journal of Comparative Pathology 80: 259.

Oxender, W.D., Newman, L.E. and Morrow, D.A. 1973. Factors influencing
    dairy calf mortality in Michigan. Journal of the American Veterinary
    Medical Association 162: 458-460.

Penhale, W.J., Logan, E.F., Selman, E., Fisher, E.W. and McEwan, A.D. 1973.
    Observations on the absorption of colostral immunoglobulins by the
    neonatal calf and their significance on colibacillosis. Annales de
    Recherches Vétérinaires 4: 223-233.

Selman, I.E., McEwan, A.D. and Fisher, E.W. 1971a. Studies on dairy calves
    allowed to suckle their dams fixed times post partum. Research in
    Veterinary Science 12: 1-6.

Selman, I.E., McEwan, A.D. and Fisher, E.W. 1971b. Absorption of immune
    lactoglobulin by newborn dairy calves. Attempts to produce consistent
    immune lactoglobulin absorptions in newborn dairy calves using
    standardised methods of colostrum feeding and management. Research
    in Veterinary Science 12: 205-218.

Selman, I.E., de la Fuente, G.H., Fisher, E.W. and McEwan, A.D. 1971c. The
    serum immune globulin concentrations of newborn dairy heifer calves:
    a farm survey. Veterinary Record 88: 460-464.

Selman, I.E. 1973. The absorption of colostral globulins by newborn calves.
    Annales de Recherches Vétérinaires 4: 213-221.

Smith, H., Williams, O., Neil, J.A. and Simmons, E.J. 1967. The immune
    globulin content of the serum of calves in England. Veterinary
    Record 80: 664-6.

Thomas, L.H. and Swann, R.G. 1973. Influence of colostrum on the incidence
    of calf pneumonia. Veterinary Record 92: 454-455.

Williams, M.R., Spooner, R.L. and Thomas, L.H. 1975. Quantitative studies on
    bovine immunoglobulins. Veterinary Record 96: 81-84.

DISCUSSION

R.W.J. Plenderleith _(UK)_

Has Dr. Fallon any information on the quality of colostrum
as to the storage? Can we store it at all, and if so, for how
long?

R.J. Fallon _(Ireland)_

In my experiments I did not store the colostrum longer than
4 days. It also appears to be quite effective to freeze colostrum and then feed it to the newborn calf. There are numerous
studies where colostrum was stored for feeding purposes. In
England they have fed colostrum that was cold-stored for up to
8 weeks. The colostrum underwent fermentation for 7 to 10 days
and then remained quite stable. This product was quite acceptable to the majority of calves. I do not know if the Ig fraction
was affected by the period of storage.

H. Thornberry _(Ireland)_

May I refer, Mr. Plenderleith, to Dr. Roy's book where he
states a period of 6 weeks for colostrum storage, at -25° to -35°C.

J. Brolund Larsen _(Denmark)_

Dr. Hammer, how can we utilise your information in practice?
More and more calves are shipped around Europe, coming to a new
environment within the first two weeks of age. What can we do,
or what should we advise farmers to do in order to stimulate the
immune mechanism within the calf?

Secondly, what do you think about Peterson's old work in
Minnesota?

D.K. Hammer _(West Germany)_

In reply to your first question, it emerges from data so far
available that, in some cases, the maternal antibody to pathogenic micro-organisms might be associated with IgG subclasses which
cannot be transferred to the offspring via colostrum. Under
these conditions protection is not afforded to the offspring and

consequently the newborn is susceptible to infections. Therefore wherever an increase of passive immunity to infections in the newborn is needed by vaccination of the dam, appropriate vaccines should be selected which efficiently produce antibodies which can be transferred to the colostrum.

In this context it should be recognised that successful bacterial immunisation can be afforded with purified capsular polysaccharide vaccines. There is a clear indication for their usefulness in preventing pneumococcal pneumonia and meningococcal meningitis. Thus a single dose of 50 µg of purified pneumococcal polysaccharide gave rise to type-specific antibody response, amounting in some cases to 50 kg antibody protein during an 8-year observation period. For too long the capsular antigens of gram-negative bacilli have been neglected as well as their possible importance in the stimulation of protective immunity of these infections. It seems to me that capsular antigens coupled to a more positively charged carrier would be good candidates to select IgG 1 class antibodies capable of being transferred to the colostrum.

There is another approach to this question of protective immunity due to capsular antigens. It takes advantage of the widespread occurrence of immunological cross-reactions between capsular antigens of pathogenic bacteria and those of the non-pathogenic bacteria of both the respiratory and the intestinal tract. It is possible that cross-protective immunity to pathogens can be stimulated by either spontaneous or deliberate colonisa-tion of either the respiratory or the intestinal tract with these non-pathogenic bacteria.

There is a second aspect of affording protective immunity which concerns cell-mediated mechanisms as another line of defence. Cell-mediated immunity represented by T cells is as-sumed to be the major mechanism for destroying micro-organisms that replicate intracellularly and thus are protected from serum antibodies. It is now recognised that the thymus produces a family of polypeptides that play an important role in the matur-ation, differentiation and function of T cells. One of these peptides, termed thymosin, with a molecular weight of less than 4,000, can act to reconstitute immune functions in certain thymus-deprived and/or immuno-deprived individuals. Clinical trials suggest that thymosin is effective in increasing T cell numbers and normalising immune function in children with thymus-dependent immuno-deficiency diseases.

It should be recognised that in the newborn calf insufficient T cell maturation may cause a decrease in the response to antigens. It is possible that the response is affected by suppressor T cells which are present in excess in the newborn to inhibit normal B-cell function. Thus treating newborn calves with thymosin would be an effective way to induce T cell differentiation and to enhance immunological function.

I shall close with the remark that the immune response is controlled by certain genes (Ir genes). Since certain diseases might also be due to the influence of these genes there might be good reasons to produce resistance to infections by selective breeding of cattle.

In reply to your second question, I don't believe too much in Dr. Peterson's work because of the results obtained 15 years ago using the same procedure to immunise cows through the teat canal via the dry mammary gland. There are two points which have to be considered. The first one is that the local antibody production, if any, is very small, and I doubt very much that this is a protective antibody. Secondly, if there is some kind of latent infection in the mammary gland, a tremendous inflammation of the udder would follow the application of the antigen in the mammary gland, and I think no farmer would accept this.

K. O'Farrell (Ireland)
I would like to put a question to Dr. Fallon. Could you give any recommendations for the farmer who buys calves? Should he have a routine Zinc Sulphate Turbidity Test carried out on them to assess their status? And a second question: do vaccines play any role in the prevention of mortality in bought-in calves and if so, at what stage should they be given?

R.J. Fallon

The Zinc Sulphate Turbidity Test should be used as a routine
test on bought-in calves. When the calf producer knows that a
test can be used to determine the Ig status of his calves he will
be more careful to ensure that his calves get colostrum. In
Ireland a cooperative which purchases 1,000 - 1,500 calves
annually from farms took blood samples from the calves prior to
purchase. They explained to the farmers that they were assessing
the amount of colostrum fed to each calf. They found that as the
purchasing season progressed, the percentage of calves with low
Ig levels decreased. This clearly demonstrates that, if the
farmer realises that he is getting paid less for calves with low
blood-serum Ig levels, he will make sure that all his calves are
given an opportunity to obtain colostrum.

In relation to purchased calves and vaccination, I would
prefer not to comment on the use of vaccines against virus pneumonia. From a practical point of view we have found that a particular Salmonella vaccine was quite effective. We had no control group of calves but when the vaccine was used the incidence
of Salmonella was less than 1%. When the vaccine was not used
the incidence of Salmonella infection increased to 10%.

H. Thornberry

At the Compton seminar we had a big discussion on this subject and Dr. William Smith, an expert in the field of salmonellosis, mentioned that the vaccination for Salmonella should
be done a fortnight before the animal is shipped. Even just
prior to the journey it is of some use. However, if the animal
hadn't received colostrum it would be contra-indicated.

D.K. Hammer

Recently some efforts have been made to bring about hybridisation of antigens of pathogenic bacteria on an apathogenic
background. That means germs that colonised normally in the
intestine were hybridised with Typhimurium or Salmonella enter-
itidis capsular antigens. This might be a very effective way
of local immunisation to resist infections.

STATUS, NUTRITION AND MANAGEMENT OF THE NEWBORN CALF, PART 2

Chairman: H. Thornberry

MANAGEMENT OF THE NEWBORN CALF: AN ATTEMPT AT AN ECONOMIC ANALYSIS*

O. Aalund

The Royal Veterinary and Agricultural University, Büllowsvej 13, 1870 Copenhagen V, Denmark.

ABSTRACT

The economic aspects of calf mortality may be analysed by the use of a model developed by Martin and Wiggins (1973). By means of the model the economic impact of variations in mortality and morbidity may be evaluated.

On the basis of information from a large Danish dairy herd, life tables have been calculated for the calves. By means of the model the net returns were calculated under conditions prevailing in the herd and comparison was made with net returns assuming a 50% reduction of mortality. Furthermore, the model was modified to suit the calf pneumonia complex. Net returns under the situation prevailing in the herd as well as under various forms of control strategies are compared.

---

* The following is an excerpt from a full report by Claus Willadsen, J. Gjøl Christensen and Ole Aalund to be published on the economic aspects of respiratory disease in a large dairy herd.

1. THE SITUATION IN DENMARK

The situation prevailing in Denmark in regard to livestock
health is dominated by the fact that diseases of monocausal
nature, eg tuberculosis and brucellosis, have been eliminated
so that diseases of multicausal nature now account for the health
problems. The mono- and the multicausal diseases require quite
different strategies in order to be properly counteracted.
While the monocausal diseases may effectively be controlled via
classical and rather simple procedures the multicausal disorders require a concerted approach in several areas. Effective
health management in calf production depends on the proper
administration of the following elements:

1) Surveillance.

2) Investigation and evaluation of the problems, including
economic assessment.

3) Intervention to control and restore health.

2. ECONOMICS AND LIVESTOCK DISEASE

The benefits that may flow from the control of diseases
are often questioned, especially by livestock owners but in
many cases also by veterinarians. Therefore, need obviously
exists for methods which make it possible with a high degree
of accuracy to estimate the amount of capital it would be
economically justifiable to spend on control measures against
diseases in livestock enterprises in order to achieve a
specified degree of improvement in the health status and, at
the same time, be fairly certain that a reasonable benefit would
arise from the extra capital invested. Hence, establishment of
disease models, ie mathematical formulae, by which it becomes
possible to forecast the economic consequences of intervention,
seems indispensable.

3. LIFE TABLES

Calf health/disease data from a large Danish dairy herd were accumulated over a 3-year period 1973/76. For analytical purposes the information was arranged in life tables. The combined mortality data may be seen in Table 1. Several other life tables were composed to study morbidity and mortality, ie in relation to sex of the calf and in relation to parity. The life tables cover the interval from day zero to the end of the 6th month of life.

TABLE 1

LIFE TABLE: CALF MORTALITY, FEMALE AND MALE CALVES COMBINED

Period	A	B	C	D
1st week of life	903	834	69	0.0769
2nd week of life	834	822	12	0.0149
3rd week of life	822	806	16	0.0197
4th week of life	806	792	14	0.0174
2nd month of life	792	750	42	0.0537
3rd - 6th months of life	750	709	41	0.0560

A. Number of calves at the beginning of the period

B. Number of calves at the end of the period

C. Number of calves dying during the period

D. Probability of death during the period.

The overall calf mortality during the entire span of the life tables was 20.5% for the two sexes combined (Table 1). The comparable figures for male and female calves were 25.8% and 17.3% respectively. The Chi^2 test showed that the difference was significant ($P < 0.001$).

Parity was associated with mortality of the calf ($P < 0.05$). There was a gradual drop in mortality from the highest figure at parity 1 over an intermediate figure at parity 2 to the lowest figure at parity 3. The sex-linked differences in calf mortality were constant at parities 1, 2 and 3.

4. THE DISEASE MODEL

The economic aspects of calf mortality may be analysed using a model which employs the principles of a model developed by Martin and Wiggins (1973). At the beginning of each period specified in the life table the calf may be subject to one of the following fates:

1) Survival, no sale

2) Survival, sale

3) Death.

The probability of the calf surviving, being sold or dying can be calculated from surveillance data. Each outcome during each period is associated with a specific economic factor.

In the present study the mortality model was modified in order to study the economic aspects of respiratory disease in calves (Figure 1). The model differentiates between the following outcomes for each calf during each period in the life table:

$\bar{B}D(R)$: The calf dies from respiratory disease for which it was never treated. The model assumes that the calf dies at the middle of the period. The calves in real life presumably die throughout the period.

$BD(R)$: The calf is attacked by respiratory diseases in the centre of the period. The calf is treated, but dies a certain number of days later. In the present material they died on the average 26 days later. Included in this category are calves treated for respiratory disease and dying from other causes. This latter group is a minority of only 1% of the total number of calves in the present material.

$\bar{B}\bar{D}(R)S$: The calf never having had respiratory disease, is sold. The sale is assumed to take place at the

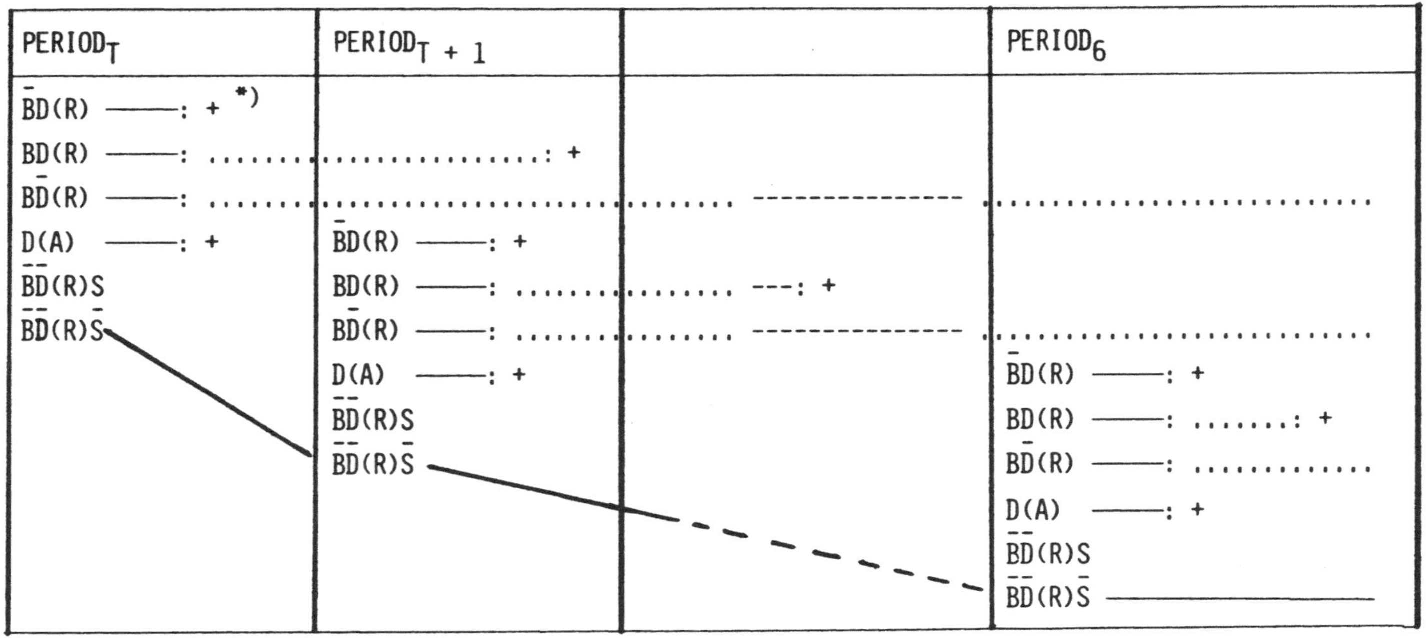

*): DEATH.   - FOR EXPLANATION OF OTHER SYMBOLS, REFER TO THE TEXT.

Fig. 1. Model for analysis of the economic impact of respiratory diseases in a population of calves during the first six months of life. Possible outcomes for a calf during each period.

beginning of a period.

$\overline{BD}$(R): The calf is attacked by respiratory disease, is treated and survives up to the end of the 6th month. Sick calves are assumed not to be marketable.

$\overline{BD}$(R)$\overline{S}$: The calf remains in the herd in a disease-free state up to the end of a given period in the life table.

D(A): The calf dies from diseases other than respiratory diseases. Death is assumed to take place at the middle of a given period. These calves may have been treated and the expenses incurred are included in 'smaller items', see later under 'costs'.

The periods employed in the life tables are the following:

Period No. 1: First week of life

Period No. 2: Second week of life

Period No. 3: Third week of life

Period No. 4: Fourth week of life

Period No. 5: Second month of life

Period No. 6: Third to sixth months of life.

The economic expectation (E) for different outcomes in different periods may now be calculated on the basis of probabilities for different fates of the calf (Table 2). The probabilities in Table 2 are absolute probabilities calculated on the basis of the conditional probabilities in the life tables. Statistically the probabilities in Table 2 fractionate the calf into different fates (Figure 2). After a period in the life table has elapsed the residual probability ($\overline{BD}$(R)$\overline{S}$) is distributed into the fate categories of the following period by multiplying it with the relevant probabilities from the life table. Thus, after a period has elapsed the sum of the residual probability ($\overline{BD}$(R)$\overline{S}$) for that period and all the other probabilities up to that point equals 1.00. The economic

TABLE 2

ABSOLUTE PROBABILITIES OF EVENTS

	$\overline{B}D(R)$	$BD(R)$	$B\overline{D}(R)$	$\overline{B}\overline{D}(R)S$	$D(A)$	$\overline{B}\overline{D}(R)\overline{S}$	Sum of rows
1st week of life	0.0050	0.0040	0.0070	0.0350	0.0710	0.8770	1.0000
2nd week of life	0.0026	0.0079	0.0175	0.0360	0.0114	0.8025	0.8770
3rd week of life	0.0008	0.0088	0.0209	0.0032	0.0088	0.7600	0.8025
4th week of life	0.0030	0.0122	0.0205	0.0068	0.0068	0.7106	0.7600
2nd month of life	0.0099	0.0242	0.0583	0.0085	0.0064	0.6026	0.7106
3rd – 6th months of life	0.0054	0.0090	0.0163	0.0054	0.0066	0.5598	0.6026
Sum of columns	0.0267	0.0661	0.1405	0.0949	0.1110		0.9990 *)

*) Sum of row plus residual probability 0.5598 ($\overline{B}\overline{D}(R)\overline{S}$)

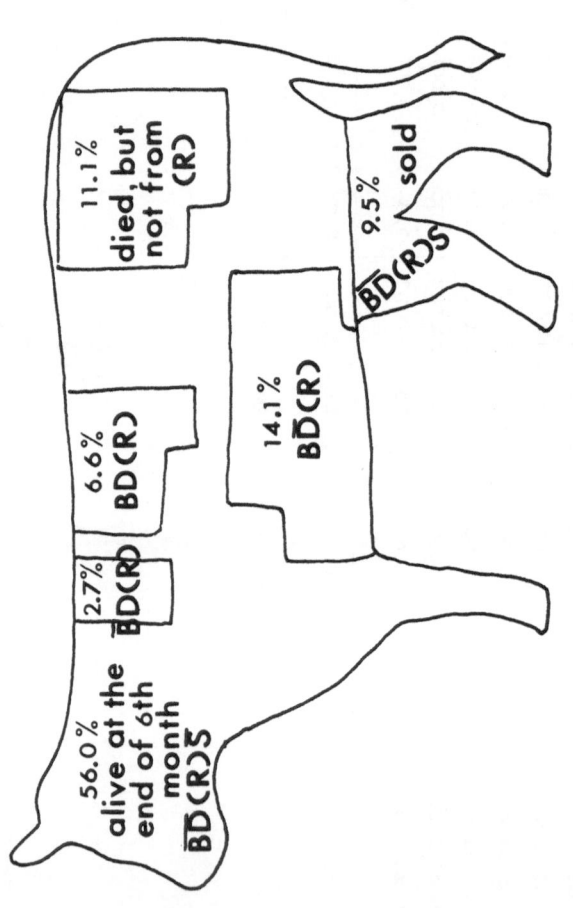

Fig. 2. Calf fractionated into its fates, see Table 2. The situation shown in the figure illustrates the balance at the end of the 6th month.

parameters to be employed include sale price and production costs. The calculation presented here is the balance up to the end of the 6th month.

The total economic expectation is the monetary profit that can be expected to flow from each calf entering the model at the beginning of the first period specified by the model and under the conditions specified by the parameters of the model. In other words, the model provides an estimate of the economic net return per calf alive at the beginning of the study period, when calculated at the end of the study period, under the conditions prevailing within the herd. These conditions are specified not only by the disease situation, ie the probabilities for the different fates of the calf, but are supposed to encompass all sources of costs and incomes arising in connection with the enterprise.

The sale price and costs used in calculations of the economic expectation are listed below:

4.1. Sale price:

S_T: The sale price at the beginning of each period for calves remaining free of respiratory ailments up to beginning of the period. The sale price is calculated as the sum of a basic price of Danish crowns (D.kr.) 328.00 and a weight value of D.kr. 9.04 per kg body weight. Assuming a body weight of 41.0 kg at birth and a linear weight gain of 146.5 kg up to the end of the 6th month the sale prices at birth and at the end of the 6th month are D.kr. 328.00 + 9.04 x 41.0 = D.kr. 698.64 (S_1) and D.kr. 328.00 + 9.04 x (41.0 + 146.5) = D.kr. 2023.00 respectively.

S_D: The sale price at the end of the 6th month for calves that have been attacked by respiratory disease but which have survived up to the end of the 6th month. These calves are assumed to have a weight gain up to the end of the 6th month which is 15.0 kg less than their non-attacked counterparts. S_D = D.kr. 328.00

$$+ 9.04 \times (41.0 + 131.5) = \text{D.kr. } 1887.40.$$

4.2. Costs:

C_T: Total costs for a disease-free calf during period T. The figure includes fixed and variable costs. Total costs comprise the following items:

C_{Fx}: Fixed costs which include depreciation on buildings and equipment plus the base price of the calf. The base price of the calf equals the sale price right after birth.

C_{FT}: Feeding costs for a calf during period T.

C_{LT}: Labour costs for a calf during period T. The calculation is based on the assumption of 2.2 minutes per day per calf at D.kr. 30.00 per h.

C_{AT}: Cost during period T of various smaller items, ie bedding, veterinary expenses except costs of treatment for respiratory diseases; D.kr. 0.15 per day per calf.

I: Interest 11% pa on base price of the calf plus 0.5 x (variable costs). Variable costs include C_F + C_L + C_A + (C_M + C_X, see later).

In connection with respiratory disease the following additional variable costs become relevant:

C_M: Cost of treatment against respiratory disease. Animals that were treated and subsequently died received on average 2.4 treatments at D.kr. 30.00 per treatment = D.kr. 72.00. Calves that were treated and subsequently survived up to the end of the 6th month received on average 1.8 treatments at D.kr. 30.00 per treatment = D.kr. 54.00.

C_{XT}: Cost of feed, labour and 'smaller items' for calves which are initially treated for respiratory diseases during period T (at the period's midpoint) until they die, in the present study 26 days later. Costs incurred during the disease period have been calculated using the costs applying to disease-free calves during the same period.

The expected economic net return (E) of the different fates to which a calf is fractionated (Table 2) is calculated by means of the following formulae:

E: E in the formulae stands for expectation, ie the expected economic return.

Q: Q in the formulae stands for the absolute probability of the calf becoming subject to the specific event.

$E_{\overline{BD}(R)}$:

For each period E has the following composition:

$$E_{T(\overline{BD}(R))} = - Q_{T(\overline{BD}(R))} \times (B + C_{T/2} + \sum_{t=1}^{T-1} C_t)$$

Hence:

$$E_{\overline{BD}(R)} = - \sum_{t=1}^{6} \left[Q_T(\overline{BD}(R)) \times (B + C_{T/2} + \sum_{t=1}^{T-1} C_t) \right]$$

$E_{BD(R)}$:

For each period E has the following composition:

$$E_{T(BD(R))} = - Q_{T(BD(R))} \times (B + C_{T/2} + C_M + C_{XT} + \sum_{t=1}^{T-1} C_t)$$

Hence:

$$E_{BD(R)} = - \sum_{t=1}^{6} \left[Q_{T(BD(R))} \times (B + C_{T/2} + C_M + C_{XT} + \sum_{t=1}^{T-1} C_t) \right]$$

$E_{\overline{BD}(R)S}$:

For each period E has the following composition:

$$E_{T(\overline{BD}(R)S)} = Q_{T(\overline{BD}(R)S)} \times (S_T - B - \sum_{t=1}^{T-1} C_t)$$

Hence:

$$E_{\overline{BD}(R)S} = \sum_{t=1}^{6} \left[Q_{T(\overline{BD}(R)S)} \times (S_T - B - \sum_{t=1}^{T-1} C_t) \right]$$

The expectations, ie the expected average economic net return per single liveborn calf at the end of the 6th month after its birth, are presented in Table 3. At the end of the 6th month the economic balance for calves at this farm was minus D.kr. 10.74. This figure would have become plus D.kr. 27.64 or plus D.kr. 66.04 if the frequency of respiratory disease had been 25% or 50% lower respectively. If the mortality rate among treated calves had been 50% lower, the net balance per calf would have been D.kr. 50.57 at the end of the 6th month.

Death, including death among calves treated for respiratory disease, for the period included in this study accounted for a loss per calf of D.kr. 187.47 which was only partially counter-balanced by an economic return of D.kr. 176.73 giving a difference of minus D.kr. 10.74.

It would be relevant now to ask the question: 'How much capital would it be economically sound to invest in each single liveborn calf in order to reach a specified target?'. Referring to the example above with 50% lower morbidity the calculated improvement in economic net return was D.kr. 66.04 plus D.kr. 10.74 = D.kr. 76.78. How much of this money would be available for investment to achieve a 50% lower morbidity with the following assumptions?

1. The livestock operator should be no worse off under the change than he is at present, ie the increase in expectation minus interest would be channelled back into the enter-prise (the 'break-even point').

2. A 100% profit is demanded from all extra investments that will have to be made by means of money borrowed on the free market at 15% per annum.

The answer to the question is D.kr. 37.01.

TABLE 3

ECONOMIC IMPACT IN DANISH CROWNS (D.kr.) OF EACH FATE DURING THE PERIODS: ACTUAL SITUATION ON THE FARM

	$\bar{B}\bar{D}(R)$	$BD(R)$	$D(A)$	$B\bar{D}(R)$	$\bar{B}\bar{D}(R)S$	$\bar{B}\bar{D}(R)\bar{S}$	Sum of row
1st week of life	-4.01	-3.99	-52.17	–	-0.70	–	
2nd week of life	-1.66	-7.28	- 8.32	–	0.42	–	
3rd week of life	-0.86	-8.68	- 6.91	–	0.12	–	
4th week of life	-2.70	-12.42	- 5.40	–	0.42	–	
2nd month of life	-9.06	-27.47	- 6.04	–	0.74	–	
3rd - 6 months of life	-7.49	-14.04	- 8.98	–	0.79	–	
Sum of columns	-25.78	-73.87	-87.82	13.68	1.78	161.27	-10.74

5. CONCLUSION

The model presented here is convenient to apply on individual livestock operations. The model makes it possible to prepare fairly accurate economic forecasts on the basis of probabilities for the fate of the calf, and selected sale prices, costs and investment policies.

REFERENCE

Martin, S.W. and Wiggins,A.D. 1973. A model of the economic costs of dairy calf mortality. American Journal of Veterinary Research. 34: 1027-1031.

TREATMENT OF THE NEWBORN CALF

P. Larvor

Laboratoire des Maladies Métaboliques, Centre de Recherches
Zootechniques et Vétérinaires de Theix, 63110 Beaumont, France.

ABSTRACT

The scientific evidence for the use of various treatments (etiological or symptomatic) in newborn calf diarrhoea is reviewed. From these considerations it is possible to summarise the main tendencies of scour therapy. Rehydration is the most important point of the treatment; various ancillary treatments are discussed and a therapeutic scheme is advocated.

1. ETIOLOGICAL TREATMENT

Three types of theory concerning the etiology of scours in
the newborn calf are current: the theory of infection with its
bacteriological and viral interpretations, the theory of
immunological defect, and the theory of congenital deficiency.
They can be combined, which means that everybody has his own
explanation. My personal view is that, starting from deficient
nutrition of the dam and poor management of the newborn calf,
there is a defect in the transmission of immunity which allows
various more-or-less pathogenic germs to colonise the digestive
tract and eventually to enter the circulation. I must admit
that there is no clear-cut demonstration of this view, and in
this field as in religious matters, all opinions must be
respected.

1.1. Effect of antibacterial agents

It is obvious that successful treatment of diarrhoea with
anti-infectious agents would offer some evidence for the view
that bacteria are involved as causative organisms. Many trials
have been carried out in order to demonstrate an activity of
antibiotics or chemotherapeutic agents, with varying results.
In animals submitted to a complete treatment (including intensive
rehydration), the antibiotics or chemotherapeutic agents are
generally ineffective (Fisher and de la Fuente, 1971; Radostits
et al., 1975) or of doubtful value (Dalton et al., 1960;
Fayet, 1975). However, in less controlled conditions, many
reports have been published of good results with such substances
(see in Roy, 1970; Fey, 1972; Dinse, 1974; Bartos et al., 1974,
1975). These discrepancies might be explained by various
reasons:

1.1.1. The acquired resistance of bacteria to the antibacterial
agents can confuse the picture. For some authors, the
efficacy of the classical antibiotics in calf diarrhoea
decreases from year to year (Roy, 1970; Fey, 1972). When
faecal swab cultures are tested for sensitivity to
antibiotics, resistance is often observed for most

classical drugs (Mylrea, 1968; Fisher and de la Fuente, 1971; Fayet, 1975). Furthermore, most bacteriologists have little confidence in the validity of the resistance tests performed on faecal flora, and attempts have been made to use bacteria isolated from the higher parts of the gut (Glantz et al., 1974). Results were disappointing; the substances which were active on isolated *Escherichia coli or Salmonella newport* (sulfachlorpyridazine and gentamycin) were unable to cure the disease!

1.1.2. The blood immunoglobulin levels of the calves can influence the effect of the antibiotic therapy. Fisher and de la Fuente (1971), commenting on the results of Dalton et al. (1960) suggest that oxytetracyclin was beneficial only when the immunological status of calves was rather good.

1.1.3. Many authors insist on the difference between the cases of purely enteric infections and the cases of septicaemia. But septicaemia seems to occur only in colostrum-deprived calves (Smith and Halls, 1968; Schoenaers and Kaekenbeeck, 1975) and appears to be infrequent in practice. This is confirmed by the high therapeutic value of rehydration alone (see later), which could not be explained if septicaemia cases were frequent.

1.1.4. Adverse conditions of treatment (poor rehydration, for instance) as it often occurs in clinical practice, might be a reason for a non-specific positive effect of anti- biotics; it is well known that germ-free animals are more resistant than conventional ones to various aggressions (for instance enteric viral infection; Mebus et al., 1973).

1.1.5. Finally, in the reports recording positive results with antibiotics, control animals are often absent, and a high percentage of recovery is by no mean a proof of the activity of the treatment, because the percentage of spontaneous recovery can be very high in some outbreaks of diarrhoea.

1.2. Effect of immunoglobulins

The preventive effect of colostral immunoglobulins on calf diarrhoea is well documented, and it has been shown that the protective amounts of immunoglobulins are 45g in colibacillosis (McEwan et al., 1970; Fisher et al., 1975); when this is compared with the concentration of injectable preparations of commercial gammaglobulins (from blood or colostral origin) it is not surprising that their clinical use is disappointing: the cost of these Igs is such that it does not permit the injection of significant amounts.

The use of oral treatment of ill calves with a pool of colostrum (frozen or fermented) certainly deserves our attention, but until now there is no detailed report on this subject.

2. TREATMENT OF THE DIGESTIVE TRACT LESIONS

2.1. Feeding the diarrhoeic calves

When treating diarrhoeic animals, some people prefer to take them off feed, in order to allow the gut to repair its lesions and to decrease the proliferation of intestinal bacteria. Some authors prefer to feed them to avoid the adverse effects of fasting on metabolism (acidosis) and on the resistance of the calves.

The comparative tests performed either in mild scouring resulting from previous overfeeding (McLean and Baily, 1972) or in serious diarrhoea with heavy losses (Radostits et al., 1975) do not show any difference in the results of treatment with or without starvation. Advantages and disadvantages of both attitudes seem to be balanced.

In the future, it would certainly be rewarding to intensify the studies on what a milk replacer specially designed for sick calves should be.

2.2. Astringents and adsorbents

Numerous astringents and adsorbents have been advocated as

ancillary treatments of diarrhoea, for instance kaolin (Radostits, 1965), bentonite (Bartos and Habrda, 1974), barium sulphate, bismuth subnitrate (Fayet, unpublished data). These substances have not been thoroughly investigated for their efficacy in the treatment of calf scours nor have they been compared in extensive field trials, but their innocuity and low cost, together with the logic of this kind of treatment, commend their systematic use.

2.3. Parasympatholytics

Parasympatholytics have sometimes been advocated in order to reduce the so-called intestinal hypermotility of scours (Link et al., 1974). The demonstration that, during newborn calf scours, there is no hypermotility but on the contrary an inertia of the digestive tract (Dardillat and Ruckebusch, 1973; Dardillat, 1975), negates the reasoning behind such treatment.

3. TREATMENT OF THE SUBSEQUENT METABOLIC DISTURBANCES

3.1. Dehydration

The clinical evidence of dehydration in scouring calves has led many workers to infuse diarrhoeic calves intravenously with an electrolyte solution. The first attempts were performed on the basis of experiments derived from the biochemical study of the human neonate (infant toxicosis), and often with the same solutions, such as Darrow's (Watt, 1967) or other isotonic fluids. The results were rather good, in spite of the fact that blood biochemistry and body fluid distribution are different in the diarrhoeic infant and calf. The main characteristics of calf dehydration are:

- Acidosis (McSherry and Grinyer, 1954a and b; Fisher, 1965; Fayet, 1968a).

- Hyponatraemia and hyperkalaemia (Roy et al., 1959; Dalton et al., 1965; Fisher, 1975; Fayet, 1968a; Lewis and Phillips, 1973).

- Hypo-osmolarity of blood plasma (Fayet, 1971; Melichar
  and Masek, 1971),

- Extracellular deyhydration with intracellular hyper-
  hydration (Fayet, 1968b, 1971; Phillips et al., 1971;
  Phillips and Lewis, 1973).

This is different in some points from infant dehydration,
which is hypertonic and concerns both intra- and extracellular
compartments. This led to the use of new types of solution
characterised by their hyperosmolarity and their high content
in sodium bicarbonate, infused in large amounts (one litre or
more), the infusion being often repeated at 1 or 2 days
intervals.

Rehydration is now generally considered as the basis for
the therapy of diarrhoea in calves, in intensive care units
as well as in clinical practice (Melichar and Masek, 1971;
Tennant et al., 1972; Radostits, 1973; Wettstein et al., 1973;
Berchtold et al., 1974; Rašková et al., 1976; Verter et al.,
1976). If properly applied, it is as effective alone as in
conjunction with antibiotics (Radostits et al., 1975; Rašková
et al., 1976). It is also effective in the treatment of
experimental colibacillosis (Massip and D'Ieteren, 1976).

The main problem for a correct application of this therapy
in practice is the difficulty of repeating the intervention as
frequently as needed. A possible answer to this difficulty is
oral rehydration, which cannot replace the intravenous route
in case of emergency, but can follow it when the dehydration is
not too acute. Several formulae have been advocated (Hamm and
Hicks, 1975; Fayet, 1975), and seem to give good results.

In a few cases, hypertonic dehydration may be observed,
not clinically different from the hypotonic, except for the
hyperthermia observed in hypertonic dehydrated calves (Fayet,
1975).

3.2. Catabolism

Catabolism is intense in scouring calves, and the blood levels of urea and phosphate are good indicators of the prognosis (Fayet, 1975, 1977). The compensation of this catabolism cannot be attained with hexoses only, because they become toxic at large doses (Fayet et al., 1977). Solutions used for human parenteral nutrition are too expensive and it would be useful to undertake research in this field.

4. CONCLUSIONS

From the preceding considerations, it is possible to summarise the main lines of scour therapy.

Besides the usual recommendations concerning the housing of the animals in a dry mild place, the core of the treatment will be an intensive intravenous rehydration with a large volume of a hypertonic alkaline solution (for instance 15 g sodium bicarbonate per litre of distilled water, at a dose of 60 ml per kg body weight, with a maximal speed of infusion of 1 litre in 20 min). If the calf has not recovered on the second day, it is possible to renew the treatment, with only 40 ml/kg body weight. In most cases, the initial intravenous rehydration can be followed by oral rehydration (sodium bicarbonate 10g, glucose 10g, water 1 litre: 1.5 litres three times a day). It is not necessary to prevent the calves from feeding. It might be useful to try colostrum feeding.

Ancillary treatment will be the distribution of a classical antidiarrhoeic mixture (adsorbent and slightly astringent), and possibly the use of an antibacterial agent. The choice of the antibiotic or chemotherapeutic drug will preferably be on the basis of the empirical knowledge of the situation in the concerned area. Vitamins are useful in moderate doses, but the use of corticotherapy is questionable.

Such treatment can often save calves, even in a moribund condition, but in some instances it is able only to lengthen for a long period the survival of the calf without really curing the disease. The fatal issue occurs then with the exhaustion of the calf or the weariness of the clinician.

This shows clearly that several factors escape our comprehension and will need further research.

REFERENCES

Bartos, J. and Habrda, J. 1974. Bentonite in the prevention and treatment
of diarrhoea in newborn calves. Veterinárni medicina, Brno. 19, 707-716.

Bartos, J., Maiková, M. and Habrda, J. 1975. Value of streptomycin furazol-
idone combination in the treatment of diarrhoea in newborn calves.
Veterinárni medicina, Brno. 20, 91-99.

Bartos, J., Poláková, M. and Habrda, J. 1974. Efficacy of chlortetracycline
and nitrofurantoin alone or combined, in calf diarrhoea. Veterinárni
medicina, Brno. 19, 21-30.

Berchtold, M., Weiss, G. and Thaller, M. 1974. Infusionstherapie bei
Kalbern unter spezieller Berücksichtigung der Acidose. Deutsche
Tierärtzliche Wochenschrift 81, 279-281.

Dalton, R.G., Fisher, E.W. and McIntyre, W.I.M. 1960. Antibiotics and
calf diarrhoea. Veterinary Record 72, 1186-1194.

Dalton, R.G., Fisher, E.W. and McIntyre, W.I.M. 1965. Changes in blood
chemistry, body weight and haematocrit of calves affected with neonatal
diarrhoea. British Veterinary Journal 121, 34-41.

Dardillat, C. 1975. Gastrointestinal motility in calf neonatal diseases.
In: 'Perinatal ill-health in calves'. Proceedings of the first
seminar on Pathology of the CEC, Sept. 22nd-24th, 1975, Compton.
Edited by J.M. Rutter. Published by Commission of the European
Communities, Brussels, Belgium.

Dardillat, C. and Ruckebush, Y. 1973. Aspects fonctionnels de la jonction
gastro-duodénale chez le veau nouveau-né. Annales de Recherches,
Vétérinaire 4, 31-56.

Dinse, P. 1974. Die Anwendung von Gentamycin in der Kälberpraxis. Deutsche
Tierärtzliche Wochenschrift 81, 614-615.

Fayet, J.C. 1968a. Recherches sur le métabolisme hydrominéral chez le veau
normal ou en état de diarrhée. 2. L'ionogramme plasmatique et le pH
sanguin. Recherche Vétérinaire (1) 109-115.

Fayet, J.C. 1968b. Recherches sur le métabolisme hydrominéral chez le veau
normal ou en état de diarrhée. 3. Les compartiments liquidiens.
Recherche Vétérinaire (1) 117-126.

Fayet, J.C. 1971. Plasma and fecal osmolarity, water kinetics and body
fluid compartments in neonatal calves with diarrhoea. British
Veterinary Journal 127, 37-44.

Fayet, J.C. 1975. A hospital unit for calves with neonatal diarrhoea. In: 'Perinatal ill-health in calves'. Proceedings of the first seminar on Pathology of the CEC, Sept 22nd-24th, 1975, Compton. Edited by J.M. Rutter. Published by Commission of the European Communities, Brussels, Belgium.

Fayet, J.C. 1977. In press.

Fayet, J.C., Renouf, C., Michel, M.C. and Overwater, J. 1977. Effect of intravenous infusion of glucose and/or fructose on the composition of blood plasma and the clinical response of the calf. Annales de Recherches, Vétérinaire 8 (2) in press.

Fey, H. 1972. Colibacillosis in calves. Hans Huber publisher, Bern, Switzerland.

Fisher, E.W. 1975. Death in neonatal calf diarrhoea. British Veterinary Journal 121, 132-138.

Fisher, E.W. and de la Fuente, G.H. 1971. Antibiotics and calf diarrhoea. The effect of serum immune globulin concentration. Veterinary Record 89, 579-582.

Fisher, E.W. and Martinez, A.A. 1975. Serum immune globulins in calves receiving E. coli endotoxin. Veterinary Record 96, 527-528.

Glantz, P.J., Kradel, D.C. and Seward, S.A. 1974. Escherichia coli and Salmonella newport in calves: efficacy of prophylactic and therapeutic treatment. Veterinary Medicine and Small Animal Clinician 69, 77-82.

Hamm, D. and Hicks, W.J. 1975. A new oral electrolyte in calf scours therapy. Veterinary Medicine and Small Animal Clinician 70, 279-282.

Lewis, L.D. and Phillips, R.W. 1973. Diarrheic induced changes in intracellular and extracellular ion concentrations in neonatal calves. Annales de Recherches, Vétérinaire 4, 99-111.

Link, R.P., Horn, J. and Joshi, H.C. 1974. Effect of isopropamide iodide on gastrointestinal motility in ruminants. Indian Veterinary Journal 51, 60-62.

McEwan, A.D., Fisher, E.W. and Selman, I.E. 1970. Observations of the immune globulin levels of neonatal calves and their relationship to disease. Journal of Comparative Pathology 80, 259-267.

McLean, D.M. and Baily, L.F. 1972. The effectiveness of three treatments for scouring in calves. Australian Veterinary Journal 48, 336-338.

McSherry, B.J. and Grinyer, I. 1954a. The pH values, carbon dioxide content and the levels of sodium, potassium, calcium, chloride and inorganic phosphorus in the blood serum of normal cattle. American Journal of Veterinary Research 15, 509-510.

McSherry, B.J. and Grinyer, I. 1954b. Disturbances in acid-base balance and electrolyte in calf diarrhoea and their treatment. A report of eighteen cases. American Journal of Veterinary Research 15, 535-541.

Massip, A. and D'Ieteren, G. 1976. Exemple de rehydration chez un veau atteint de diarrhée. Annales de Médicine Vétérinaire.

Melichar, B. and Masek, J. 1971. Biochemical aspects of rehydration therapy in diarrhoeic calves. Acta Veterinaria, Brno. 40, Supp. 2, 117-123.

Mebus, C.A., Stair, E.L., Rhodes, M.B., Underdahl, N.R. and Twiehaus, M.J. 1973. Calf diarrhoea of viral etiology. Annales de Recherches, Vétérinaire 4, 71-78.

Mylrea, P.J. 1968. Passage of antibiotics through the digestive tract of normal and scouring calves and their effect upon the bacterial flora. Research in Veterinary Science 9, 5-13.

Phillips, R.W. and Lewis, L.D. 1973. Viral induced changes in intestinal transport and resultant body fluid alterations in neonatal calves. Annales de Recherches, Vétérinaire 4, 87-98.

Phillips, R.W., Lewis, L.D. and Knox, K.L. 1971. Alterations in body water turnover and distribution in neonatal calves with acute diarrhoea. Annals of the New York Academy of Science 176, 231-243.

Radostits, O.M. 1975. Clinical management of neonatal diarrhoea in calves with special reference to pathogenesis and diagnosis. Journal of the American Veterinary Medical Association 147, 1367-1376.

Radostits, O.M. 1973. Clinical management of diarrhoea in calves. Bovine Practitioner (8) 20-24.

Radostits, O.M., Rhodes, C.S., Mitchell, M.E., Spotswood, T.P. and Wenkoff, M.S. 1975. A clinical evaluation of antimicrobial agents and temporary starvation in the treatment of acute undifferentiated diarrhoea in newborn calves. Canadian Veterinary Journal 16, 219-227.

Rašková, H., Sechser, T., Vaněček, J., Polák, L., Treu, M., Mužik, J., Sklenář, V., Rabas, O., Raška, K., Matějowská, D. and Matějowská, V. 1976. Neonatal *Escherichia coli* infection in calves. 1. Appraisal of rehydration. Zentrablatt for Veterinärmedizin 23B, 131-142.

Roy, J.H.B. 1970. The Calf. 3rd Edition, 2nd Vol. Iliffe Publisher, London.

Roy, J.H.B., Shillam, K.W.G., Hawkins, G.M., Lang, J.M. and Ingram, P.L. 1959. The effect of white scours on the sodium and potassium concentration in the serum of newborn calves. British Journal of Nutrition 13, 219-226.

Schoenaers, F. and Kaekenbeeck, A. 1975. Enteric colibacillosis in calves.
    Aetiology and experimental reproduction. In: 'Perinatal ill-health
    in calves'. Proceedings of the first seminar on Pathology of the
    CEC, Sept. 22nd-24th, 1975, Compton. Edited by J.M. Rutter.
    Published by Commission of the European Communities, Brussels, Belgium.

Smith, H.W. and Halls, S. 1968. The experimental infection of calves with
    bacteriemia-producing strains of *Escherichia coli*: the influence of
    colostrum. Journal of Medical Microbiology 1, 61-78.

Tennant, B., Harrold, D. and Reina-Guerra, M. 1972. Physiologic and
    metabolic factors in the pathogenesis of neonatal enteric infections
    in calves. Journal of American Veterinary Medical Association 161,
    993-1007.

Verter, W., Totziski, G. and Liebaug, E. 1976. Experimentelle
    Untersuchungen zur Pathophysiologie und Therapie der Dehydration des
    Kalbes. Monatshefte Veterinärmedizin 31, 254-259.

Watt, J.G. 1967. Fluid therapy for dehydration in calves. Journal of
    American Veterinary Medical Association 150, 742-750.

Wettstein, R., Gerber, H. and Schneider, E. 1973. Praktische Behandlung
    schwerer Durchfälle des neugeborenen Kalbes. Schweizer Archiv für
    Tierheilkunde 115, 453-464.

ACIDOSIS AND CLINICAL STATE IN DEPRESSED CALVES

K. Walser and H. Maurer-Schweizer

Gynäkologische und Ambulatorische Tierklinik der Universität
München, D-8000 München, Germany.

ABSTRACT

*Even in normal delivery the foetus comes into a hypoxic state which
is followed by pre-pathological acidosis in the newborn at birth. Recovery
to normal values of blood gases and acid-base balance is completed
spontaneously within the first 24 h of life.*

*In depressed calves (asphyxia) there is at birth an evident combined
respiratory-metabolic acidosis in the pathological range. In comparison
with normal calves there are significant deviations in pH values, base
excess, standard bicarbonate and actual bicarbonate. Carbon dioxide
tensions are significantly higher. The time needed for adaptation is
prolonged. A definite correlation can be demonstrated between the
clinical state and pH value and the metabolic parameters. Therefore an
infusion of a suitable buffer solution is of great benefit in the early
treatment of neonatal acidosis.*

When using the term asphyxia one should be aware that its current clinical use is etymologically incorrect. The Greek word "asphyxia" means "absence of pulse". In modern clinical usage, on the other hand, asphyxia means a state of suffocation in the newborn which is characterised by absence of or reduced breathing, affected function of heart and circulation, and lowered functions of the central nervous system (motility, muscle tone, reflexes). In human medicine it was already suggested years ago that the term asphyxia be replaced by terms like 'depressed newborns', 'hypoxia', or even more accurately 'acidosis during parturition'. Hypoxia is certainly the cause of this dangerous condition but cell death is ultimately caused by a shift of the acid-base balance in blood and tissues to acidosis.

For understanding these facts it is necessary to recall the physiological state of the foetus and the physiological stresses on foetus and newborn during normal delivery. There are disturbances in the utero-placental circulation caused by contractions of the uterus during labour. This may result in a lack of oxygen to the foetus. So even during undisturbed parturition the foetus can get into hypoxia. At the same time, by this disturbance in the circulation the transfer of carbon dioxide to the mother is reduced which means an increase of carbon dioxide in foetal blood.

Following the theory of Saling (1966) which is generally accepted today the foetus responds to hypoxia by an oxygen-conserving adaptation of its circulation. That means all organs not essential for intra-uterine life (lungs, spleen, thymus, muscles and skin, gastro-intestinal tract, possibly also liver, kidneys and adrenal glands) are supplied with a minimum of blood. Oxygen is spared for essential organs like heart and brain (Figure 1). By this centralisation of circulation and by reduced total oxygen consumption, oxygen tension in the blood can be maintained within physiological limits for some time. This oxygen-conserving adaptation, however, means anaerobic glycolysis in all tissues with minimum blood supply.

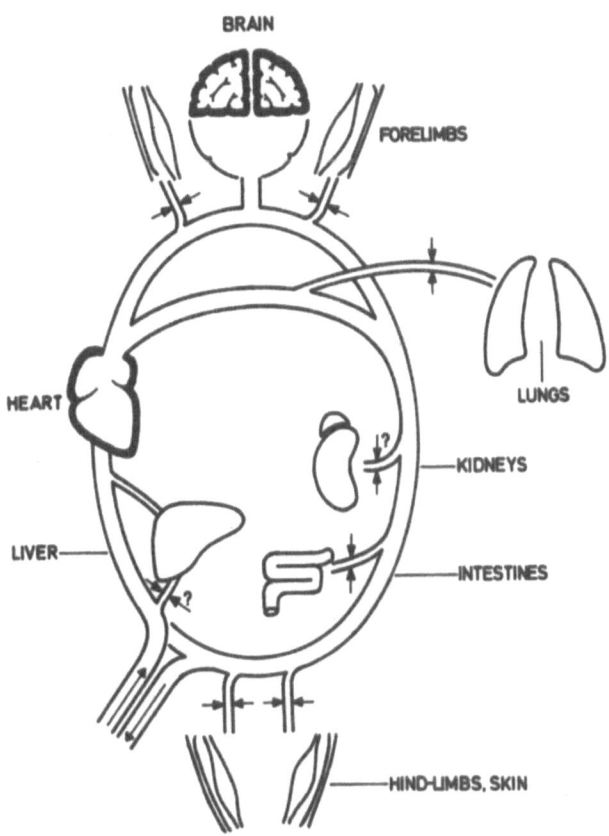

Fig. 1. Oxygen conserving adaptation (from Stoll, 1970).

Under normoxic conditions glucose as main energy source is reduced to pyruvate via the citric-acid cycle (Figure 2). The first step to pyruvate is anaerobic, the second step consists of oxidation of pyruvate via the citric-acid cycle to the final products CO_2 and H_2O. During oxygen shortage glucose can only be metabolised anaerobically to pyruvate which is mostly reduced to lactic acid. Energy output in this process is small but it is sufficient to maintain metabolism for some time. Anaerobic glycolysis however has a great disadvantage; although

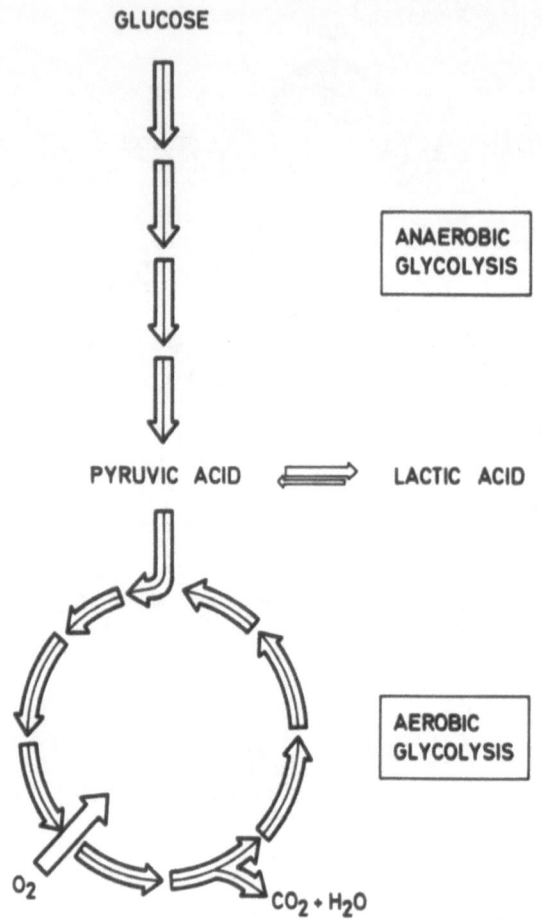

Fig. 2. Glycolysis (from Stoll, 1970).

energy production is reduced the carbohydrate reserves are
rapidly exhausted and metabolic acidosis develops by accumulation
of acid metabolites (lactic acid).

At birth the foetus therefore suffers from a respiratory
as well as a metabolic acidosis (Figure 3). It is the degree
of acidosis that finally determines whether the foetus lives
or dies. Vital cell functions cannot take place in a severe
acidosis; at a pH value of 6.70 in blood foetal life ends.

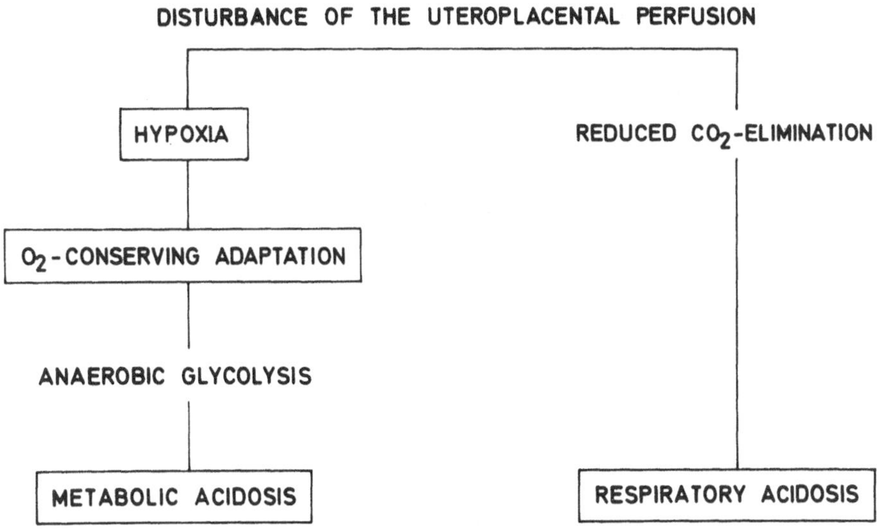

DISTURBANCE OF THE UTEROPLACENTAL PERFUSION

```
                    ┌──────────────────────────┐
                    │                          │
        ┌───────────┴──────┐                   │
        │     HYPOXIA      │        REDUCED CO₂-ELIMINATION
        └────────┬─────────┘                   │
                 │                             │
  ┌──────────────┴─────────┐                   │
  │ O₂-CONSERVING ADAPTATION│                  │
  └───────────┬────────────┘                   │
              │                                │
      ANAEROBIC GLYCOLYSIS                     │
              │                                │
  ┌───────────┴────────────┐   ┌───────────────┴──────────┐
  │   METABOLIC ACIDOSIS   │   │   RESPIRATORY ACIDOSIS   │
  └────────────────────────┘   └──────────────────────────┘
```

Fig. 3. Origin of acidosis.

Against this regression into pathological acidosis there is the organism's regulatory system of chemical buffering in the blood. Bicarbonate is the most important buffer. It reacts with excess H^+ ions from lactic acid setting free carbonic acid which is disintegrated into H_2O and CO_2. Moreover there are other buffering systems such as haemoglobin, plasma proteins and phosphate buffers.

Values important for the degree of acidosis - as for example pH value, blood gases like oxygen and carbon dioxide, oxygen saturation, buffer bases and bicarbonate, a possible base-deficit caused by consumption of buffers - can be measured in the newborn's blood. This gives exact information about the actual condition of the newborn.

From previous investigations quite a lot is known about blood gases and acid-base balance of newborn calves (Moore, 1969; Mülling et al., 1972; Ammann et al., 1974; Maurer-Schweizer et al., 1977). Even normal calves are born in a preacidotic state after undisturbed delivery. At birth there is a slight combined respiratory-metabolic acidosis. Even in

healthy calves with the start of spontaneous regular breathing
the situation grows worse during the first ten minutes post
partum. We can give the following explanation. After the start
of breathing the oxygen-conserving adaptation of the foetus
is stopped. Now lactic acid previously accumulated in the
minimum supplied tissues is released into circulation and
stresses acid-base balance. During the following hours
spontaneous normalisation of acid-base balance and blood gases
takes place. So this is the typical sequence of events:

1) a slight combined respiratory-metabolic acidosis during
parturition,

2) a post partal acid maximum 10 - 15 minutes later,

3) a gradual adaptation within the next few hours.

This corresponds in general with the situation of newborns
in human and other species (Figures 4, 5 and 6; Maurer-Schweizer
et al., 1977a).

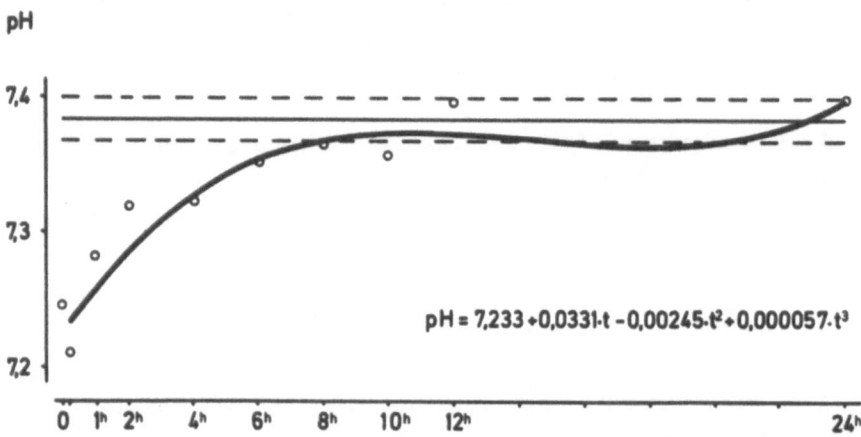

$$pH = 7.233 + 0.0331 \cdot t - 0.00245 \cdot t^2 + 0.000057 \cdot t^3$$

Fig. 4. pH-values in normal calves.

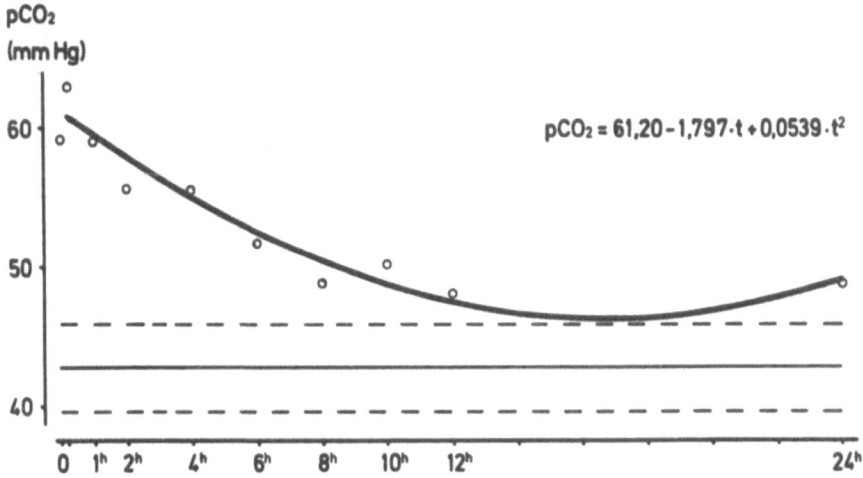

Fig. 5.   Carbon dioxide tension in normal calves.

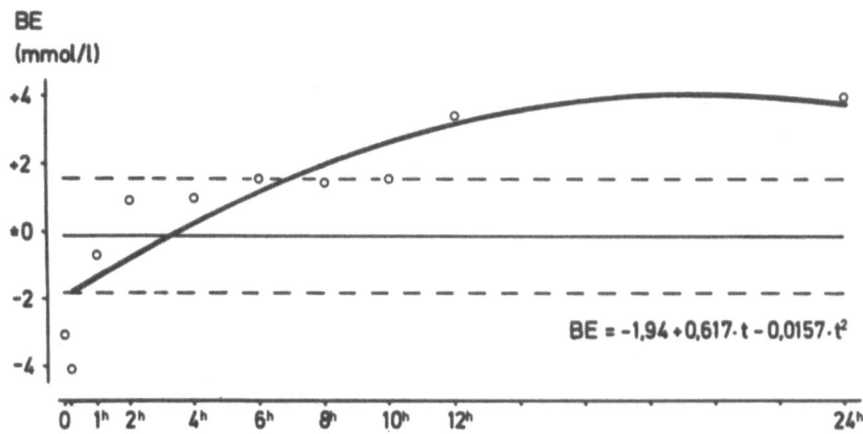

Fig. 6.   Base excess in normal calves.

In depressed calves there are typical signs of acidosis at birth mainly of the metabolic type. In our own investigations (Maurer-Schweizer and Walser, 1977) we found a pH of 7.08; pCO_2 of 73.3 mm Hg; base excess (BE) of -10.6 mmol/l (Figures 7, 8 and 9). Within the first ten minutes metabolic acidosis increases (pH=6.99; BE= 15.5 mmol/l; acute HCO_3^-=17.3 mmol/l). Metabolic parameters are returning to physiological values during the next hours but with a clear delay compared to healthy calves. Carbon dioxide tension and oxygen saturation have not reached standard values even 24 h post partum.

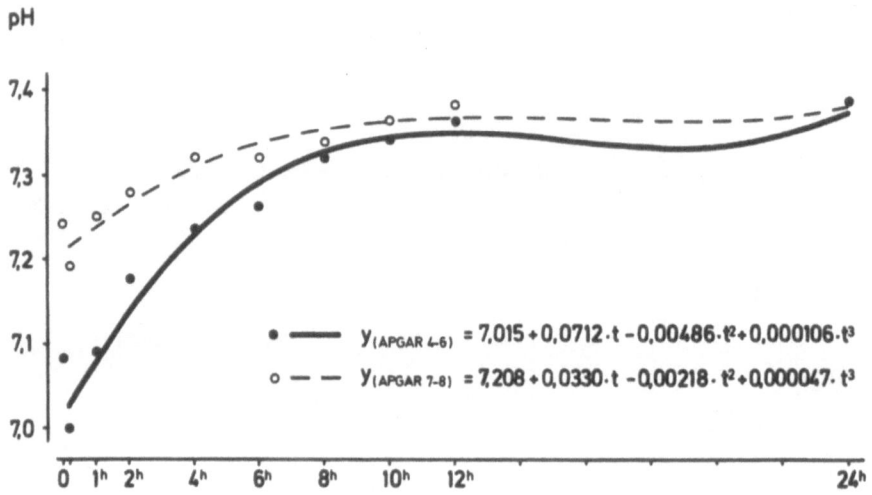

Fig. 7. pH-values in depressed and in normal calves.

There are strong correlations between clinical state and degree of metabolic acidosis (ph value, BE, standard bicarbonate, acute HCO_3^-) whereas such a relation to pCO_2, pO_2 and SO_2 cannot be found (Figure 10).

Regarding these patho-physiological events it is clear what has to be done therapeutically to overcome dangerous acidosis at birth. Proceeding from ideal conditions the programme for first treatment (so-called immediate reanimation)

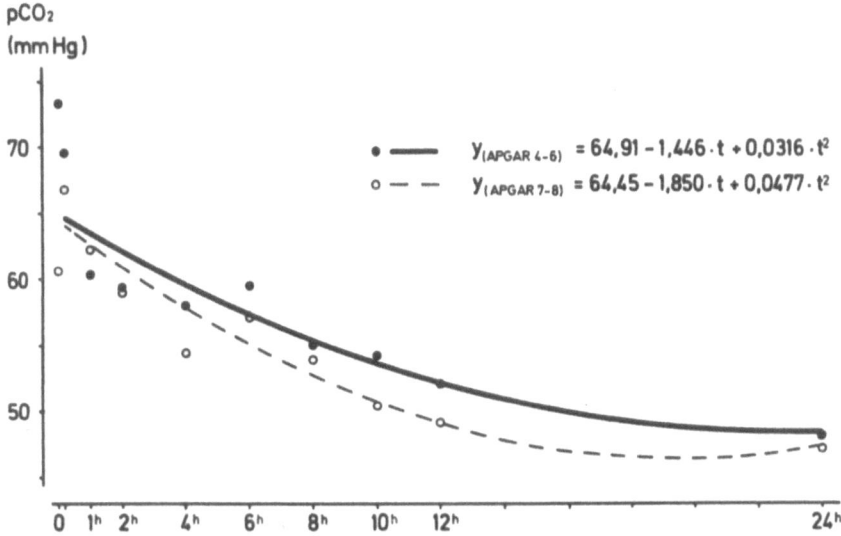

Fig. 8. Carbon dioxide tension in depressed and in normal calves.

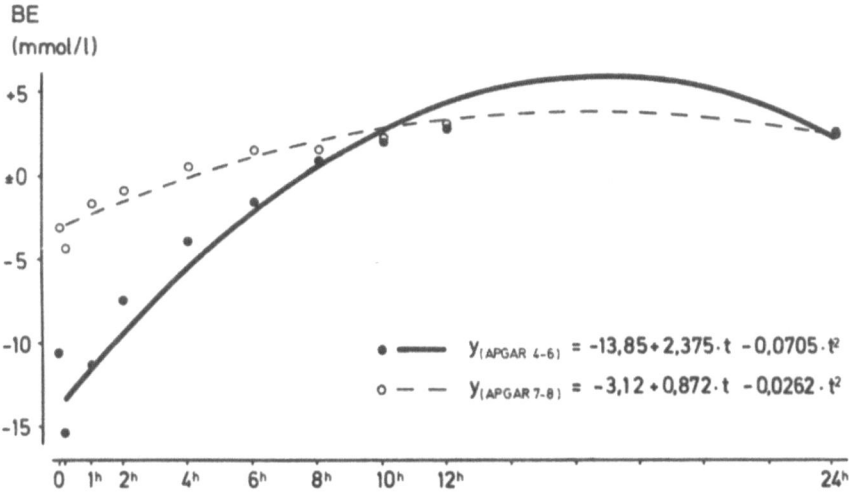

Fig. 9. Base excess in depressed and in normal calves.

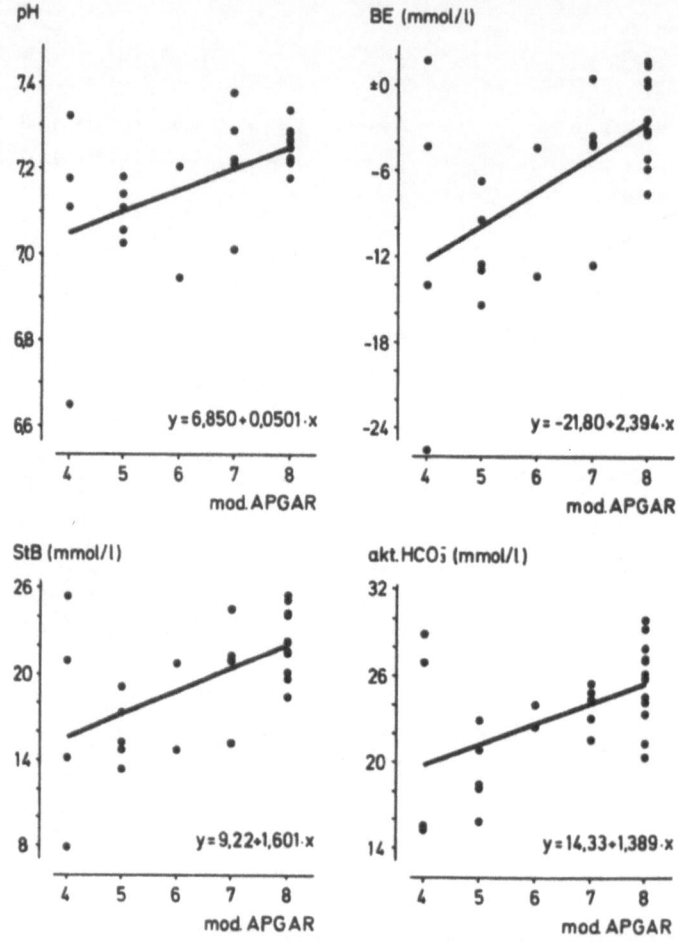

Fig. 10. Relationship between clinical state and metabolic acidosis.

of depressed newborns has to consist of:

1) ensuring clear air passages,

2) mechanical ventilation,

3) administration of buffer solutions,

4) if necessary, additional medication.

The injection of buffer solutions is a new therapy in veterinary medicine compared with former treatment schedules.

In human medicine today, blind buffering is applied even before mechanical ventilation in special cases. This can easily be understood if you remember once more the enormous importance of a severe acidosis in depressed newborns.

The first aim must be to stop the dangerous acidosis by filling up exhausted buffer reserves as quickly as possible. There are two buffers that can be used.

1)   Sodium-bicarbonate buffer 8.4%

Sodium bicarbonate is the most important biological buffer. Before application, however, it must be ensured that respiration is undisturbed. Otherwise carbon dioxide which is set free by buffering with bicarbonate cannot be eliminated by respiration. In dyspnoeic newborn there is therefore possibly a further increase in acidity.

2)   Tris buffer 7% (THAM)

This buffer's action is independent of lung ventilation. Tris reacts with carbonic acid as well as with non-volatile acids, for instance lactic acid. In contrast to buffering with bicarbonate there is no additional charge for the organism by carbon dioxide after injection of Tris. Buffered Tris-H+ is excreted by the kidneys.

Consequently the following should be done to overcome acidosis:

1) immediate injection of Tris buffer,

2) after start of spontaneous breathing (if necessary), injection of sodium bicarbonate.

In practice, of course, the necessary quantity of buffer solution cannot be determined by measured values but must be chosen according to clinical experience.

As far as we know today about 5 - 7 ml/kg body weight of the above mentioned buffer solutions should be given intravenously as a continuous infusion. At the same time the exhausted carbohydrate reserves should be filled up by infusion of 5 - 10% glucose solution.

Concerning mechanical ventilation concessions must be made to the restricted technical and personnel possibilities in a veterinarian's practice at least in the country.

Of course endotracheal ventilation after intubation is the most reliable and effective method. For newborn animals masks are usually employed but they have a large dead space. Some of the commercially offered respiration units can be used with oxygen as well as with air. If this is possible pure oxygen should be applied first; later, if necessary, mechanical ventilation can be continued with air.

In practice we possibly cannot do without the old methods of stimulating respiration by physical stimuli (such as cold water, rubbing etc.). The same applies to rhythmical thoracic compression and expansion.

Additional medication has totally changed in recent years. According to scientific findings from human medicine we do not use central analeptics any longer. These drugs are not necessary in mild asphyxia and they are of no use or even contra-indicated in severe asphyxia as has been shown by experimental results.

Today we use only vasolidators which remove pulmonary vasoconstriction during acidosis. Blood supply to the lungs is improved and alveolar gas exchange is increased by this method.

REFERENCES

Ammann, H., Berchtold, M. and Schneider, F. 1974. Blutgas- und Säuren-Basen-Verhältnisse bei normalen und asphyktischen Kälbern. Berliner und Münchener Tierärztliche Wochenschrift 87: 66 - 68.

Maurer-Schweizer, H., Wilhelm, U. and Walser, K. 1977a. Blutgase und Säure-Basen-Haushalt bei lebensfrischen Kälbern in den ersten 24 Lebensstunden. Berliner und Münchener Tierärztliche Wochenschrift (in press).

Maurer-Schweizer, H., Wilhelm, U. and Walser, K. 1977b. Blutgas- und Säure-Basen-Verhaltnisse bei lebensfrischen Kaiserschnittkälbern in den ersten 24 Lebensstunden. Berliner und Münchener Tierärztliche Wochenschrift (in press).

Maurer-Schweizer, H. and Walser, K. 1977. Azidose und klinischer Zustand bei asphyktischen Kälbern. Berliner und Münchener Tierärztliche Wochenschrift (in press).

Moore, W.E. 1969. Acid-base and electrolyte changes in normal calves during the neonatal period. American Journal of Veterinary Research 30: 1133-1138.

Mülling, M., Henning, H.J. and Marcks, Ch. 1972. Aktuelle pH-Werte im Blut neugeborener Kälber. Tierärztliche Umschau 27: 180 - 181.

Saling, E. 1966. Das Kind im Bereich der Geburtschilfe. Georg Thieme Verlag, Stuttgart.

Stoll, W. 1970. Die Reanimation des Neugeborenen. Der Kinderarzt 18: 10 -13.

DISCUSSION

B. Hoffmann *(West Germany)*

I have a question for Dr. Walser. I have heard the expression 'early respiratory distress syndrome' being used for human problems. Is this the same situation you were talking about? And if so, I have heard that this symptom would be related to dysfunctions of the newborn's adrenal and it was Liggins who used corticoids for treatment. Would this also apply to the newborn calf, and do you have any indications whether the problem you have discussed has some genetic background?

K. Walser *(West Germany)*

In human medicine, distressed newborns are individuals who are not able to blow up and to ventilate their lungs; they are close to suffocation. I think this problem is not the problem we see in calves but perhaps it may occur after induced parturition. Concerning the use of corticoid in calves, I personally have no experience, but I think that there is no use in practice. I cannot say whether there are any genetic factors.

M.A.N. Taverne *(Netherlands)*

I have two questions for Dr. Walser. Do you have data on changes in the blood parameters determined during parturition? Did you, for example, try to take blood samples from the calves at the moment the cervix is partly dilated?

The other question is this: you showed us a nice correlation between these different parameters in the blood and what you called the clinical state. I read from your paper that you used a modified Apgar score. What is the score, and on what criteria is it based?

K. Walser

Firstly, we have no data on blood gases or pH values during parturition but Berchthold in Zürich has shown that the values just before parturition are similar to those immediately after birth.

Now to the clinical state. We use our score, modified after the so-called Apgar score in human medicine. Apgar is a Canadian human obstetrician and she was the first to make a score for a quick assessment of the newborn. As you have seen from the slides, we call a calf normal with an Apgar score of 7 or 8; and we call it mildly or moderately asphyctic with an Apgar score of 4 to 6.

M.A.N. Taverne

Perhaps quite important, and directly related to this seminar is, whether you could - as you did for the pig - establish any correlations between the course of parturition, expressed for example in the duration of the expulsive phase, and the parameters measured in the blood of the newborn calves. This could perhaps enable us to predict the status of the newborn.

K. Walser

No, we have not yet tried to establish such relationships. We divided our material into calvings lasting for three hours and those lasting three to six hours. The bursting of the amnion was taken as the beginning of calving. There were no statistical differences between the two periods. In piglets, however, we see that the values decrease after prolonged expulsion.

GENERAL DISCUSSION WITH PANEL OF SESSION CHAIRMEN

SURVEY OF THE PRESENT SITUATION IN THE EEC

I.L. Mason

My general impression is that there are really two problems. One is the increase in calving problems as a result of concentrating on large beef breeds, and the other is the occurrence of calving problems and calf mortality due to pathological situations.

I think the answer to the first problem must be found in genetic terms. We must select the breeds and the sires which give the easiest calving. I think this will pay off in the future because the production system is bound to become less intensive. When looking far enough ahead, and when we realise what has to be produced for proper feeding of cattle under intensive management conditions, we must realise that we will encounter another difficulty, especially in developing countries where management and food supply are poor. So we must consider beef production on a more extensive scale and this will mean selecting breeds which will calve easily and produce sufficient milk. We may even look favourably on some of the disappearing, locally-adapted breeds which still remain in Europe, at least around the Mediterranean shores. I can't see any future in continuing to breed for large calves, producing calving difficulties and trying to get round them by using these unnatural methods like inducing parturition. I feel strongly about this, especially because of the situation in poor countries. The unfortunate thing is that so many people from these countries come to the rich countries to be trained, and to learn about embryo transplants or induction of parturition, or synchronisation of oestrus. They go back and try these things in their own countries but of course what they need is simply to feed their animals better, keep them under more hygienic conditions, and to treat them better.

We are dealing with a very delicate biological mechanism; we know a lot about it, but I am sure that there is a great deal we don't know. We have to learn from the animal itself, to select the animals which are performing naturally instead of treating them in this rather rough way which to me is similar to knocking them on the head to cure a headache.

Now the second problem. I think it is quite different when we are dealing with a highly selected population of animals which calve well, and whose calves remain alive. Obviously there are cases which are difficult to deal with. These are the cases which people like Mr. Plenderleith and Dr. Grunert have the privilege to deal with, and they have given us very good advice on what is the best treatment for these difficult cases. I think we must keep the two things separate.

GENETIC FACTORS AND BREEDING FOR CALVING PERFORMANCE.   Part I.

R.D. Politiek *(The Netherlands)*

This session gave a very good survey of the breeding aspects of the calving performance in beef breeds and crosses.   Dr. Hanset and Dr. Jaudrain described the consequences on calving problems of selection for double muscling.   An increasing per-- centage of Caesarians in the Belgian Blue-White is accepted by the breeders.   The high price of double-muscle calves compensates for the extra costs.

Crossbreeding experiments in the Federal Republic of Germany (Drs. Langholz, Diehl and Pabst) and also in Denmark (Dr. Liboriussen) and Canada (Dr. Berg) have demonstrated the great differences between sire breeds, in their influence on calving difficulty and the viability of the crossbred offspring.   Large framed beef breeds as a rule give heavier calves with a better growth potential but with more calf losses than do small beef breeds.   Marked differences between crossbred progeny groups, however, indicate a chance of reducing the disadvantages.   There- fore sire evaluation and selection can give special 'beef-sire lines' for cows and for heifers.

Sessions 1 and 2 make it very clear that breeding for calv- ing performance is of economic importance in many countries.   A careful and well defined recording method for calving data, a correct analysis of these data and a construction of an index for selection on ease of calving are important.

My proposal is to ask a small international working group to develop recommendations on the basis of present knowledge and experience in order to reach a more uniform and coordinated system of sampling, evaluation and selection.   This report should be published by the EEC and also be offered for discussion in a combined session of the Cattle and Genetic Commissions of the EAAP.

In the EAAP Commission a working group prepared a report on sire evaluation (Zürich, 1976) and on cow evaluation (Brussels, 1977).

Proposals

EEC Working group: Sire evaluation for dystocia

Members: To be chosen by the Commission in Brussels

Task: Recommendation and report on how to evaluate sires for dystocia, including:

Sampling of calving data
definitions
accuracy (numbers)
sampling methods
production systems - Dairy/Beef/Crosses

Evaluation methods
how to analyse, corrections, BLUP
selection index
combination of parameters
optimisation
selection index
presentation of breeding value for dystocia
effect of sire on calves
effect of sire on daughters

Meetings: Four days at Wageningen, Netherlands. Discussion on recommendations, first draft of report. Visit to AI centre of Dr. v. Dieten (birth registration of calves since 1955). Meeting facilities in the new building of the Department of Animal Science (Zodiac).

If necessary a second meeting. Final report.

GENETIC FACTORS AND BREEDING FOR CALVING PERFORMANCE.   Part 2.

R. Hanset *(Belgium)*

    Dr. Bar-Anan gave an account of his thorough study on the
environmental and genetic components of difficult calving and
perinatal mortality in the Israeli-Friesian breed. He showed
that real genetic differences exist between sires for the
direct effect as well as for the maternal effect, at least when
these effects are evaluated on heifers, as mates or daughters.

    The standard deviations of the sire means were high com-
pared with their respective averages which means that there was
scope for selection. A long term selection programme was not
recommended but rather a strategy consisting of nominated matings
of a large proportion of the heifers with bulls proven for easy
calving and low perinatal mortality on heifers. This procedure
brought along an immediate and important reduction to such an
extent that this result could be explained by the action of but
a few major genes.

    Dr. Foulley considered the problem for the French beef
breeds. After reviewing our knowledge on genetic variability of
the paternal and maternal components, he studied two situations:

    1. the selection of terminal sire lines;
    2. the production of breeding females with beef breeds,
       in pure and crossbreeding.

    With respect to selection of the terminal sire lines, Dr.
Foulley suggested utilising more objective selection criteria,
such as selection indices, with restriction on birth weight.
But here complications arise because of the non-linear relationship between birth weight and calving problems. For producing
breeding females, selection of bulls ought to consider simultaneously direct and maternal effects. In beef breeds, selection
goals tend to be incompatible with each other (eg low birth weight
and high growth rate); therefore compromises have to be found. Dr.
Foulley suggests selection for economic traits with appropriate
restrictions for traits involved in calving difficulties, calf
mortality and maintenance costs.

Dr. Osterkorn and co-workers studied the relationship between performance test data and calving performance of test bulls, on a very large material of almost two million calvings. They showed that selection of bulls on weight-for-age had a negative influence on the rate of veterinary aid and rate of stillbirth.

Dr. Philipsson presented the results of a very important study based on Swedish data. He analysed the effects of environmental factors such as season, age of the cow, nutritional status of the heifer. He confirmed that genetic variability exists for calving performance and calf mortality, mainly at first calving. He gave estimates of genetic correlations between the traits, for the direct as well as for the maternal effects. He studied, by model calculations, the efficiency of different selection schemes, and recommended the selection of bulls on the basis of daughter group results.

Dr. Politiek reported on a study on dystocia in Red-and-White cattle in the Netherlands. He demonstrated the effects of age at first calving and season on incidence of stillbirth, of age and sex on the incidence of dystocia. He considered the possibility of evaluating the calving performance of bulls in mating with heifers from data on matings with cows. He showed that a score combining dystocia and birth weight could be a good indicator of stillbirth percentage in heifers. He recommended a preliminary sire evaluation for the direct effect on 500-750 calf births in second calvers. Bulls so tested could then be judged as suitable, or not, for matings with heifers. The heifers obtained by this preliminary test could be used for testing for maternal effects on daughter groups.

Dr. Schlote reported on investigations still in progress on the relationship between body measurements and weight taken on the dam (heifer) and on the calf, and calving difficulties in German Fleckvieh cattle. Up to now he has data on 180 heifer-calf pairs, and only preliminary conclusions can be drawn. The final results of such a study will be of great interest.

Dr. Ménissier presented the preliminary results of cross-breeding experiments with several European dual-purpose and beef breeds. He showed that the effects of the paternal breed on calving difficulties were primarily related to the size (birth weight) of the calves rather than to their conformation.

From this session it appears that:

1. Genetic variability does exist and can be exploited either for nominated matings of proven bulls with heifers or in long term selection programmes;

2. A antagonism between the classical selection criteria in dual purpose and beef breeds (rate of gain - conformation - mature weight) and calving ability has been underlined, the double-muscled being the most extreme example. The solution will probably be a kind of compromise between these conflicting characteristics. Selection for an aggregate genotype will include several interrelated criteria with the appropriate restrictions. Nevertheless, the larger the number of bulls put on the test, the higher the chance of detecting outstanding males transmitting acceptable combinations of negatively correlated characteristics.

3. A better cooperation in the detection of good bulls concerning the relevant traits could be obtained if the different countries had a uniform way of collecting, analysing and presenting the progeny-test data.

PHYSIOLOGICAL ASPECTS OF PARTURITION

M. Bosc *(France)*

Labour, in the cow, can be considered as an explosive process. The first change in the pattern of the electromyographic activity occurs 16 hours before the expulsion of the calf. This electromyographic activity has no definite direction in each uterine horn but it is related to the mechanical activity and also to cervical dilatation. This activity can be modified during labour by epinephrin secreted by the adrenals but not by the sympathetic innervation of the uterus. In fact, in the cow there is no special report about this innervation, or about the parasympathetic. In this particular field we have to use analogies with other animals.

Labour is closely related in the cow to the decline in progesterone, which is mainly produced by the corpus luteum. This production is under foetal control and we do not know the precise connection between this foetal signal and the regression of the corpus luteum. Evidence for this foetal signal is now well documented but the future steps will include further studies of the relationship between adrenals, hypothalamus and hypophysis.

Much attention has been paid to oestrogen production by the foeto-placental unit. These steroids certainly have a permissive role in regard to the regression of corpus luteum function and also to the preparation of the genital tract for birth. As has been emphasised, oestrogens reflect the foetal genotype and could be used for genetic or clinical studies related to calving problems and to viability of the young calf. If the endocrinological studies have often emphasised the role of the placenta, the complete absence of work on the mechanism of its delivery is noteworthy. Special attention should be paid in the future to settling this question which involves the techniques of calving induction and the prevention of possible infections of the genital tract.

INDUCED PARTURITION

H. Karg (West Germany)

The session dealt with many technical details and with the effects of using hormone injections to induce parturition. There were four complementary papers, covering the different points of view on possibilities and limitations of this bio-technical procedure. All four papers gave excellent reviews and summaries, and I do not intend to repeat them in detail. I just want to make some suggestions on indications for inducing parturition, and compounds to be used.

The papers agreed that restriction of gestation length, and hence the birth weight and size of the calf, would be one way to avoid the main causes of dystocia. This measure is accept-able, for instance, for young heifers, as in the baby calf pro-gramme mentioned by Dr. Gravert. But the discussion made it clear that artificial termination of gestation should only be accepted as an interim approach in breeding programmes since otherwise the danger of negative genetic selection for dystocia may arise. Further indications were mentioned, such as timing of parturition as a management tool, for instance to synchronise animals going on pasture or perhaps to help the workers involved in handling the offspring. The third indication, medical term-ination, is of course without question. Hormonal compounds used successfully up to now are short-acting corticosteroids and prostaglandins; attempts with oestrogens, which were sucessful in the ewe, were failures in the cow.

The principal obvious setback in all these measures is the incidence of retained placenta. It ranges from 50 to 100%, depending on how far the treatment is before normal term. If we apply this method we have to accept this incidence and hence we have to be aware that management and health conditions may have to be adjusted. Unfortunately, nobody could show any pro-gress or further attempts to solve this problem by scientific methods. But it also became clear that the situation in the animal is otherwise not seriously impaired, as the immune globulin status of the calf is normal, and milk yield was

affected only during the first 100 days and not for the full
lactation period. Also fertility seems not to be impaired.
Hence this hormonal method of termination of pregnancy is at
least better than Caesarean section.

NUTRITION AND MANAGEMENT OF THE DAM IN RELATION TO CALVING
PROBLEMS.   Part 1

R. Bar-Anan *(Israel)*

In the session on feeding and management of the dam there
was concensus that any new developments of economic importance
such as early calving, crossing with oversized bulls and in-
dustrialisation, may increase calf mortality when not supplemented by old fashioned care. All three speakers emphasised
that the heifer should be neither too thin nor too fat.

Dr. Brolund Larsen suggested that young heifer calvings
could increase beef production considerably but attention should
be paid, by breeding or feeding, to proper growth of the heifer
dam without unduly increasing the size of the calf at birth.

Dr. Lowman suggested that special attention at calving
should be given to dams that go beyond the normal gestation
length. Under range conditions calvings have to be restricted
within a short season so that proper care and feeding conditions can be provided for the suckler dam.

Dr. Oxender indicated that in calf rearing there is no
magic panacea. Prevent disease, attend parturition, assure
that the calf gets its colostrum and that the scouring calf
does not get dehydrated.

Summing up, it seems that modern cattle breeding introduces
new stresses, which can be minimised by old established know-how.

NUTRITION AND MANAGEMENT OF THE DAM IN RELATION TO CALVING PROBLEMS
Part 2

D. Smidt *(West Germany)*

The main concern of this session was what the farmer can do,
by himself and in cooperation with his veterinary service, to
avoid dystocia and high calf mortality rates. Papers were given
by Dr. Drennan, Dr. Sejrsen, Professor Farries and Professor
Grunert. Precalving feeding and management seem to have certain
impacts on these traits, but are different in their quantitative
importance. This is probably due to some principal differences
between beef and dairy cattle.

In beef heifers, at least, high planes of nutrition in
late pregnancy increased the incidence of calving problems
depending, however, on the age of the heifers. Low plane nutrition seems to have a negative effect on the viability of the
calf. Dairy cattle obviously do not show these effects, at
least not to the same extent. Only severe undernutrition was
regarded as a possible reason for decreased birth weight. It
has been demonstrated, however, that precalving nutrition can
definitely affect postpartum health and fertility in high-
producing dairy cows. The discussion focussed on the effects
of different nutritional levels on milk production and milk composition. This reflects very well the main orientation in defining nutrient requirements according to production.

It may be advisable to give more consideration to the impacts of precalving nutrition on metabolism, and consequently
on the health of the animals with special emphasis on their
fertility.

Findings and statements concerning the condition in which
heifers should be at the time of calving - thin or fat - turned
out to be somewhat contradictory. This could be due to breed
differences, but possibly also to problems in evaluating the
condition of the animal as well as the consequences in terms of
the course of parturition. It was suggested that the influence
of breeds should be studied in more detail.

The housing of calving animals - dairy cows - did not seem
to have striking effects on calving problems, except for the
positive effect of calving pens. One may conclude that calving
pens should be available as part of the general parturition
management.

Seasonal effects on calving performance and calf mortality
(a higher incidence in winter than in summer) may be due partly
to seasonal differences in the exercise of the animals, but also
to effects on birth weight and gestation length.

The paper on clinical aspects demonstrated very nicely the
differences in tackling the problems from various angles. The
clinician focusses his attention on the single case, its diagnosis,
pathogenesis, etiology and therapy. He cannot accept general
descriptions such as 'calving difficulties', 'dystocia', 'stillbirth' etc. in his clinical terminology because each of these
terms represents numerous completely different clinical situations, each of which requires different therapeutic measures.

The breeder, however, needs a generalised description of
certain situations in order to be able to collect suitable data,
more or less neglecting diagnostic accuracy. The necessity for
both approaches should be mutually understood, which may not
always be the case.

Breeding and management measures can improve the general
situation regarding calving problems and mortality, whereas the
clinician has to handle the problem cases remaining, and his
preventive measures are certainly more of a specific nature.

This meeting was a good opportunity to clarify the positions
and attitudes of both sides.

NUTRITION AND MANAGEMENT OF THE NEWBORN CALF. Part 1.

<u>H. Thornberry</u> *(Ireland)*

This session demonstrated very clearly that it is extremely important to look at the problem of calf mortality from aspects other than animal breeding, since better management of the calf can also be an important factor in reducing calf mortality.

In my review I stressed the point that there is a 9 - 13% mortality of calves in Europe. Of these, 50 - 62% are perinatal deaths in some member countries. In the UK (in the years prior to 1975) there was a 9.4% mortality in calves. This represents a loss of £66 million annually.

One of the major reasons for calf losses seemed to be *E. coli* infections but it was thought that to attribute 60% of calf mortality to *E. coli* was not completely accurate. Also, there seemed to be doubt as to whether dehydration is such an important factor in these deaths as is generally supposed. Undoubtedly, viruses play some part in the disease and may often be the primary pathogens. They are very resistant to environmental factors. The problem of salmonellosis could be solved in some countries but still remains a problem in others (Denmark and the Netherlands). In most cases, the disease is passed on to the calf via the dam's excretions, and from the calf to its companions. As a preventive medical treatment, vaccination is useful if administered preferably about two weeks before the calf is due to travel. Even shortly before a journey it is of some value. In order to raise the threshold for infections, colostrum should be given to the calf in the first six hours which is a very critical period in the newborn. However, 'severely' heated milk powder substitutes are the main cause of calf diarrhoea, especially if they contain non-milk proteins. The maximal transfer of immune globulins is by natural means, namely, suckling the dam or by bucket feeding. Deficiency of minerals in the dam's diet leads to the calves being weak at birth, with a subsequent high mortality.

The breed of the sire is important. Heavier birth weights, though commercially desirable, are a major factor in the complex interrelationships which result in dystocia and calf mortality.

Dr. Hammer discussed some basic immune mechanisms in the newborn calf. The transmission of Igs is entirely postnatal through the colostrum and via the intestinal epithelium. The humoral antibodies acquired by the newborn calf are effective only against organisms replicating outside cells and in neutral-ising bacterial toxins. There are four principal types of cellular interactions playing important roles in the regulation of immune responses. These are apparently controlled by genes. It is hoped that geneticists will be able to make practical use of these findings.

During the discussion it became obvious that there are probably three other basic approaches from the immunological point of view to fight calf mortality. These are:

a) to use highly effective antigens, for example by coupling to appropriate carriers;

b) to make use of the 'cross-reactivity' of apathogenic germs and to introduce 'hybridisation'. These 'hybrids' could then be inoculated into the intestinal tract of the calf which might be a very effective way to resist in-fections; and,

c) to induce early maturation of the immune-response system by using thymosine, a hormone isolated from the calf thymus but so far never used in the calf.

All three approaches look very promising, and their further development should be pushed forward much more rapidly.

Dr. Fallon reported about a more practical approach to the effects of different methods of feeding colostrum on calf blood-serum immunoglobulin levels.

A survey conducted on 1 250 calves from various sources over a period of 4 years shows a positive relationship between

absorbed colostral Igs and calf survival. The method used was
the Zinc Sulphate Turbidity Test (ZST). It was shown that 40%
of the calves had insufficient Ig. This was mainly due to
failure to make sufficient colostrum available during the first
6 hours after birth.

Three experiments were carried out with a view to examining
factors which might affect Ig levels in the calf:

1) feeding method;

2) feeding frequency;

3) feeding in presence of dam;

4) feeding temperature.

Feeding by nipple or pail did not significantly affect intake or Ig status. One feed did not provide the calf with an
adequate intake to provide sufficient Ig protection. Feeding
in the present of the dam had a significant effect and increased
the Ig absorption by the calf. There was no significant difference in Ig levels between feeding at body temperature ($38^{\circ}C$) and
room temperature ($14^{\circ}C$).

In answer to a question, it was stated that colostrum can
be stored at -18° to $-25^{\circ}C$ without reducing its protective value
for at least 6 months, provided it is frozen in small amounts
(0.6 litre = 1 pint cartons). During the discussion it was also
stated that the biggest effect of routinely applying the ZST
test is that, after a while, the farmer begins to feel controlled
and thus takes better care of his calves, especially in that he
feeds them adequate levels of colostrum which expresses itself
in higher Ig levels.

NUTRITION AND MANAGEMENT OF THE NEWBORN CALF.   Part 2.

W. Oxender *(USA)*

A computer programme can be a valuable aid for predicting
the economic costs and benefits of calf mortality and morbidity.
This programme can be developed from calf mortality and morbidity
data from the local area. The economic costs and benefits will
vary with prevailing market prices. Diseases of the digestive
system are the major clinical problem in neonatal calves. Enteritis and septicaemia are common, with enteritis being the major
problem in newborn calves. Correction of dehydration and metabolic acidosis appear to be the most beneficial treatments.
Antibacterial agents are usually ineffective in preventing mortality in agammaglobulinaemic calves.

Perinatal survival of calves can be improved through experienced assistance at birth. Some calves have severe respiratory acidosis following birth, particularly after difficult
deliveries. Correction of acidosis with intravenously administered fluids increases the survival rate of calves. Thus,
adequate clinical treatment can be given; it is simple and cheap
and should be considered as an important part in the programme
to decrease calf mortality.

584

SUMMARY OF THE DISCUSSION

B. Hoffmann *(West Germany)*

I would just like to make a few comments on the summaries we have heard.

I think, Mr. Mason, that you are quite right in your policy but to me it seems to be too limited. When you can drive a good car, why throw it away if somebody cannot drive it; rather educate him to run it.

I was also very much impressed by Dr. Politiek's suggestion of more meetings and his activity to arrange more sessions and panels.

I think what we have heard during this meeting was that many factors contribute to the problems we have discussed. For instance, Dr. Hammer has clearly indicated that the problem of stillbirth could also be related to the immune response of the animal. Dr. Walser has indicated that acidosis could be an important factor in causing stillbirths. We have heard that management, feeding, breeding, or breeding at a certain age, may contribute to dystocia and stillbirth. Let me put this question: how do you distinguish between these factors when talking about genetics? How can you exclude this from your programme? I would suggest that you take somebody on your panel who is not a geneticist.

H.O. Gravert *(West Germany)*

There may be a certain misunderstanding here. I think the panel group which Dr. Politiek proposed would not be solving the problem itself. What it should do is just set up some kind of uniform system between the EEC member countries for evaluating data on calving difficulties. At present we have different systems in different countries, and I think we should put all the practical experience together. That was the reason why we propose one of our co-workers from Germany who is really doing the practical work.

A. Osinga *(The Netherlands)*

I get the impression that if we select for a lower rate of stillbirth we only select against the dystocia-related stillbirth. And this stillbirth - at least in my mind - has nothing to do with the immune status of the animal. It is just the difficult birth causing the problems of inoxia, hypoxia or acidosis. Dystocia is related to this difficult position of the calf during parturition; the calf takes too long to be born so that insufficient oxygen is supplied for its metabolism. I think that is the factor that the geneticist wants to select out: stillbirth directly related to dystocia. That means that stillbirth in heifers is the main topic and not that in cows because, in cows, dystocia related stillbirth is only a minor percentage of the stillbirth rate, while it forms the major percentage of stillbirth observed in heifers. I think this is a very important point to make during the closing part of the meeting.

F. Pirchner *(West Germany)*

I don't know exactly why size in cattle is so highly valued. Partly it is artificial because feeding and fattening to fixed age and fixed weight introduce a positive correlation with efficiency. Now, due to the negative correlation of dystocia with growth rate, there are certain disadvantages connected with this, as Dr. Osterkorn said yesterday, and as Dr. Bar Anan added in his thesis. So I think if you have a terminal crossing system it doesn't matter too much. That's all right; there was a recent paper in the Journal of Animal Science where the author studied various European breeds with very high dystocia rates and calf mortality. In the end it turned out that using these large sires on small cows would outweigh all the disadvantages due to dystocia calf loss. But it is a different story if you are within a breed, and there must still be something in the size factor. Farmers ought to explain why they prefer large cattle. I think this is a generally valid statement.

R. Bar-Anan *(Israel)*

May I make some general remarks on the breeding aspects from which I think farmers would benefit.

There are two main points. One is unification. I would agree, as Dr. Politiek has suggested, with the unification of methods, recording, testing and evaluation. We have just introduced in Israel an integrated scheme with milk recording for doing so. That means new methods of recording, new methods of contemporary comparison for heifers and cow daughters and, of course, a new scheme for evaluation.

On the other hand, for breeding it seems that there should be diversification, which would solve most of the problems. I suggest that we should use bulls of smaller size, proven for easy calving, for the first two calvings, and hence for herd replacement, and use oversized bulls for later calvings (including crossbreeding).

B. Hoffmann (West Germany)

Thank you, Dr. Bar Anan. I quite agree with you that it is a question of matching the two compartments, the foetus and the dam, and we should never forget that it is the foetus which determines when it is born.

J. Brolund Larsen (Denmark)

I would like to add to the comment of Dr. Bar Anan. I will repeat the suggestion I made two years ago in Copenhagen, that smaller breeds should be used for the first calving for dairy breeds but especially for beef breeds. First oestrus is observed, depending upon the growth rate, feeding condition, environment, and so on, at about 6 to 8 months of age. Then, if a Jersey or an Aberdeen Angus bull is used, for instance on Charolais heifers, and the first calving will be easy. Use these smaller breeds as 'heifer openers' as we would like them called. We know that we don't have problems with the second calving -- only with the first.

B. Hoffmann

Thank you, Dr. Brolund Larsen. May we consider your contribution as the final scientific words of this conference.

LIST OF PARTICIPANTS

Prof. Dr. O. Aalund	Royal Veterinary and Agricultural University, Bulowsvej 13 Copenhagen Denmark
Dr. G. Averdunk	Bayer. Landesanstalt für Tierzucht Prof. Zorn Str. 6 8011 Grub, Post Poing W. Germany
Dr. R. Bar-Anan	Dept. of Animal Husbandry PO Box 7054 Hakirya 61070 Tel Aviv Israel
Prof. R.T. Berg	Dept. of Animal Science University of Alberta Edmonton, Alberta, T6G 2E3 Canada
G.J. Breslin	Commission of the European Communities Rue Alcide de Gasperi Luxembourg
Dr. M. Bosc	Station de Physiologie de la Reproduction Inst. National de la Recherche Agronomique BP 1 37 Nouzilly France
Dr. J. Brolund Larsen	National Institute of Animal Science Rolighedsvej 25 1958 Copenhagen V Denmark
Dr. M.J. Drennan	The Agricultural Institute Grange, Dunsany Co. Meath Ireland
Dr. R.J. Fallon	The Agricultural Institute Grange, Dunsany Co. Meath Ireland
Prof. Dr. E. Farries	Inst. f. Tierzucht u. Tierverhalten der FLA Braunschweig-Völkenrode 3057 Neustadt 1 W. Germany

Dr. J.L. Foulley | Station de Génétique quantitative et appliquée
Centre National de Recherches Zootechniques
Domaine de Vilvert
78350 Jouy en Josas
France

Prof. Dr. H.O. Gravert | Kronshagener Weg 7 - 9
2300 Kiel
W. Germany

Prof. Dr. E. Grunert | Klinik für Geburtshilfe und Gynäkologie des
Rindes
Bischofsholer Damm 15
3000 Hannover
W. Germany

Prof. Dr. D.K. Hammer | Max-Planck-Institut für Immunbiologie
Stübeweg 51
7800 Freiburg/Brsg.
W. Germany

Prof. Dr. R. Hanset | Chaire de Génétique et d'Economie Rurale
Faculté de Médicine Vétérinaire
Rue des Vétérinaires 45
1070 Bruxelles
Belgium

Priv. -Doz. Dr. B. Hoffmann | Südd. Versuchs- und Forschungsanstalt
für Milchwirtschaft Weihenstephan
Institut für Physiologie
Technische Universität München
8050 Freising
W. Germany

Prof. Dr. H. Karg | Südd. Versuchs- und Forschungsanstalt
für Milchwirtschaft Weihenstephan
Institut für Physiologie
Technische Universität München
8050 Freising
W. Germany

Prof. Dr. H. Kräusslich | Institut für Tierzucht,
Vererbungs- u. Konstitutionsforschung
der Universität München
Veterinärstr. 13
8000 München
W. Germany

J. Kuyl | EEC, Agr. Structures and Environment
Div. E 4: Co-ordination of Agricultural Research
Rue de la Loi 200
1049 Bruxelles
Belgium

Prof. Dr. H.J. Langholz

Institut für Tierzucht und Haustiergenetik
der Universität Göttingen
Albrecht-Thaer-Weg 1
3400 Göttingen
W. Germany

Dr. P. Larvor

Institut National de la Recherche Agronomique
Theix par St. Genès Champanelle
63110 Beaumont
France

P.L. L Hermite

EEC, Agr. Structures and Environment
Div. E 4: Co-ordination of Agricultural Research
Rue de la Loi 200
1049 Bruxelles
Belgium

Dr. B.G. Lowman

Edinburgh School of Agriculture
West Mains Road
Edinburgh EH9 3JG
UK

Dr. T. Liboriussen

Dept. Cattle Experiments
Nat. Inst. Animal Science
25 Rolighedsvej
Copenhagen V-1958
Denmark

I.L. Mason

29, Via S. Anselmo
00153 Roma
Italy

Prof. Dr. F. Ménissier

Station de Génétique quantitative et appliquée
CNRZ
78350 Jouy en Josas
France

Dr. K. O'Farrell

The Agricultural Institute
Moorepark Research Centre
Fermoy
Ireland

Dr. A. Osinga

Afd. Veeteeltwetenschap
(Dept. of Animal Husbandry)
Agricultural University
Duivendaal 5
PO Box 338
Wageningen
The Netherlands

Prof. W. Oxender

Dept. of Large Animal Surgery and Medicine
Michigan State University
East Lansing, Michigan 48824
USA

Dr. J. Philipsson

Dept. of Animal Breeding
The Swedish College of Agriculture
75007 Uppsala 7
Sweden

Prof. Dr. F. Pirchner

Lehrstuhl für Tierzucht
Techn. Universität München
8050 Weihenstephan
W. Germany

R.W.J. Plenderleith

The Rase Veterinary Centre,
Pasture Lane
Market Rasen,
Lincolnshire
UK

Prof. R.D. Politiek

Dept. of Animal Husbandry
Agricultural University
Duivendaal 5
PO Box 338
Wageningen
The Netherlands

Min. Rat Dr. J. Riemensberger

c/o Bundesministerium für Ernährung,
Landwirtschaft und Forsten
Postfach
5300 Bonn-Duisdorf
W. Germany

P.P. Rotondo

Commission of the European Communities
Rue Alcide de Gasperi
Luxembourg

Prof. Dr. M. Rüsse

Gynäkol. Tierklinik der Universität München
Veterinärstr. 13
8000 München
W. Germany

Dr. W. Schlote

Universität Hohenheim
Postfach 106 (06300)
7000 Stuttgart 70
W. Germany

Dr. J. Schmidt

Südd. Versuchs- und Forschungsanstalt
für Milchwirtschaft Weihenstephan
Institut für Physiologie
Technische Universität München
8050 Freising
W. Germany

Dr. K. Sejrsen

National Institute of Animal Science
25 Rolighedsvej
1958 Copenhagen V
Denmark

Prof. Dr. D. Smidt

Institut für Tierzucht und Tierverhalten der FAL
Brauschweig Völkenrode
3057 Neustadt 1
W. Germany

Dr. M.A.N. Taverne

Faculteit voor Diergeneeskunde RU
Kliniek voor Veterinaire Verloskunde
en Gynaecologie, 'De Uithof'
Yalelaan 7
Utrecht
The Netherlands

Dr. H. Thornberry

Veterinary Research Laboratory
Abbotstown, Castleknock
Co. Dublin
Ireland

Prof. Dr. K. Walser

Gynäkol. Tierklinik der Universität München
Veterinärstr. 13
8000 München
W. Germany

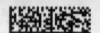